VOLUME FOUR

ADVANCES IN NEUROTOXICOLOGY

Neurotoxicity of Pesticides

VOLUME FOUR

ADVANCES IN NEUROTOXICOLOGY

Neurotoxicity of Pesticides

Edited by

MICHAEL ASCHNER

Albert Einstein College of Medicine, Bronx, NY, United States

LUCIO G. COSTA

University of Washington, Seattle, WA, United States; University of Parma, Parma PR, Italy

ELSEVIER

ACADEMIC PRESS

An imprint of Elsevier

Academic Press is an imprint of Elsevier
50 Hampshire Street, 5th Floor, Cambridge, MA 02139, United States
525 B Street, Suite 1650, San Diego, CA 92101, United States
The Boulevard, Langford Lane, Kidlington, Oxford OX5 1GB, United Kingdom
125 London Wall, London, EC2Y 5AS, United Kingdom

First edition 2020

ISBN: 978-0-12-820515-0
ISSN: 2468-7480

For information on all Academic Press publications
visit our website at https://www.elsevier.com/books-and-journals

Publisher: Zoe Kruze
Acquisitions Editor: Sam Mahfoudh
Editorial Project Manager: Peter Llewellyn
Production Project Manager: Abdulla Sait
Cover Designer: Alan Studholme

Typeset by SPi Global, India

Contents

Contributors

Arturo Anadón
Department of Pharmacology and Toxicology, Faculty of Veterinary Medicine, Universidad Complutense de Madrid, Madrid, Spain

Irma Ares
Department of Pharmacology and Toxicology, Faculty of Veterinary Medicine, Universidad Complutense de Madrid, Madrid, Spain

Jason R. Cannon
School of Health Sciences; Purdue Institute for Integrative Neuroscience, Purdue University, West Lafayette, IN, United States

Lucio G. Costa
Department of Environmental and Occupational Health Sciences, University of Washington, Seattle, WA, United States; Department of Medicine and Surgery, University of Parma, Parma, Italy

John K. Fink
Department of Neurology, University of Michigan Medical School; Ann Arbor Veterans Affairs Medical Center, Ann Arbor, MI, United States

Paul Glynn
Department of Molecular and Cell Biology, University of Leicester, Leicester, United Kingdom

Ramesh C. Gupta
Toxicology Department, Breathitt Veterinary Center, Murray State University, Hopkinsville, KY, United States

Robert B. Hufnagel
Ophthalmic Genetics and Visual Function Branch, National Eye Institute, National Institutes of Health, Bethesda, MD, United States

Vivek Lawana
School of Health Sciences; Purdue Institute for Integrative Neuroscience, Purdue University, West Lafayette, IN, United States

Galina F. Makhaeva
Institute of Physiologically Active Compounds Russian Academy of Sciences, Chernogolovka, Russia

María-Aránzazu Martínez
Department of Pharmacology and Toxicology, Faculty of Veterinary Medicine, Universidad Complutense de Madrid, Madrid, Spain

Marta Martínez
Department of Pharmacology and Toxicology, Faculty of Veterinary Medicine, Universidad Complutense de Madrid, Madrid, Spain

María-Rosa Martínez-Larrañaga
Department of Pharmacology and Toxicology, Faculty of Veterinary Medicine, Universidad Complutense de Madrid, Madrid, Spain

Rudy J. Richardson
Molecular Simulations Laboratory, Department of Environmental Health Sciences, University of Michigan; Department of Neurology; Center for Computational Medicine and Bioinformatics, University of Michigan Medical School; Michigan Institute for Computational Discovery and Engineering, University of Michigan, Ann Arbor, MI, United States

David M. Soderlund
Department of Entomology, Cornell University, Geneva, NY, United States

Sanjeeva J. Wijeyesakere
Department of Microbiology and Immunology, University of Michigan Medical School, Ann Arbor, MI, United States

Preface

The fourth volume of *Advances in Neurotoxicology* is devoted to discussions on the neurotoxicity of pesticides. The volume is not intended to provide a comprehensive coverage of the field, but rather to highlight selected aspects of pesticide neurotoxicity, particularly in areas of current interest and concern. The reviews focus on insecticides, as these compounds exert similar toxicity in target and nontarget species, though it is acknowledged that other classes of pesticides may also induce neurotoxicity.

Two chapters are devoted to organophosphorus (OP) compounds; one deals with the decade-long enigma of OP-induced delayed neuropathy and the potential underlying mechanisms, while another presents information on the neurotoxic effect of OP compounds used as chemical warfare agents. Two additional important classes of widely used insecticides are discussed, namely the pyrethroids and the neonicotinoids. Both have their share of current controversies associated with neurotoxicity. The association of pesticides with neurodegenerative diseases is still object of scientific debate, and the topic of the neurotoxicity of rotenone, a natural insecticide, in relationship to Parkinson's disease is discussed in depth. Finally, the "unusual" neurotoxicity of amitraz, a formamidine insecticide/acaricide, is also reviewed, as several cases of poisoning with this compound are still reported very year. The volume will be of interest to neurotoxicologists in general and to all scientists and regulators dealing with various aspects of pesticide toxicology.

This series fills a gap in the growing field of neurotoxicology and should be of interest to academic and industrial researchers, graduate students, and practicing toxicologists. The series Editors will be happy to entertain proposals for specific topics to be the subject of future volumes, and for guest editing.

MICHAEL ASCHNER
LUCIO G. COSTA

CHAPTER ONE

Neuropathy target esterase (NTE/PNPLA6) and organophosphorus compound-induced delayed neurotoxicity (OPIDN)☆

Rudy J. Richardson[a,b,c,d,*], John K. Fink[b,e], Paul Glynn[f], Robert B. Hufnagel[g], Galina F. Makhaeva[h], Sanjeeva J. Wijeyesakere[i,†]
[a]Molecular Simulations Laboratory, Department of Environmental Health Sciences, University of Michigan, Ann Arbor, MI, United States
[b]Department of Neurology, University of Michigan Medical School, Ann Arbor, MI, United States
[c]Center for Computational Medicine and Bioinformatics, University of Michigan Medical School, Ann Arbor, MI, United States
[d]Michigan Institute for Computational Discovery and Engineering, University of Michigan, Ann Arbor, MI, United States
[e]Ann Arbor Veterans Affairs Medical Center, Ann Arbor, MI, United States
[f]Department of Molecular and Cell Biology, University of Leicester, Leicester, United Kingdom
[g]Ophthalmic Genetics and Visual Function Branch, National Eye Institute, National Institutes of Health, Bethesda, MD, United States
[h]Institute of Physiologically Active Compounds Russian Academy of Sciences, Chernogolovka, Russia
[i]Department of Microbiology and Immunology, University of Michigan Medical School, Ann Arbor, MI, United States
*Corresponding author: e-mail address: rjrich@umich.edu

Contents

† Present address: FMC Corporation, Newark, DE 19711-3506.

☆We dedicate this chapter to the memory of a cherished mentor, colleague, and friend—Dr. Martin Keith Johnson (1930–2018)—who discovered NTE and its connection to OPIDN (Lock et al., 2019).

Advances in Neurotoxicology, Volume 4
ISSN 2468-7480
https://doi.org/10.1016/bs.ant.2020.01.001

1. Introduction

Systemic inhibition of a sufficient quantity of neuropathy target esterase (NTE) with certain organophosphorus (OP) compounds produces OP compound-induced delayed neurotoxicity (OPIDN), a distal degeneration of axons in the central nervous system (CNS) and peripheral nervous system (PNS), thereby providing a powerful model for studying a spectrum of neurodegenerative diseases (Richardson et al., 2013). Axonopathies are important medical entities in their own right (Watson and Dyck, 2015), but in addition, illnesses once considered primary neuronopathies are now thought to begin with axonal degeneration. These disorders include Alzheimer's disease (Liang et al., 2019; Stokin et al., 2005; Yuan et al., 2019),

Parkinson's disease (Burke and O'Malley, 2013; Pelzer et al., 2019; Tagliaferro et al., 2015), and motor neuron diseases such as amyotrophic lateral sclerosis (ALS) (Fischer et al., 2004; Moloney et al., 2014; Venkova et al., 2014). Moreover, conditional knockout of NTE in the mouse CNS produces vacuolation and other degenerative changes in large neurons in the hippocampus, thalamus, and cerebellum (Akassoglou et al., 2004), along with degeneration and swelling of axons in ascending and descending spinal cord tracts (Read et al., 2009). In humans, NTE mutations cause a variety of neurodegenerative conditions resulting in a range of deficits including spastic paraplegia and blindness (Hufnagel et al., 2015; Kmoch et al., 2015; Synofzik et al., 2014). Mutations in the *Drosophila* NTE orthologue SwissCheese (SWS) produce neurodegeneration characterized by vacuolization that can be partially rescued by expression of wild-type human NTE, suggesting a potential therapeutic approach for certain human neurological disorders (Sujkowski et al., 2015; Sunderhaus et al., 2019b).

This chapter defines NTE and OPIDN, presents an overview of OP compounds, provides a rationale for NTE research, and traces the history of discovery of NTE and its relationship to OPIDN. It then briefly describes subsequent studies of NTE, including practical applications of the assay; aspects of its domain structure, subcellular localization, and tissue expression; abnormalities associated with NTE mutations, knockdown, and conventional or conditional knockout; and hypothetical models to help guide future research on elucidating the role of NTE in OPIDN.

2. NTE definition: Gene, protein, and enzyme

Neuropathy target esterase (NTE), formerly called neurotoxic esterase (Johnson, 1970), is now also known as patatin-like phospholipase domain-containing protein 6 (PNPLA6), the sixth member of a nine-protein family of patatin domain lipid hydrolase proteins expressed in humans (Kienesberger et al., 2009).

NTE is encoded by the *PNPLA6* gene located on human chromosome 19p13.2 and containing 37 exons, giving rise to at least 4 splice variants of which the canonical protein sequence, UniProt isoform-4 or National Center for Biotechnology Information (NCBI) isoform-a, has 1375 amino acid residues. The *PNPLA6* gene is highly conserved evolutionarily, with orthologues in diverse species, including *Drosophila*, zebrafish, chicken, mouse, rat, and chimpanzee (NCBI, 2019; UniProt, 2019a).

The NTE protein is expressed ubiquitously in tissues and anchored in the endoplasmic reticulum (ER) of cells, with highest levels in lung, testis, spleen, brain, and kidney (NCBI, 2019). NTE has multiple roles in development, including neurite outgrowth and vasculogenesis (Moser et al., 2004). In the adult, the protein appears to be involved in axon maintenance in neurons (Read et al., 2009) and regulation of osmolarity in the kidney (Gallazzini et al., 2006).

Biochemically, NTE is a serine hydrolase with phospholipase B activity (UniProt, 2019a). As such, NTE possesses both phospholipase A1 and A2 activities, sequentially deacylating both acyl bonds of phosphatidylcholine (PtdCho), with cleavage of the *sn*-2 bond being rate-limiting, followed by rapid hydrolysis of the *sn*-1 bond to yield glycerophosphocholine (GroPCho) and 2 free fatty acids (FFA or RCOOH) (Glynn, 2013; van Tienhoven et al., 2002) (Fig. 1). Moreover, the enzymatic activity of NTE is enhanced in the presence of phospholipids—in particular, phosphatidylcholine (Atkins et al., 2002; Davis and Richardson, 1987; Pope and Padilla, 1989), and lysophospholipid hydrolysis by the catalytic domain of NTE in artificial bilayer membranes alters membrane fluidity (Greiner et al., 2010). Given its phospholipase capabilities and anchorage in the ER, it is likely that NTE plays a role in membrane lipid homeostasis (Zaccheo et al., 2004).

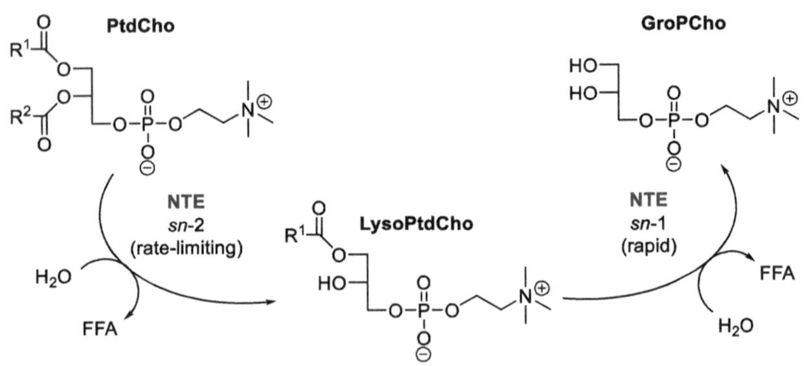

Fig. 1 Catalytic activity of NTE as a phospholipase B. The substrate, phosphatidylcholine (PtdCho), is sequentially hydrolyzed by NTE (shown in blue), first at the *sn*-2 position (rate-limiting) to yield lysophosphatidylcholine (LysoPtdCho) and a free fatty acid (FFA), followed by rapid cleavage at the *sn*-1 position to yield glycerophosphatidylcholine (GroPCho) and a FFA (Glynn, 2013; van Tienhoven et al., 2002). R^1 and R^2 represent the hydrocarbon chains of the FFA components of the phospholipid.

3. NTE and OPIDN

NTE was first identified by a systematic experimental search for the site of initiation of the delayed central-peripheral distal axonopathy (CPDA) brought about by sufficiently high single or repeated exposures to certain OP compounds—a syndrome called OPIDN (Abou-Donia, 1981; Carrington, 1989; Richardson et al., 2013; Wijeyesakere and Richardson, 2010). CPDA is a term used to describe the pattern of lesions found in OPIDN and a number of other neurodegenerative conditions, whereby distal degeneration of axons occurs in both the CNS and PNS (Moretto et al., 1987; Spencer and Schaumburg, 1977). OP compounds that can produce OPIDN are termed *neuropathic*.

Variations of the name ascribed to this chemically induced axonal degeneration include "organophosphorus," "organophosphorus ester," or "organophosphate" instead of "OP compound" and/or "neuropathy" instead of "neurotoxicity" (Abou-Donia, 1981; Cherniak, 1988; Emerick et al., 2010; Masoud and Sandhir, 2012). In addition, some authors favor "polyneuropathy" over "neuropathy" or "neurotoxicity," in which case the acronym becomes "OPIDP" (Jokanović et al., 2011, 2019; Lotti and Moretto, 2005; Monroy-Noyola et al., 2015).

In this chapter, we use OPIDN rather than OPIDP, and we understand the term to indicate "organophosphorus compound-induced delayed neurotoxicity" for two reasons. First, from a chemical standpoint, "organophosphorus" is the correct name for this generic class of chemicals. In contrast, "organophosphate" and "organophosphorus ester" refer to specific subclasses of OP compounds (Eto, 1974; IUPAC, 1978). Moreover, there are representatives from a number of other OP subclasses that can produce OPIDN (Richardson, 2010). Second, from a neurological perspective, "neuropathy" and "polyneuropathy" are rather imprecise terms that usually refer to pathology occurring in one or more peripheral nerves (Albers, 2005; Hutton et al., 2011; Karam and Dyck, 2015). However, OPIDN affects axons in ascending and descending spinal cord tracts as well as sensory and motor axons in peripheral nerves (Abou-Donia, 1981; Emerick et al., 2010; Richardson et al., 2013). Therefore, we regard "delayed neurotoxicity" within the context of exposure to neuropathic OP compounds as a better descriptor of this chemically produced condition than "delayed neuropathy" or "delayed polyneuropathy." The *delayed* aspect of OPIDN refers to a clinically quiescent period between exposure

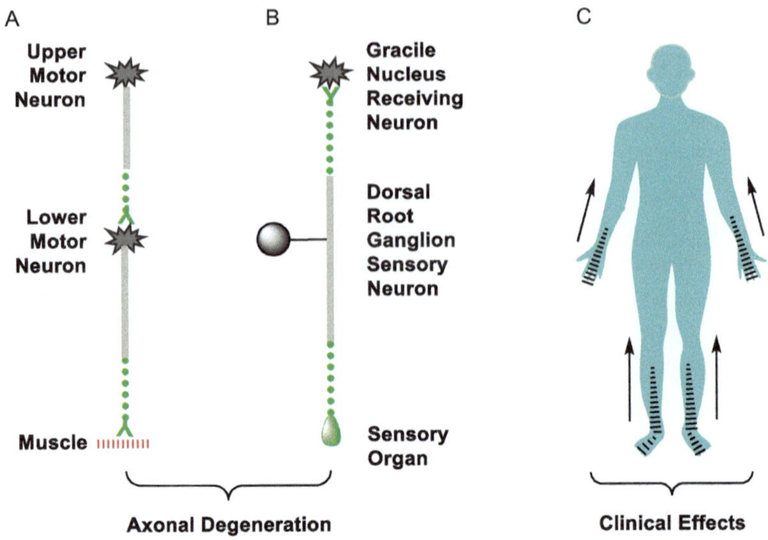

Fig. 2 Distribution of axonal lesions and sensorimotor deficits in OPIDN. (A) (Motor pathway). A descending axon from an upper motor neuron in the motor cortex of the brain passes along the corticospinal tract to synapse with a lower motor neuron in the anterior horn of the spinal cord. The lower motor neuron sends an axon through an anterior spinal root and peripheral nerve to synapse onto a muscle. (B) (Sensory pathway). A sensory neuron in the dorsal root ganglion sends the ascending branch of its bifurcated axon through a dorsal (posterior) column in the spinal cord to synapse with a receiving neuron in the gracile nucleus of the brainstem. The descending branch of the sensory neuron passes through a dorsal spinal root and peripheral nerve, terminating in a sensory organ, such as a Pacinian corpuscle (sensing changes in pressure) or a pain receptor. (C) (Clinical effects). Dashed lines depict the "stocking and glove" distribution of sensory effects (e.g., paresthesias) and motor deficits (e.g. paralysis); arrows indicate the spatial-temporal "dying back" progression from distal to proximal regions. Neurons and intact myelinated axons = gray; distal regions of degenerating axons = dotted green lines; green Y-shapes; sensory organ = green teardrop; muscle = dashed red rectangle (Abou-Donia, 1981; Bouldin and Cavanagh, 1979a,b; Davis and Richardson, 1980; Lotti and Moretto, 2005).

to a neuropathic OP compound and the appearance of clinical signs and symptoms. This hiatus is typically 8–14 days (Emerick et al., 2010), but periods of up to 4 weeks have been reported (Lotti and Moretto, 2005).

The motor and sensory pathways affected in OPIDN along with the distribution and progression of clinical signs and symptoms is shown schematically in Fig. 2. Motor effects include weakness and ataxia progressing to bilateral paralysis. Sensory manifestations comprise bilateral paresthesias such as numbness and tingling; in addition, cramping or lancinating muscle pain may be present. Histopathological findings reveal a primary distal axonopathy with accumulation of tubulovesicular elements accompanied by secondary demyelination. Degeneration preferentially occurs in long, large-diameter axons in both sensory and motor modalities. Mild cases mainly affect the lower limbs, while more severe cases can also affect the trunk and upper limbs. The clinical course in humans consists initially of a stocking and glove distribution of sensory disturbances, attenuated reflexes, and flaccid paralysis. Gradually, some peripheral nerve regeneration occurs, but owing to the persistence of lesions in spinal cord tracts, hyperreflexia and spastic paralysis supervene. Electrophysiological findings over time are consistent with the clinical interpretation of CPDA followed by some degree of peripheral nerve regeneration accompanied by lingering central nervous system involvement (Abou-Donia, 1981; Bouldin and Cavanagh, 1979a,b; Davis and Richardson, 1980; Lotti and Moretto, 2005).

Animal studies and observations of human cases have revealed the following characteristics of OPIDN and its relationship to NTE (Carrington and Abou-Donia, 1984; Glynn, 2000; Johnson, 1982, 1988; Lotti and Moretto, 2005; Moretto et al., 1991; Richardson et al., 2013, 2020; Richardson and Makhaeva, 2014):

- Within a species, adults are vulnerable to OPIDN, while juveniles are resistant.
- Within a species, adults recover incompletely from OPIDN, while juveniles tend to recover completely.
- Within a species, OPIDN preferentially affects long, large-diameter axons in spinal cord and peripheral nerves.
- Within a species, the clinical course begins with paresthesias progressing to weakness, ataxia, hyporeflexia, and flaccid paralysis; mild cases mainly affect the lower limbs; severe cases affect lower limbs, upper limbs, trunk, and the left recurrent laryngeal nerve (which is longer than the right recurrent laryngeal nerve). Over time, some peripheral nerves regenerate while spinal lesions persist resulting in hyperreflexia and spasticity.
- Among species, large animals (e.g., humans, cattle, water buffaloes, sheep, pigs, cats, chickens) tend to be more susceptible to OPIDN than small animals (e.g., rats, mice, gerbils, guinea pigs).
- Among species, large animals tend to have higher NTE specific activity than small animals.
- Among species, those resistant to OPIDN tend to have higher NTE turnover rates than those susceptible to OPIDN.
- A threshold of >70% inhibition of nervous system NTE is necessary but not sufficient for OPIDN to ensue.

- Inhibited NTE must also undergo "aging" (loss of a secondary leaving group from the OP moiety, yielding a negatively charged organophosphyl group covalently attached to the active-site serine) in order for OPIDN to occur.
- NTE inhibition in brain has been used as a convenient surrogate for NTE inhibition in spinal cord or peripheral nerve.
- The animal model of choice for OPIDN studies is the adult hen.
- In hens, clinical signs of OPIDN are delayed by ~8 days from the time of achieving >70% inhibition of brain NTE following a single or repeated dose of a neuropathic OP compound.

Table 1 summarizes chemical and genomic manipulations of NTE and their outcomes in various species. The most enigmatic and controversial aspect of NTE and its presumptive role in OPIDN is the apparent requirement for aging of the inhibited enzyme as the initiating event in the axonopathy. This theory proposes that inhibition of the catalytic activity of NTE is not enough to precipitate OPIDN and that a suitable chemical modification of the protein is also required. By this criterion, OPIDN results from combining a loss of physiological function with a gain of pathological/toxic function in NTE rather than arising from a loss of its physiological function alone. Decades of research focused on OPIDN have led toxicologists to this interim conclusion (Glynn, 2000; Richardson et al., 2013). On the other hand, as we shall see, recent studies directed toward understanding the physiological and pathogenic functions of NTE/PNPLA6 have challenged the gain of toxic function hypothesis, favoring instead the seemingly simpler idea that OPIDN is due to prolonged inactivation of the catalytic activity of the protein (loss of physiological function) (Glynn, 2007; Read et al., 2010). As we seek explanations that might reconcile the toxicological and genetic findings, we should keep two important facts in mind: (1) no drug or toxicant acts exclusively on a single target (Talevi, 2015), and (2) genetic or chemical manipulation of a given target elicits compensatory mechanisms (Schenone et al., 2013).

4. Overview of OP compounds

NTE was discovered using a chemical approach. This could not have been done without prior knowledge of the chemistry of OP compounds and some aspects of their biochemistry and toxicology. Therefore, a brief overview of these materials is in order.

Table 1 Summary of chemical and genetic manipulations of NTE and their outcomes.

Conditions	Species	NTE Status	Results	Selected References
Nonaging inhibitor—acute or chronic doses in adult	Hen	Inhibited	Loss of catalytic function	Baker et al. (1980)
	Cat		No gain of toxic function	Carrington (1989)
	Rat		No OPIDN	Johnson (1975c)
	Mouse		Protection	Richardson et al. (2013, 2020)
				Veronesi and Padilla (1985)
Aging inhibitor—acute or chronic doses in adult	Human	Inhibited + aged	Loss of catalytic function	Davis and Richardson (1980)
	Hen		Potential gain of toxic function	Howland et al. (1980)
	Cat		OPIDN (distal axonopathy in PNS and spinal cord)	Johnson (1990)
	Rat			Lapadula et al. (1985)
	Mouse			Richardson et al. (2013, 2020)
	+Others			Veronesi et al. (1991)
Mutation(s)—chronic throughout lifespan	Human	Mutated	Variable loss of catalytic function	Hein et al. (2010a,b)
				Hufnagel et al. (2015)
			Potential gain of toxic function	Kmoch et al. (2015)
	Drosophila			Rainier et al. (2008)
			Variable types and severity of diseases	Sunderhaus et al. (2019b)
				Topaloglu et al. (2014)

Continued

Table 1 Summary of chemical and genetic manipulations of NTE and their outcomes.—cont'd

Conditions	Species	NTE Status	Results	Selected References
Conventional KO (−/−) – chronic from embryonic day 0	Mouse	100% embryonic deletion	Lethality after embryonic day 7.5 due to placental failure and disrupted vasculogenesis	Moser et al. (2004) Winrow et al. (2003)
Conventional KO (−/+)—chronic from embryonic day 0	Mouse	~50% embryonic deletion	50% loss of catalytic function No gain in toxic function No apparent ill effects	Moser et al. (2004) Winrow et al. (2003)
Conditional KO (CNS)—chronic from embryonic day 11	Mouse	100% embryonic CNS deletion	100% loss of catalytic function No gain in toxic function Neurodegeneration in brain; distal axonopathy in spinal cord	Akassoglou et al. (2004) Read et al. (2009)
Knockdown (silencing)—chronic from 10 h postfertilization	Zebrafish	Variable depression of NTE expression	Various developmental abnormalities including motor neuron defects	Song et al. (2013)

We tend to think of OP compounds as synthetic products of chemical, agrochemical, or pharmaceutical industry, but many of the most important natural constituents of life are OP compounds, including DNA, RNA, phospholipids in cellular membranes, and a number of important enzyme cofactors and metabolic intermediates (Berg et al., 2012). There are even naturally occurring OP compounds found in certain bacteria and marine sponges that exhibit high potency as acetylcholinesterase (AChE) inhibitors with consequent insecticidal activity (Neumann and Peter, 1987; Petroianu, 2012). Moreover, in the broadest chemical sense, OP compounds are either naturally occurring or synthetic organic molecules containing trivalent or pentavalent phosphorus. However, this chapter focuses on synthetic substances containing pentavalent phosphorus, as these constitute the vast majority of exogenous OP compounds to which most people are likely to be exposed. In addition, the term "organophosphyl" is used to indicate a generic OP group attached to another entity such as the active site of an esterase, in recognition of the fact that the OP compound could be from any of a number of subclasses, e.g., organophosphate, organophosphonate, organophosphinate, or organophosphoramidate. Likewise, "organophosphylation" is used to indicate the generic process of adding an OP group to a protein.

Examples of some pentacovalent OP compounds are shown in Fig. 3. Note that the phosphorus atom is tetrahedral; consequently, it can be chiral, giving rise to optical isomers. For example, the nerve agent soman (GD; 3,3-dimethylbutan-2-yl methylphosphonofluoridate) has a chiral phosphorus and a chiral carbon, so that there are four possible stereoisomers. The P_SC_S isomer of soman has the highest inhibitory potency against AChE (Ordentlich et al., 1999; Sanson et al., 2009). Diisopropylphosphorofluoridate (DFP) and its analogue, *N,N'*-diisopropylphosphorodiamidic fluoride (mipafox; MIP) are aging inhibitors of NTE and can produce OPIDN. Diethyl 4-nitrophenyl phosphate

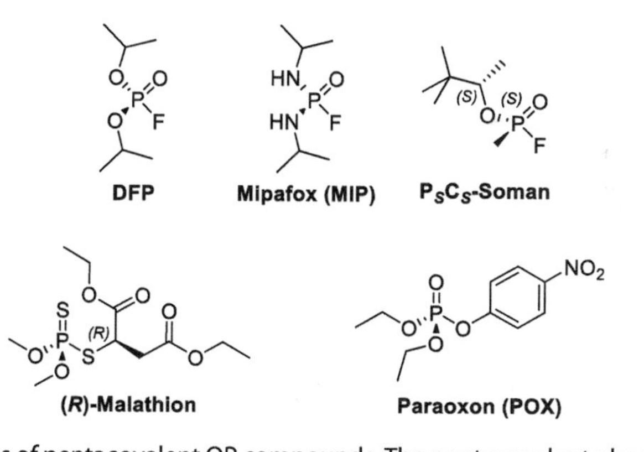

Fig. 3 Examples of pentacovalent OP compounds. The pentacovalent phosphorus atom is tetrahedral; consequently, it can be chiral, yielding optical isomers. Soman has a chiral phosphorus and a chiral carbon, yielding four stereoisomers; the most potent anti-AChE isomer (P_SC_S) is shown. Malathion has a chiral carbon; the (*R*)-isomer is shown, but the commercial insecticide is supplied as the racemic mixture. DFP and MIP are analogues; both are aging inhibitors of NTE and are neuropathic. Paraoxon (POX) is the active metabolite of the insecticide parathion; it is not an effective inhibitor of NTE and is not neuropathic. Mipafox (MIP) is employed in the NTE assay as the selective neuropathic inhibitor of NTE, whereas POX is used as the nonneuropathic inhibitor (Fig. 11) (Jianmongkol et al., 1996; Johnson, 1977; Kayyali et al., 1991; Richardson et al., 2020).

(paraoxon) is the active metabolite of the insecticide, diethyl 4-nitrophenyl phosphorothioate (parathion; POX); it is a potent anti-AChE compound but not an effective inhibitor of NTE, and it is not neuropathic (Lotti and Johnson, 1978). Diethyl 2-[(dimethoxyphosphorothioyl)sulfanyl]butanedioate (malathion) is a widely used insecticide that requires metabolic activation from the parent P=S (thion) form to the active P=O (oxon) form in order to be an effective inhibitor of AChE. Malathion has a chiral carbon; the (*R*)-isomer is shown here, but the commercial insecticide is supplied as the racemic mixture. Neither malathion nor its active metabolite, diethyl 2-(dimethoxyphosphorylsulfanyl)butanedioate, (malaoxon) is an NTE inhibitor and neither compound is neuropathic (Jianmongkol et al., 1996; Lotti and Moretto, 2005).

4.1 OP compounds: Uses and exposures

OP compounds are employed in myriad forms as flame retardants, fuel additives, hydraulic fluids, lubricants, pesticides, pharmaceuticals, and plasticizers (Davis and Richardson, 1980; Richardson et al., 2020). However, they are perhaps best known in the US for their use as insecticides in domestic and agricultural applications (CDC, 2019a). OP insecticides are also recognized throughout the world for the part they play in the control of vector-borne diseases such as African trypanosomiasis, Chagas disease, dengue, leishmaniasis, lymphatic filariasis, and malaria (WHO, 2011). Recently, OP insecticides have been pressed into increased service as a component of integrated vector management amid growing concern over the spread of mosquito-borne viruses such as Eastern Equine Encephalitis, West Nile, and Zika (CDC, 2019b; USEPA, 2019; WHO, 2016).

A consequence of widespread use of OP compounds is that some degree of exposure is inevitable throughout the population, thus prompting the need to understand potential adverse health outcomes so that proper risk assessments can be done and weighed against the societal benefits of using these products. Episodically increased exposures through such well-intended measures as vector control can generate controversy and alarm (Reuters, 2016), heightening the need for trustworthy data on health effects that can be effectively communicated to populations potentially at risk. Public apprehension surrounding exposures to OP compounds is understandable in view of the notoriety of highly toxic OP nerve agents such as (*RS*)-propan-2-yl methylphosphonofluoridate (GD; sarin), which have been used as chemical weapons in a number of conflicts (Jamal, 1995; Page, 2003), reportedly including the Syrian civil war (Brooks et al., 2018; John et al., 2018). Moreover, these compounds have been used against civilian populations as agents of chemical terrorism (Nozaki et al., 1995; Okamura et al., 1996; Suzuki et al., 1995; Worek et al., 2016), and they continue to pose such a threat (Richardson et al., 2020). However, it is reassuring to note that nonoccupational exposures to low-level environmental levels of OP insecticides have not thus far been conclusively linked to adverse neurological outcomes in humans (Reiss et al., 2015).

4.2 Interactions of OP inhibitors with serine esterases

Fig. 4 shows the interactions of a generic OP inhibitor of serine esterases with a target enzyme. Except for the formation of an aged enzyme, the steps in the process are analogous to hydrolysis of a carboxylic acid ester substrate. The initial interaction is reversible, whereby the inhibitor forms a Michaelis-type complex with the enzyme, shown in blue as "EOH" to emphasize the

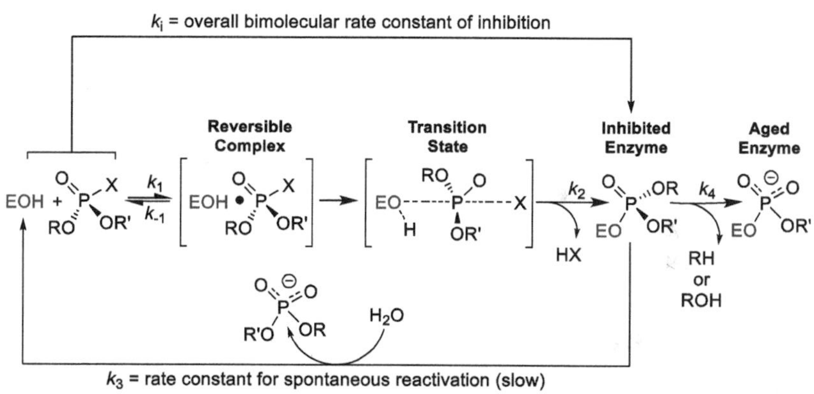

Fig. 4 Interactions of an OP inhibitor with a serine esterase. The esterase (blue) is depicted as EOH to emphasize the active site serine hydroxyl group that engages in a nucleophilic attack on the electropositive phosphorus atom, displacing the leaving group, X, and becoming organophosphylated. Kinetic rate constants: k_1, k_{-1} = formation and dissociation, respectively, of the reversible OP-EOH complex; k_2 = phosphylation; k_3 = spontaneous reactivation; k_4 = aging; k_i = overall bimolecular rate constant of inhibition. The tetrahedral OP inhibitor progresses through a trigonal bipyramidal transition state, which collapses to a tetrahedral species covalently attached to the active site serine residue, yielding inhibited enzyme, which can undergo spontaneous reactivation or aging. The aged enzyme is intractable to reactivation, even by powerful nucleophiles such as oximes (Richardson, 1992; Richardson et al., 2020). R and R′ are substituted or unsubstituted alkyl or aryl groups that can be the same or different. X is a primary leaving group such as fluorine or 4-nitrophenoxyl.

serine hydroxyl nucleophile that attacks the electropositive phosphorus atom in the OP compound. The rate constants k_1 and k_{-1} indicate the forward and reverse reactions of formation and dissociation, respectively, of the reversible complex (Richardson, 1992; Richardson et al., 2020).

On its way to covalently inhibiting the enzyme, the OP compound morphs from a tetrahedral geometry to a trigonal bipyramidal transition state. Covalent acylation (phosphylation) occurs as the leaving group, X, is rapidly displaced while the transition state collapses back to a tetrahedral geometry; this step is characterized by the rate constant of phosphylation, k_2, and yields the covalently inhibited enzyme. If the OP inhibitor were chiral, in the absence of pseudorotation, inhibition would be expected to result in inversion of chirality at phosphorus (Eto, 1997), although the (*R*) or (*S*) designation could change depending on the assignment of group priorities according to the Cahn-Ingold-Prelog (CIP) rules (Carbonell et al., 2013; Epstein et al., 2009).

The dissociation constant of the reversible EOH-OP complex can be approximated by Eq. (1).

$$K_{\mathrm{d}} \cong (k_{-1}/k_1) \tag{1}$$

This simplification is justified for most OP inhibitors of esterases because k_3 and k_4 are often orders of magnitude smaller than k_2, and k_2 is usually much smaller than k_{-1}. K_{d} is a Michaelis-type equilibrium constant in molar units that represents the concentration required to achieve 50% of the maximum rate of production of the acylated (phosphylated) enzyme (see Fig. 6 below). K_{d} is also regarded as a measure of the affinity of an OP inhibitor for binding to the active site of the enzyme. Thus, low values of K_{d} correspond to high affinity, and high values of K_{d} correspond to low affinity (Aldridge and Reiner, 1972; Main, 1980; Richardson et al., 2020).

The overall progress of the reaction from enzyme (EOH) and inhibitor (AX) to phosphylated (inhibited) enzyme (EOA) with expulsion of the primary leaving group (X) is characterized by the bimolecular rate constant of inhibition, k_{i}. This important measure of inhibitory potency is determined by measuring the activity remaining as a function of time of preincubation of the enzyme with various concentrations of inhibitor, where $[AX] > 10$ $[EOH]$. The ester substrate is added after the preincubation interval for a further incubation period to determine the residual activity (Richardson, 1992; Richardson et al., 2020). When $[AX] << K_{\mathrm{d}}$, k_{i} is given by Eq. (2):

$$k_{\mathrm{i}} = k_2/K_{\mathrm{d}} \tag{2}$$

It is important to realize that k_i is a composite of the acylation (phosphylation) rate constant with units of min^{-1} and the dissociation equilibrium constant, K_d, with units of M. Therefore, k_i has the units of a second-order (bimolecular) rate constant, M^{-1} min^{-1}.

It is straightforward to determine k_i when pseudo-first-order kinetics is obtained. These conditions commonly occur when the concentration of the Michaelis-type reversible EOH-AX complex is low, k_2 is high, $k_3 << k_2$, and $[AX] > 10[EOH]$ (Richardson, 1992; Richardson et al., 2020). In this event, Eq. (3) applies:

$$\ln(v/v_0) = k_2[AX]t/([AX] + K_d) \quad (3)$$

where v is the velocity (i.e., rate) of enzymatic hydrolysis of its ester substrate at time $= t$, and v_0 is the rate at time zero. Substituting (% *activity*/100) for (v/v_0), letting $k' = k_2[AX]/([AX] + K_d)$, and rearranging yields Eq. (4):

$$\ln(\%activity) = -k't + \ln(100) \quad (4)$$

Accordingly, plots of ln(% *activity*) versus t will be straight lines with slopes $= -k'$ and y-intercept $= \ln(100) \approx 4.605$, as shown in Fig. 5A. These are the primary kinetic plots. Moreover, the experimentally determined dependence of ln(% *activity*) on the preincubation time (t) and inhibitor concentration $[AX]$ is given by Eq. (5):

$$\ln(\%activity) = -k_i[AX]t + \ln(100) \quad (5)$$

Setting Eqs. (4) and (5) equal to each other gives Eq. (6):

$$k' = k_i[AX] \quad (6)$$

A plot of $-k'$ vs. $[AX]$ will be a straight line with slope $= k_i$, as shown in Fig. 5B. This is the secondary kinetic plot. The k_i value is a measure of the overall inhibitory potency of a given compound (such as a direct-acting OP inhibitor) against a particular serine esterase, such as AChE or NTE. Again, it is important to realize that k_i is a composite quantity that includes K_d, an indication of the affinity of the inhibitor for the enzyme, and k_2, the rate of organophosphylation of the enzyme by the inhibitor (Eq. 2), so that the units of k_i are $[AX]^{-1}$ t^{-1}. The k_i obtained for chlorpyrifos methyl oxon (CPMO) against hen brain AChE (microsomal fraction, pH 7.6 phosphate buffer, 37 °C) is $1.09 \times 10^7 M^{-1} min^{-1}$ (Kropp and Richardson, 2003). Note that the extrapolated primary plots exhibited a small amount of zero-time inhibition, indicating a low concentration of a reversible Michaelis-type complex.

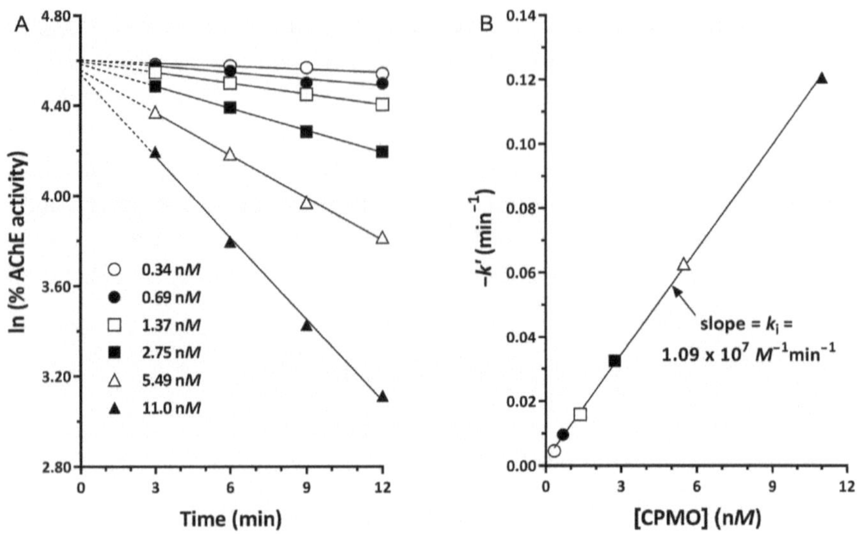

Fig. 5 Kinetics of inhibition of hen brain AChE by chlorpyrifos methyl oxon (CPMO). (A) Primary kinetic plots of ln(% activity remaining) after preincubating AChE with various concentrations of CPMO for timed intervals before adding substrate. Note that the extrapolated lines show a small amount of zero-time inhibition, indicating the presence of a low concentration of a reversible Michaelis-type complex (B) Secondary kinetic plot of –slopes ($-k'$) of the primary plots vs. CPMO concentration; the slope of this line is the bimolecular rate constant of inhibition, $k_i = 1.09 \times 10^7\ \mathrm{M}^{-1}\ \mathrm{min}^{-1}$ (Kropp and Richardson, 2003).

Substituting a percent activity of interest into Eq. (5) gives the inhibitor concentration at a given time of preincubation with enzyme that would yield the particular percent activity. For example, for $[AX]_{50} = IC_{50}$, the inhibitor concentration required to produce 50% inhibition of the enzyme at a given time, t, of preincubation of enzyme and inhibitor at defined conditions of pH, temperature, and ionic strength before adding substrate, yields Eq. (7):

$$IC_{50} = \ln(2)/k_i t \cong 0.693/k_i t \tag{7}$$

Note from Eq. (7) that k_i and IC_{50} are reciprocally related, and that IC_{50} is time dependent. It is valid to calculate an IC_{50} from a k_i value when pseudo-first-order kinetics is obtained, but it is not valid to calculate a k_i from an experimentally determined fixed-time IC_{50}, because the IC_{50} alone contains no information about the kinetics of the reaction. Moreover, if inhibitory potency is assessed by measuring fixed-time IC_{50} values directly, it is essential to report the time of preincubation along with the concentration, because the IC_{50} decreases as the preincubation time increases. For example, the 20-min IC_{50} for CPMO against hen brain microsomal AChE at pH 7.6 and 37 °C is $0.693/[(1.09 \times 10^7 M^{-1} min^{-1})(20 min)] = 3.18 nM$. However, if we were to choose $t = 40$ min, the 40-min IC_{50} decreases to 1.59 nM, giving an unwarranted impression of greater potency. It is preferable to assess inhibitory potency of an OP compound against an esterase kinetically rather than by fixed-time IC_{50} values, but the latter can be easier to conceptualize (Richardson, 1992; Thompson and Richardson, 2004).

For even greater ease of conceptualization, the pIC_{50} is recommended (Navre, 2019). After converting the IC_{50} to M units, $pIC_{50} = -\log IC_{50}$. For example, $IC_{50} = 3.18 nM = 3.18 \times 10^{-9} M$; $-\log(3.18 \times 10^{-9}) = pIC_{50} = 8.50$. The pIC_{50} has some advantages over the IC_{50}. For example, because IC_{50} values in M units require exponents, it is less cumbersome to present pIC_{50} data. Moreover, pIC_{50} values are more intuitive than IC_{50} data for conveying relative potencies—the pIC_{50} is directly related to inhibitory potency. In addition, when concentrations are equally spaced on a log scale, pIC_{50} confidence intervals and standard errors are symmetrical on a log scale (GraphPad, 2016a). However, as with the IC_{50}, the time of preincubation should be stated along with the pIC_{50}. Overall, it is preferable to use the pIC_{50} rather than the IC_{50} as a fixed-time measure of inhibitory potency, but the kinetically determined bimolecular rate constant of inhibition, k_i, is preferable to either fixed-time value (Kropp and Richardson, 2003; Mortensen et al., 1998; Strelow, 2017).

Under certain conditions, the K_d and k_2 components of k_i can be determined. For example, if the primary kinetic plots do not pass through the origin or if the secondary plot is not linear, there may be an appreciable concentration of a Michaelis-type complex. Consequently, the K_d term must be explicitly included. Combining Eqs. (3) and (4) yields Eq. (8):

$$k' = \ln(v_0/v) = k_2/(1 + K_d/[AX]) \quad (8)$$

Eq. (8) has the same form as the Michaelis-Menten equation, and a plot of k' vs. $[AX]$ is hyperbolic, as shown in Fig. 6A. The hyperbolic plot can be linearized as shown in Fig. 6B, but double-reciprocal plots magnify errors; therefore, values of k_2 and K_d should ideally be determined using nonlinear regression on Eq. (8) (GraphPad, 2016b; Ritchie and Prvan, 1996). In this example, $k_2 = 1.0 min^{-1}$ and $K_d = 3.0 \times 10^{-6} M$; therefore, $k_i = k_2/K_d = 1.0 min^{-1}/(3.0 \times 10^{-6})M = 3.3 \times 10^5 M^{-1} min^{-1}$.

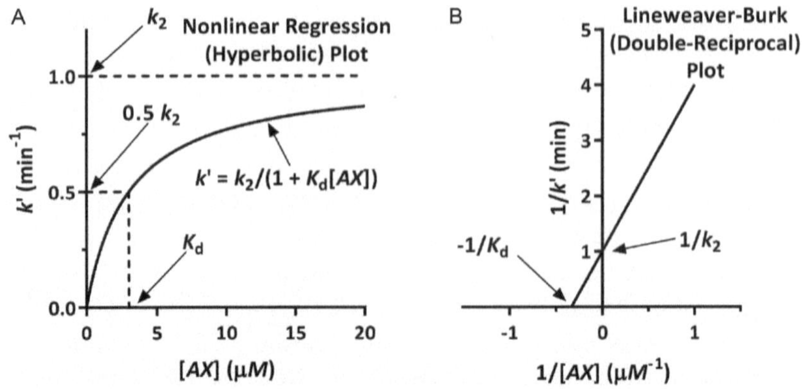

Fig. 6 Plots of Eq. (8) to determine k_2 and K_d values. (A) Direct nonlinear (hyperbolic) plot. (B) Linearized Lineweaver-Burk (double-reciprocal) plot. In this hypothetical example, $K_d = 3.0$ μM and $k_2 = 1.0\,\text{min}^{-1}$; therefore, $k_i = k_2/K_d = 3.3 \times 10^5\ \text{M}^{-1}\ \text{min}^{-1}$ (Main, 1980).

Whenever possible, it is good to determine k_2 and K_d, as these values provide information about the rate of phosphylation and the affinity of the inhibitor for the enzyme, respectively, whereas k_i alone provides information only about the overall potency of the inhibitor for the enzyme (Aldridge and Reiner, 1972; Main, 1980; Richardson et al., 2020).

Spontaneous reactivation is described by a pseudo-first-order rate constant, k_3, because the concentration of the reactivator, H_2O, is essentially constant and in great excess. The spontaneous reactivation rate constant is determined by preincubating the enzyme with a concentration of inhibitor that produces >90% inhibition within a short interval. Inhibitor is then removed or diluted to prevent ongoing inhibition, and the return of enzyme activity is measured in aliquots at timed intervals. A plot of ln(%activity) vs. time yields a straight line with slope $= -k_3$. The corresponding half-life can be calculated via Eq. (9):

$$t_{½} = \ln(2)/k_3 \cong 0.693/k_3 \tag{9}$$

AChE phosphylated by most OP compounds reactivates quite slowly. A notable exception is dimethylphosphoryl-AChE, which reactivates much faster than the diethylphosphorylated enzyme. For example, rat dimethylphosphoryl-AChE $k_3 = 5.7 \times 10^{-3}\ \text{min}^{-1}$ ($t_{1/2} = 2\,\text{h}$), whereas rat diethylphosphoryl-AChE $k_3 = 2.5 \times 10^{-4}\ \text{min}^{-1}$ ($t_{1/2} = 44\,\text{h}$). Human diethylphosphoryl-AChE reactivates at a similar rate as the rat enzyme ($k_3 = 2.0 \times 10^{-4}\ \text{min}^{-1}$; $t_{1/2} = 58\,\text{h}$). Reactivation can be accelerated by strong nucleophiles such as certain oximes (Fig. 8). (Main, 1980; Richardson, 1992; Thompson and Richardson, 2004).

Serine esterases inhibited by OP compounds containing certain labile linkages, such as P–O–R or P–NH–R, can undergo a further reaction (aging), whereby a net loss of the R-group occurs, yielding a negatively charged organophosphyl group still covalently attached to the active-site serine residue of the enzyme. The rate constant for aging is denoted by k_4. Depending on the esterase and the inhibitor, aging can proceed via an S_N1 mechanism (e.g., unimolecular scission of an R–O bond to yield RH) or an S_N2 process (e.g., nucleophilic displacement of RO– by H_2O to yield ROH) (Richardson et al., 2020). If the original inhibitor contains a chiral phosphorus atom, the aged species produced by either mechanism becomes achiral, because the negative charge is delocalized over two oxygen atoms, rendering them equivalent. The charged phosphyl group is stabilized by hydrogen bonding to the oxyanion hole of the enzyme, and the negative charge repels nucleophilic reactivators (Masson and Nachon, 2017). The result is that the aged enzyme is intractable to reactivation. Thus, k_4 can be determined by rapidly inactivating the enzyme via preincubation with an OP inhibitor, removing or diluting the inhibitor to prevent ongoing inhibition, taking aliquots at timed intervals for reactivation by adding fluoride ion or an oxime, and measuring the residual activity (Richardson, 1992).

As aging progresses, the enzyme becomes increasingly resistant to reactivation. Because the process is either first-order or pseudo-first-order, a plot of ln(% activity restored) vs. t is linear, with slope $= -k_4$. As with reactivation, the half-life may be calculated using Eq. (9), substituting k_4 for k_3. For example, k_4 for human AChE inhibited by DFP is 2.5×10^{-3} min^{-1}; $t_{1/2} = 4.6$ h. In contrast, k_4 for human AChE inhibited by paraoxon k_4 is 2.8×10^{-4} min^{-1}; $t_{1/2} = 41$ h. The diisopropylphosphorylated enzyme ages faster than the diethylphosphorylated enzyme because branched alkyl chains favor stabilization of the carbocation produced by S_N1 displacement (Main, 1980). This phenomenon has been exploited in the design of nerve agents such as soman, which has a highly branched ageable pinacolyl group, yielding an aging $t_{1/2}$ of approximately 2 min (Shafferman et al., 1996). Because aged enzyme cannot be reactivated by oximes, rapid aging following exposure to nerve agents precludes treatment with these antidotes (Thompson and Richardson, 2004).

4.3 Inhibition of AChE and cholinergic toxicity

The previous section dealt with interactions of OP inhibitors of serine esterases in general, although many examples were drawn from OP-AChE interactions in particular. Therefore, this section will discuss AChE inhibition and aging only briefly, mainly in the context of toxicity.

Exposure to sufficiently high doses of OP insecticides or nerve agents produces cholinergic toxicity via inhibition of AChE in the central and/or peripheral nervous systems (Richardson, 2010; Thompson and Richardson, 2004). Such anti-AChE compounds are termed *cholinergic*. Because the physiological function of AChE is to terminate the action of the neurotransmitter, acetylcholine (ACh), in cholinergic synapses throughout the central and peripheral nervous systems, a relatively high level of inhibition of AChE can result in an accumulation of excess ACh. Consequently, postsynaptic muscarinic and nicotinic ACh receptors undergo hyperstimulation followed by fatigue, ultimately resulting in muscle paralysis, central depression of respiration, coma, and death.

4.4 Aging of OP-inhibited AChE

As noted above regarding OP interactions with serine esterases, AChE inhibited by OP compounds containing linkages such as P–O–R or P–NH–R can undergo aging, a postinhibitory reaction whereby a net loss of an R-group occurs, yielding a negatively charged organophosphyl group still covalently attached to the active-site serine residue of the enzyme. For AChE inhibition, this secondary reaction does not change the type or severity of toxicity—it only alters the kind of antidotal therapy that can be used. Whether or not aging occurs with inhibited AChE, the outcome of a sufficient level of inhibition is the same—cholinergic neurotoxicity. Aging of phosphylated AChE changes only the types of treatments that are available (Richardson, 2010). These distinctions are illustrated in Fig. 7.

To counteract the excess ACh resulting from inhibition or inhibition plus aging of AChE, atropine is given to attenuate muscarinic effects. If the dose of OP compound were high enough to cause convulsions or seizures, anti-convulsants such as benzodiazepines are also administered. In addition, if aging of inhibited AChE has not occurred, the oxime 2-[(hydroxyimino)methyl]-1-methylpyridin-1-ium (2-pyridine aldoxime methyl; 2-PAM) is given to reactivate the inhibited enzyme (Fig. 8) (Thompson and Richardson, 2004). However, if aging has occurred, 2-PAM is ineffective for two reasons: the negative charge of the aged phosphyl group repels the nucleophilic oxime, and the aged phosphyl group is strongly stabilized by hydrogen bonding to a region of the active site of the enzyme called the oxyanion hole (Masson and Nachon, 2017;

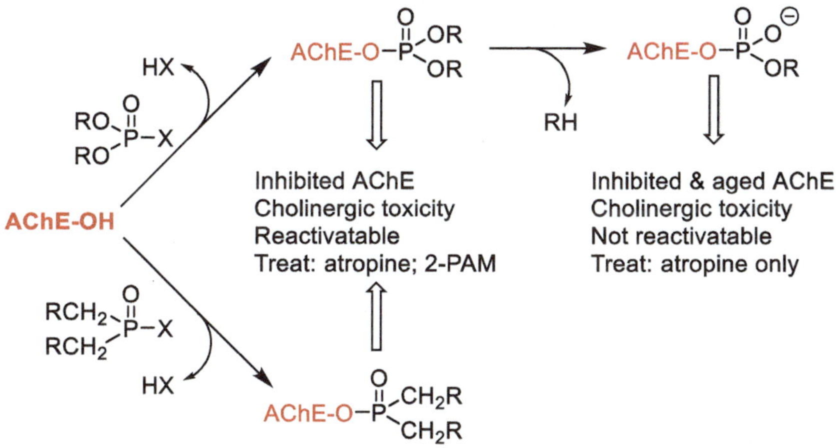

Fig. 7 Toxicological consequences of AChE inhibition and aging. The upper reaction of AChE (shown in red) with an organophosphate first yields an inhibited enzyme resulting in cholinergic toxicity, treatable with atropine. Before the inhibited enzyme ages, reactivation is possible, so that 2-PAM can also be administered (Fig. 8). After aging occurs, AChE is still inhibited but cannot be reactivated; therefore, 2-PAM will be ineffective and treatment is limited to atropine alone (although anticonvulsants could be given if convulsions were present). The lower reaction with an organophosphinate yields an inhibited enzyme resulting in cholinergic toxicity. Aging cannot occur because of the stability of the carbon-phosphorus bonds; both atropine and 2-PAM can be given as treatments. Note that aging of AChE is depicted here as proceeding via S_N1 scission, yielding RH (Richardson and Makhaeva, 2014). R = substituted or unsubstituted aryl or alkyl groups.

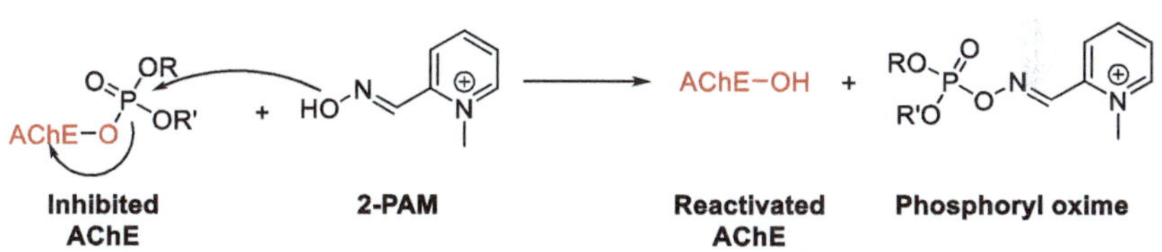

Fig. 8 Reactivation of inhibited AChE by 2-PAM. The oxime oxygen of 2-PAM attacks the phosphorus atom of the unaged organophosphoryl moiety that is covalently attached to the active-site serine of AChE, displacing reactivated AChE and producing a phosphoryl oxime compound. 2-PAM is ineffective if the inhibited enzyme has undergone the aging reaction (Thompson and Richardson, 2004). R, R′ = substituted or unsubstituted alkyl or aryl groups that may be the same or different.

Quinn et al., 2017). However, recent reports suggest that it might be possible to "resurrect" aged AChE using quinone methide precursors (QMPs) to realkylate the aged enzyme to render it reactivatable; the same compounds can also serve as reactivators (Franjesivic et al., 2019; Zhuang et al., 2018).

4.5 Inhibition and aging of NTE and OPIDN

Unlike cholinergic toxicity, which only depends upon inhibition of a critical amount of nervous system AChE, OPIDN appears to require inhibition of at least 70% of nervous system NTE along with aging of the inhibited enzyme (Richardson and Makhaeva, 2014). As noted above, OP compounds that can produce OPIDN are said to be neuropathic. The consequences of inhibiting NTE with an aging vs. a nonaging inhibitor are shown in Fig. 9.

It is possible for OPIDN to arise insidiously, because some neuropathic compounds are potent NTE inhibitors but poor AChE inhibitors. Accordingly, such compounds can trigger axonal degeneration well below the doses required to elicit cholinergic toxicity (Richardson, 2010; Richardson et al., 2020).

The aging process of inhibited NTE is similar to that for inhibited AChE—a postinhibitory loss of a secondary leaving group from the OP moiety resulting in the formation of a negatively charged organophosphyl group covalently attached to the active-site serine residue of the enzyme.

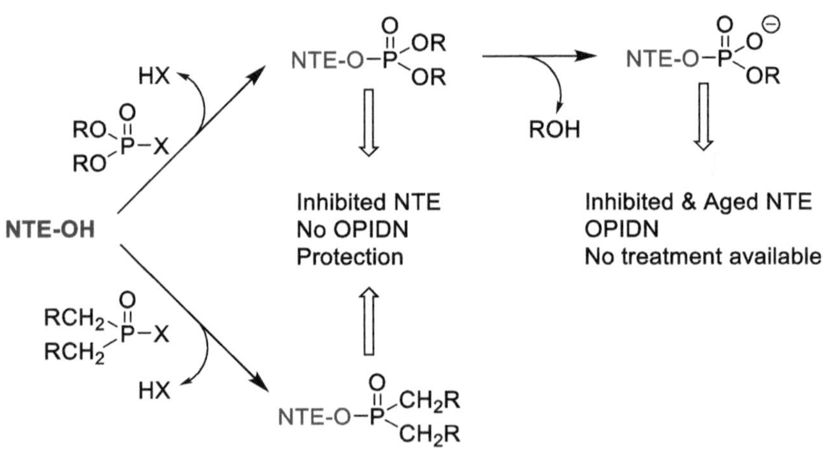

Fig. 9 Toxicological consequences of NTE inhibition and aging. The upper reaction of NTE (shown in blue) with an organophosphate yields inhibited enzyme, but OPIDN does not result until after aging occurs. Currently, there is no specific treatment for OPIDN. The lower reaction with an organophosphinate produces inhibited enzyme. Aging cannot occur because of the stability of the carbon-phosphorus bonds; there is no OPIDN, but there is protection against a subsequent exposure to a neuropathic OP compound. Note that aging of NTE is thought to proceed via an S_N2 reaction with H_2O, so that the R-group is displaced as ROH. Moreover, for most neuropathic NTE inhibitors, aging occurs within minutes, precluding postinhibitory treatment with reactivators (Clothier and Johnson, 1979, 1980; Kropp et al., 2004; Richardson and Makhaeva, 2014). R=substituted or unsubstituted alkyl or aryl groups; they may be the same or different.

However, rather than proceeding via an S_N1 mechanism, aging of NTE inhibited by neuropathic OP compounds is thought to occur by an S_N2 mechanism for organophosphates and organophosphonates, with aging $t_{1/2}$ values ranging from <1 to 10 min for compounds with R-groups that are straight-chain (e.g., *n*-butyl) or "slightly branched" (e.g., isopropyl) (Clothier and Johnson, 1979, 1980). NTE inhibited by the phosphoramidate mipafox appears to age instantaneously, possibly by deprotonation (Kropp et al., 2004). When the ageable R-group on organophospylated NTE is the highly branched pinacolyl moiety, the aging $t_{1/2}$ lengthens to 10 h, lending support to the proposed S_N2 mechanism for aging of organophosphylated NTE. Given these relatively rapid rates of aging of NTE inhibited by neuropathic OP compounds, aging is not the rate-limiting step in the delay of 1–4 weeks between inhibition of NTE and the onset of clinical signs of OPIDN. Likewise, inhibition of NTE is not the rate-limiting step, given that inhibition in experimental animals occurs within hours after dosing (Wijeyesakere and Richardson, 2010).

4.6 "Jake Leg" epidemic

While sporadic cases of OPIDN have been reported since the nineteenth century, the largest epidemic took place in the United States in the early 1930s, toward the end of the era of Prohibition, when the production, importation, transportation, and sale of alcoholic beverages was banned (Davis and Richardson, 1980). During this period, an extract of Jamaican ginger (colloquially called "Jake") was used widely as a means to obtain ethyl alcohol that was presumably potable. However, certain stocks of this extract were adulterated with solvents that contained tris(2-methylphenyl) phosphate (tri-*ortho*-cresyl phosphate; TOCP, also known as tri-*ortho*-tolyl phosphate, TOTP), which is metabolized to the neuropathic NTE inhibitor, 2-(*ortho*-cresyl)-4*H*-1,2,3-benzodioxaphosphoran-2-one (CBDP) (Fig. 10) (Carletti et al., 2013; Reinen et al., 2015).

Consumption of these TOCP-contaminated mixtures throughout the country resulted in a series of large-scale outbreaks of paralysis known by various names, including "Jake Leg," "Jake Walk, and "Ginger Paralysis" (Morgan and Tulloss, 1976; Smith et al., 1930a,b). The number of afflicted individuals has never been accurately determined, although figures as high as 50,000 have been cited (Morgan, 1982). A bulletin from the US Veteran's Administration states that there were 4837 officially reported cases

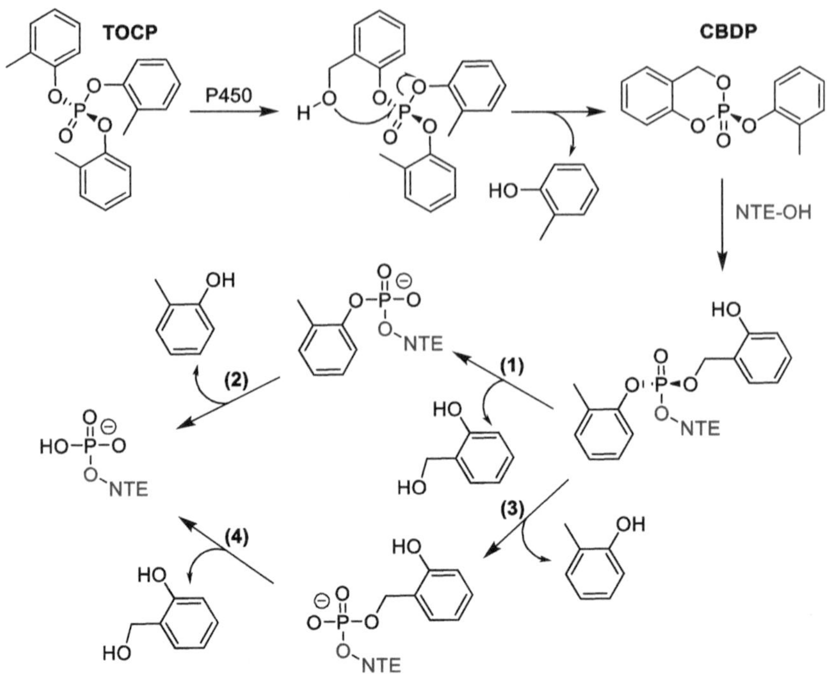

Fig. 10 Formation of CBDP from TOCP and its NTE inhibition and proposed alternative aging pathways. TOCP is metabolized by cytochromes P450 to a hydroxylated intermediate that undergoes internal cyclization with elimination of a cresyl group to form the active NTE inhibitor, CBDP. Inhibited NTE could then undergo aging via proposed alternative pathways with net loss of a saligenin (1) or cresyl (3) group. The aging reaction could stop after the first elimination or proceed with sequential loss of cresyl (2) or saligenin (4) to yield a final aged product with a phosphate group covalently attached to the NTE active site serine residue. (Carletti et al., 2013; Reinen et al., 2015). The (*S*)-isomer of CBDP is shown, as this enantiomer reacts faster with the catalytic domain of NTE (Wu and Casida, 1994); however, it is not yet known if the initial CBDP-NTE adduct has retained, inverted, or racemized stereochemistry. The aged products are achiral owing to delocalization of the negative charge between two equivalent oxygen atoms. NTE is shown in blue.

(Bevis, 1936), but the same article indicates that according to estimates from the Bureau of Prohibition, the true number was at least 20,000. Since then, smaller outbreaks of OPIDN following exposure to TOCP (usually in contaminated cooking oil) have been reported in several countries, including Germany (during World War II) (Karczmar, 2007), Morocco (1959) (Glynn, 2003; Smith and Spalding, 1959) and Sri Lanka (1977–1978) (Senanayake, 1981).

5. Rationale for NTE research

In recent times, reports of OPIDN cases worldwide have been infrequent and usually associated with failed suicide attempts involving ingestion of OP insecticides, although accidental and occupational exposures have also been involved (Jokanović et al., 2019; Lotti and Moretto, 2005). The global decline in OPIDN cases has been due in part to increasingly stringent regulation of OP compounds and the use of the NTE assay to screen OP compounds for their neuropathic potential.

Nevertheless, continued study of neuropathic OP compounds, OPIDN, and NTE is warranted for several reasons. For example, OP compounds continue to be employed in numerous applications, and their use as flame retardants is increasing owing to the phase-out of other classes of these agents, such as the polybrominated diphenyl ethers (PBDEs) (Ospina et al., 2018). Moreover, anticholinesterase OP compounds remain as important agents of chemical warfare and terrorism (Figueiredo et al., 2018), and neuropathic OP compounds pose a latent threat in this regard (Richardson et al., 2020). Exposures to OP compounds also continue to be implicated in the etiology of various conditions for which solid research is needed to resolve issues of causation vs. correlation amid complexity and controversy. These disorders include aerotoxic syndrome (Carletti et al., 2011; Hageman et al., 2019), Gulf War Illness (Michalovicz et al., 2019), thyroid and other endocrine disruptions (Yang et al., 2020), and Type 2 diabetes and obesity (Czajka et al., 2019). However, arguably the most compelling justifications for NTE research are embodied in the intriguing scientific questions concerning the mechanisms of axonal degeneration following changes in the structure and/or function of NTE brought about by chemical exposures and/or genetic mutations (Hufnagel et al., 2015; Kmoch et al., 2015; Read et al., 2009; Richardson et al., 2013; Sunderhaus et al., 2019a,b).

6. History of NTE

6.1 Importance of history

In 1810, Goethe wisely asserted, "… *die Geschichte der Wissenschaft die Wissenschaft selbst sei.*" ("… the history of a science is the science itself.") (Goethe, 1810), and in 1905, Santayana famously observed, "Those who cannot remember the past are condemned to repeat it" (Santayana, 1905). Accordingly, to understand the science of NTE and to avoid reinventing what has already been devised, it is prudent to retrace the story of its discovery and conceptual evolution. This story began with some clever chemical thinking about an important and intriguing toxicological problem long before the application of modern molecular biology and bioinformatics gave us our current view of this fascinating protein. Moreover, there is still much to be learned about NTE, and in the spirit of "… what's past is prologue …." (Shakespeare, 1623), its history is likely to provide valuable clues that can help us draw the blueprint for solving more of its mysteries in the future.

6.2 M.K. Johnson and the MRC toxicology unit

M.K. Johnson discovered NTE through work carried out during 1966–1969 in the Biochemical Mechanisms Section of the Medical Research Council (MRC) Toxicology Unit in Carshalton, England. Johnson was trained as a chemist and accustomed to solving intractable problems. Before coming to the MRC, his former mentor would repeatedly exhort him with the challenge, "Johnson, this won't work—but you try it!" (M.K. Johnson, personal communication). Thus forearmed, Johnson was accustomed to undertaking experiments that others would consider impossibly difficult and fruitless.

6.3 MRC leadership and intellectual climate

When Johnson joined the MRC Toxicology Unit, it was primed with the most favorable leadership and attitudes imaginable to help ensure the success of the work he was to undertake. The Unit was under the directorship of John Barnes, a physician committed to fostering mechanistic research who, among many other notable achievements, had conducted the first toxicology test on penicillin shortly before the onset of WWII (Witschi, 2002). Johnson reported directly to W.N. Aldridge, a highly respected biochemist in charge of the Biochemical Mechanisms Section. At this time, Aldridge was formulating the unifying thesis for much of his life's work—the notion that enzyme inhibitors resembled substrates. Aldridge had already done pioneering work on understanding the acute cholinergic effects of OP compounds by studying the kinetics of their binding to AChE. Somewhat later, in 1972, Aldridge collaborated with Elsa Reiner, a visiting scientist from the Institute for Medical Research in Zagreb, to produce the book, *Enzyme Inhibitors as Substrates: Interactions of Esterases with Esters of Organophosphorus and Carbamic Acids* (Aldridge and Reiner, 1972), which is now regarded as one of the great classics of mechanistic toxicology (Witschi and Lock,

2001). Aldridge kept his staff heading in the right direction for making important discoveries with the continual reminders, "What question are you asking?" and "Go after the anomalies — expected answers seldom tell us anything interesting; pursuing the unexpected will lead to the breakthroughs" (W. N. Aldridge, personal communication).

6.4 MRC philosophy

Barnes and Aldridge established the guiding philosophy of the MRC Toxicology Unit that defined the way its scientists would choose their research problems. There were two main criteria. First, the problem had to have been referred to the Toxicology Unit as a matter of practical public health or medical importance, usually arising from occupational exposures. Second, the problem had to be of intrinsic scientific interest and to show promise as something whose intensive and unrestricted study would yield mechanistic insights, thus advancing the science of toxicology (Witschi and Lock, 2001). These principles were to become official MRC doctrine by 1972, as evidenced by Johnson's quotation from the *MRC Handbook* in a review article on OPIDN:

"The aim of the (Toxicology) Unit is to establish the basic concepts necessary for explaining at a molecular level the mechanisms of toxicity and of chemical and physical injury" (Johnson, 1975b).

Several of the scientists who came to work at the Toxicology Unit had been associated during WWII with the Chemical Defence Establishment at Porton Down, where one of the chief concerns was to understand the acute cholinergic toxicity of OP nerve agents in order to develop prophylactic drugs or antidotes.

6.5 Mipafox incident

Although the delayed neurotoxicity of OP compounds had been well known since the major episodes of the early 1930s, this aspect of OP toxicity had been associated with a single noncholinergic compound, tri-*o*-cresyl phosphate (TOCP) (Bevis, 1936; Smith et al., 1930a,b). Then, in 1951 at Fisons PLC, three workers who were involved with the pilot plant production of a new insecticide candidate called mipafox presented with acute cholinergic toxicity (Bidstrup et al., 1953). Two of the three affected individuals were admitted to hospital for treatment and ultimately discharged. Subsequently, the two patients developed a delayed neurotoxicity that closely resembled poisoning by TOCP.

The mipafox incident demonstrated that an OP compound that was structurally dissimilar to TOCP or its active metabolite CBDP could produce OP delayed neurotoxicity (Figs. 3 and 10). The episode likewise raised concern that this form of insidious and permanently debilitating toxicity could extend to other OP compounds—either already in use or to be developed in the future. Here was a perfect problem to be taken on by the MRC Toxicology Unit. It had arisen from an industrial exposure, it was clearly of medical and public health importance, and it was an intriguing scientific mystery whose investigation held the promise of yielding mechanistic insights into neurotoxicity and neurodegenerative disease. Consequently, research was started at the Unit to investigate the mechanism of OP delayed neurotoxicity.

6.6 Discovery and naming of NTE

Barnes, together with Aldridge and E. Poulsen, laid the foundation for Johnson's ultimate discovery by initiating the search for a suitable target in hen brain (Aldridge, 1964; Aldridge and Barnes, 1966a,b; Poulsen and Aldridge, 1964).

By this time, earlier work had established adult hens as the vulnerable animal model of choice for OPIDN studies (Smith et al., 1930b). Rats, mice, and immature chickens appeared to be resistant; adult roosters were vulnerable, but less readily available and far less docile than hens. AChE and butyrylcholinesterase (BChE) had also been ruled out as potential targets for OPIDN (Cavanagh, 1954; Davison, 1953; Smith and Lillie, 1931).

Poulsen and Aldridge (1964) reasoned that neuropathic OP compounds should covalently bind to an unknown serine esterase in hen brain capable of hydrolyzing substrates such as phenyl phenylacetate (PPA) and phenyl phenylpropionate (PPP), whose structures somewhat resembled that of CBDP, the cyclic saligenin phosphate active metabolite of TOCP (Fig. 10). Initially, such enzyme activities were found, but a clear correlation between their inhibition and delayed neurotoxicity was lacking. In retrospect, these early experiments had come tantalizingly close to defining a relevant esterase activity, but they were hampered by the relative lack of sensitivity of the Warburg manometric assay method that required measuring small volumes of CO_2 gas liberated from a bicarbonate buffer through reaction of the acids produced by enzymatic cleavage of ester substrates. The true signal had been lost in the noise. Years later, Poulsen lamented his narrow miss in identifying NTE, but he was glad for Johnson's success (E. Poulsen, personal communication).

Undaunted by these initial failures, Johnson was convinced that the theory of selective binding was correct. Using a sensitive radiometric approach employing [^{32}P]-diisopropylphosphorofluoridate (DFP), a radiolabeled neuropathic analogue of mipafox, he was able to identify a portion of labeled sites in hen brain tissue that was blocked by neuropathic mipafox, but not blocked by nonneuropathic tetraethylpyrophosphate (TEPP) (Aldridge et al., 1969; Johnson, 1968, 1969a). Furthermore, he was able to quantify the decrease in radiolabeling of the phosphorylation site both in vitro by preincubating hen brain homogenates with neuropathic OP compounds and ex vivo in brain homogenates from hens that had been dosed with neuropathic OP compounds.

Johnson then returned to Aldridge's theory that the site identified by differential incubation with selective inhibitors and radiolabeling was likely a serine esterase. First, he was able to show that radiolabeling of this site could be blocked by preincubating hen brain homogenates with the carboxylic acid ester, PPA. Further study of the PPA hydrolase activity using a colorimetric assay showed that about 10% was resistant to inhibition by nonneuropathic OP compounds such as TEPP or paraoxon and that a portion of this residual activity was sensitive to inhibition by the neuropathic compound, mipafox (Johnson, 1969b,c). It was this relatively small proportion of mipafox-sensitive paraoxon-resistant activity that came to be known first as "neurotoxic esterase" (Johnson, 1973, 1974) and later as "neuropathy target esterase" (Johnson et al., 1985).

Among the journals that were circulated to Johnson's desk from the MRC library was *Analytical Biochemistry*, reflecting his deep appreciation for the fact that while scientific advances required dreamy qualities such as a lively imagination, practical progress depended critically on methods. For example, he was fond of telling his coworkers in the lab that despite the astonishing success of the Lowry protein assay (Lowry et al., 1951) in being the most highly cited scientific paper of all time (Aksnes et al., 2019), there was still no perfect method for quantifying proteins in crude mixtures. Johnson predicted that fame if not fortune awaited the investigator who could devise—not a better mousetrap—but a better procedure than the Lowry method for assaying protein content in biological samples.

Having lived through the privations in Britain imposed by WWII, Johnson was also keen to economize whenever possible. He made good use of his admiration and broad knowledge of techniques along with his desire to get the most out of sparse resources in at least two ways. First, recognizing that energetic electrons such as those emitted from ^{32}P emit visible light when passing through a dielectric medium such as water (the Cherenkov Effect) (Cherenkov, 1934), he devised a method for measuring ^{32}P radiation using a standard scintillation counter and the simplifying economy of not requiring addition of a chemical scintillant or elaborate sample preparation (Johnson, 1969d). Second, in order to simplify and increase the sensitivity of esterase assays compared to the cumbersome and relatively insensitive Warburg manometric method, Johnson sought to develop a colorimetric assay. Realizing that his substrate of choice was PPA, which would be hydrolyzed by esterases to phenylacetic acid and phenol, he reached back to 1946 to retrieve a report on a quantitative colorimetric approach for determining concentrations of phenolic fungicides in fabrics (Gottlieb and Marsh, 1946) and adapted it for use in his esterase assays. The substrate was ultimately changed from PPA to phenyl valerate (PV), but the principle of the assay remained the same (see Fig. 11 below) (Johnson, 1977; Kayyali et al., 1991).

It is interesting to note that in Johnson's first published account—in abstract form—of this as-yet unnamed esterase activity, he inadvertently put together the right words but with a different meaning: "The phenyl esters of 1-phenylacetic acid (PPA) and 2-phenylpropionic acid (PPP) are structurally analogous to the *neurotoxic esterase* inhibitor phenylsaligenin phosphate" (Johnson, 1968; emphasis added). In this abstract, Johnson was referring to the fact that the esterase inhibitor was "neurotoxic," a term used then in context to indicate that a compound was neuropathic, i.e., capable of producing OPIDN, but it appears that he may have at that time unconsciously christened NTE with its initial name, "neurotoxic esterase." In his first full paper referring to neurotoxic esterase (Johnson, 1974), Johnson maintained a provisional status to the name by always putting quotation marks around it. Soon thereafter, at the urging of one of us (RJR), who was then a postdoc in his lab, Johnson became convinced that his new entity was fully fledged and allowed its name to stand undiminished by, as he would say, "inverted commas" (Johnson, 1975a,b,c).

During the next few years, in private correspondence and conversations at scientific conferences, a number of scientists expressed confusion over the name of Johnson's enzyme, saying that the nomenclature suggested that the esterase itself was neurotoxic. This seemed untenable under the premise that the protein surely must have a normal physiological function and would therefore not be neurotoxic in its own right. Johnson acquiesced to this logic and began referring to NTE as "neuropathy target esterase" (Johnson et al., 1985), a new name that seemed more descriptive of its role as the esterase site

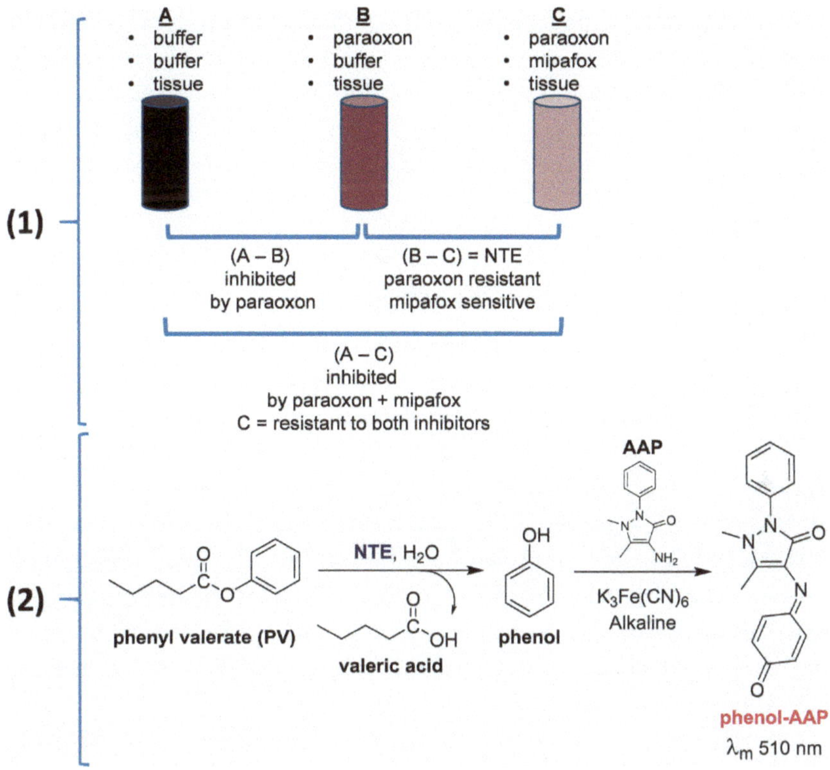

Fig. 11 Principle of the NTE assay. (1) Tissue aliquots are preincubated for a timed interval at a set temperature and pH with (A) buffer only, (B) paraoxon (nonneuropathic), and (C) paraoxon (nonneuropathic) + mipafox (neuropathic). (2) At the end of the preincubation interval, substrate (PV) is added for a timed incubation period, during which NTE and other esterases hydrolyze PV to valeric acid and phenol. The colorimetric reagent 4-aminoantipyrine (AAP) is then added along with $K_3Fe(CN)_6$ to form the plum-red phenol-AAP chromophore. Absorbances (A, B, and C) are read at 486 nm and converted to molar units of phenol formed by calibration against a phenol standard. NTE activity is proportional to (A–C) – (A–B) = (B–C). Absorbance A is not required; it is included here for illustrative purposes. The blank is self-correcting, but it is good practice to include a tissue blank to check for nonenzymatic generation of phenol and a reagent blank to check for decomposition of the substrate into phenol and valeric acid (Johnson, 1977; Kayyali et al., 1991). Structures of the neuropathic inhibitor mipafox and the nonneuropathic inhibitor paraoxon are shown in Fig. 3. The structure of the phenol-AAP chromophore was deduced correctly by Gottlieb and Marsh (1946), and later confirmed by P.F. Jones and K. E., Johnson using proton NMR (Jones and Johnson, 1973). The AAP-phenol reaction has come to be widely used in numerous assays employing colorimetric or electrochemical endpoints (Varadaraju et al., 2018).

that was attacked by a neuropathic OP compound to produce axonopathy. Moreover, the refreshed moniker preserved the same set of initials that had originally been used for the protein. To paraphrase the proclamation of continuity echoed down the centuries by the British (and loyal subjects of other kingdoms) at the concerted passing and accession of their monarchs—NTE is dead! Long live NTE!

6.7 Biochemical lesion and pathogenesis

Sir Rudolph Peters was a renowned biochemist at Oxford who led the effort during WWII to develop British Anti-Lewisite (BAL), an effective antidote against the organoarsenic chemical warfare agent, lewisite (Peters et al., 1945). His subsequent work showing that the toxicity of fluoroacetate was due to its biotransformation to fluorocitrate generated his concept of "lethal synthesis," a powerful idea that had a major influence on toxicological thinking. Equally potent was his notion of the "biochemical lesion," a term he coined in 1931 to refer to the initial molecular perturbation in a chain of events leading to eventual tissue degeneration before any histopathological changes could be detected (Gavrilescu and Peters, 1931; Peters, 1963, 1969).

The concept of lethal synthesis and the biochemical lesion clearly had an impact on Johnson, who launched the theory of "neurotoxic esterase" with his seminal paper titled, "The primary *biochemical lesion* leading to the delayed neurotoxic effects of some organophosphorus esters" (Johnson, 1974; emphasis added). In this publication, Johnson recounted the key observation that not all inhibitors of NTE produced OPIDN. However, the nonneuropathic inhibitors were not biologically inert—they protected chickens (adult hens) against OPIDN from subsequent challenge by neuropathic NTE inhibitors (Johnson and Lauwerys, 1969). The critical molecular difference between the two types of inhibitors was that nonneuropathic protective compounds were incapable of forming aged NTE, whereas neuropathic OP compounds were capable of producing aged NTE—a protein covalently linked to a negatively charged organophosphyl group. Thus, here was a perfect example of lethal synthesis giving rise to a biochemical lesion.

Johnson and colleagues established that the concerted molecular reactions of inhibition and aging of NTE can take place within hours of dosing, well before histological lesions or clinical signs of OPIDN appear, typically 2–3 weeks later (Clothier and Johnson, 1979, 1980; Johnson, 1975a,b,c, 1982, 1990; Makhaeva et al., 2007). At present, this specific chemical modification of NTE is still the earliest known stage in the pathogenesis of OPIDN and thereby regarded as the initiating event. Nevertheless, we have yet to elucidate the steps involved in the ensuing axonal degeneration and to determine how organophosphylation and aging of NTE trigger these degenerative changes.

However, important clues about pathogenesis were uncovered by J.B. Cavanagh, a distinguished neuropathologist at the Institute of Neurology in London who worked closely with scientists at the MRC Toxicology Unit throughout his career. Cavanagh and coworkers helped establish that OPIDN was a primary axonopathy (with demyelination a secondary process) especially affecting long axons of large diameter in peripheral nerves and spinal cord tracts. In addition, they determined that AChE and BChE were not involved, and observed that neuropathic OP compounds acted like chemical scalpels to produce a "chemical transection" of the axon causing the region distal to the initial lesion to undergo Wallerian degeneration. In addition, Cavanagh noted that the lesions resembled those produced by thiamine deficiency, which was thought to decrease energy metabolism, leading to degeneration of distal axons—the regions of neurons most distant from their cell bodies (Bouldin and Cavanagh, 1979a,b; Cavanagh, 1954; Cavanagh and Holland, 1961). We will return later to these observations in a suggested model of OPIDN as Wallerian degeneration.

6.8 Passing the torch

Just as Johnson had been inspired by the scientists around him, he had the gift of instilling others with a fascination for NTE research. He frequently attended international conferences and made visits to other laboratories around the world, taking a keen interest in the many people he met. These activities resulted in his attracting visiting workers to the MRC Toxicology Unit who would later take up NTE research in their home institutions. In addition, he would make a point of introducing one scientist to another, thereby engendering new and fruitful collaborations in the field.

7. Applications of NTE

7.1 NTE Assay and neuropathic potential

Since its inception (Johnson, 1969b), the NTE assay has enjoyed a number of relatively minor improvements (Correll and Ehrich, 1991; Johnson, 1977;

Johnson and Richardson, 1983; Kayyali et al., 1991). However, it is still based on differential inhibition of the hydrolase activity toward an artificial ester substrate (e.g., phenyl valerate) by neuropathic (e.g., mipafox) and nonneuropathic (e.g., paraoxon) inhibitors, and detecting the difference in product formation (e.g., phenol production) colorimetrically or electrochemically (Sigolaeva et al., 2001). A differential assay is required in cells or tissues containing esterases in addition to NTE because inhibition by both paraoxon and mipafox still leaves a residual hydrolase activity that is resistant to both inhibitors. The principle of the NTE assay is shown in Fig. 11.

The NTE assay made it possible to screen OP compounds for a quantitative assessment of their neuropathic potential using in vitro preparations or ex vivo following dosing in experimental animals. It was soon discovered that there was an apparent threshold of ~70% inhibition (and presumably concomitant aging) of nervous system NTE activity in order to precipitate OPIDN (Johnson, 1970).

Moreover, the 70% NTE inhibition threshold was found to apply to chronic dosing as well as to an acute dose regimen. It had been shown in several independent experiments carried out between 1932 and 1975 that there was a dose threshold for OPIDN induced by TOCP given daily for 26 to 140 days (Barnes, 1975; Henschler, 1958; Smith et al., 1932). Using the potent neuropathic OP compound, mono-*o*-cresyl diphenyl phosphate (MOCP), Lotti and Johnson (1980) found that daily dosing at 2.5 mg/kg/day for 8 weeks resulted in a chronic inhibition of brain and spinal cord NTE of 60% and 45%, respectively, with no clinical signs of OPIDN. However, OPIDN could be precipitated by superimposing a single high dose of 50 mg/kg or by increasing the repeated dose to 5 mg/kg/day, which resulted in elevating NTE inhibition to >80%. These experiments also demonstrated that the effects of inhibiting NTE below the threshold were not cumulative.

For those compounds that have anti-AChE as well as anti-NTE activity, it is possible to calculate the relative inhibitory potency (*RIP*), defined as the ratio of bimolecular rate constants of inhibition (k_i) against each enzyme. Alternatively, the *RIP* can be defined using fixed-time IC_{50} values (concentration of inhibitor required to inhibit enzyme activity by 50% at a given time of preincubation of the inhibitor with the enzyme). Because the IC_{50} and k_i are reciprocally (see Eq. 7), the RIP based on IC_{50} values is the reciprocal of the ratio based on k_i values, as shown in Eq. (10).

$$\begin{aligned} RIP\ (cholinergictoxicity) &= k_i(AChE)/k_i(NTE) \\ &= IC_{50}(NTE)/IC_{50}(AChE) \end{aligned} \tag{10}$$

Defined in this way, if *RIP (cholinergic toxicity)* > 1, then the compound is more likely to produce cholinergic toxicity and death than to produce OPIDN (Kropp and Richardson, 2003; Lotti and Johnson, 1978; Richardson, 1992).

Note that some studies have preferred to emphasize the neuropathic potential of an OP compound and to define the *RIP* with the potency against NTE in the numerator, as in Eq. (11).

$$RIP(OPIDN) = k_i(NTE)/k_i(AChE) = IC_{50}(AChE)/IC_{50}(NTE) \quad (11)$$

Defined in this reciprocal manner, if *RIP(OPIDN)* > 1, then the compound is more likely to produce OPIDN than cholinergic toxicity (Makhaeva et al., 2014).

7.2 Structure-activity relationships (SARs)

In the earlier years of NTE research at the MRC, much of the important SAR data that was to be accumulated on NTE inhibitors was made possible by chemical industries supplying Johnson's lab with hundreds of OP compounds for NTE testing. This work was undertaken based upon the understanding that the data could be freely published in the open literature (Davis et al., 1985; Johnson, 1975a,b,c, 1988). The results also made it possible to quantify differences in potency among NTE inhibitors, thus giving rise to quantitative SAR (QSAR) data.

This SAR and QSAR information proved quite useful in making predictions about the neuropathic potential of a given compound. For example, the results of a hen brain NTE assay 24–48 hours after dosing enabled the following conclusions to be drawn:

1. If the compound failed to produce >70% inhibition, it would not cause OPIDN.
2. If the compound produced >70% inhibition, it would cause OPIDN if it belonged to a structural class (Type A) capable of aging (such as phosphates, phosphonates, or phosphoramidates).
3. If the compound produced >70% inhibition, it would not cause OPIDN if it belonged to a structural class (Type B) incapable of aging (such as carbamates, sulfonyl fluorides, or phosphinates).

The generic structures of Type A and Type B NTE inhibitors are shown in Fig. 12.

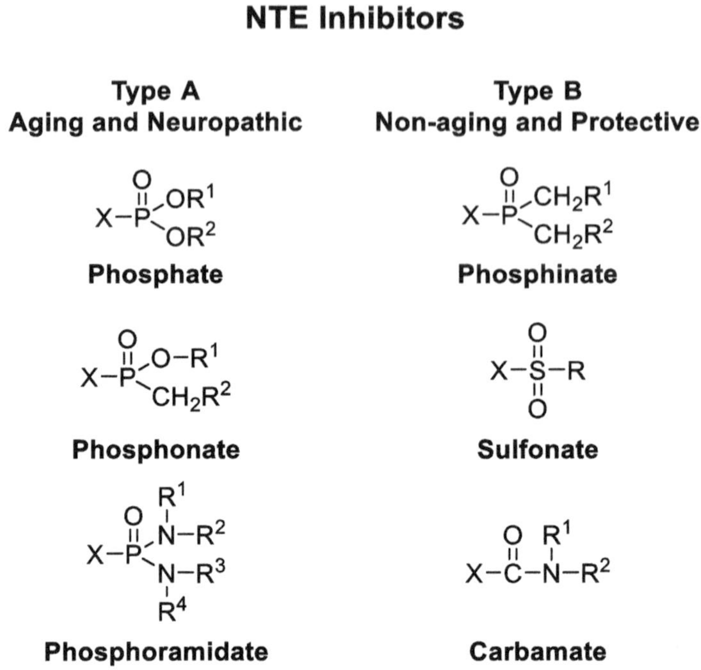

Fig. 12 Generic structures of class A and B NTE inhibitors. Type A inhibitors include phosphates, phosphonates and phosphoramidates. Type A inhibitors are capable of aging, and inhibition of >70% of nervous system NTE results in OPIDN. Type B inhibitors include phosphinates, sulfonates, and carbamates. Type B inhibitors cannot age, and inhibition of >70% of nervous system NTE does not result in OPIDN; however, inhibition of >30% of nervous system NTE with a nonaging inhibitor confers protection against OPIDN from subsequent administration of an otherwise neuropathic dose of a Type A inhibitor. The R-groups are substituted or unsubstituted alkyl or aryl moieties, and for phosphoramidates and carbamates, R can also be hydrogen. X is the primary leaving group in each case.

7.3 Protection (prophylaxis) against OPIDN

Although the extensive SAR and QSAR studies on NTE inhibitors provided partial support for the theory that both inhibition and aging of NTE were required to precipitate OPIDN, the more convincing evidence came from experiments showing that pretreatment with nonageable NTE inhibitors protected against OPIDN from subsequent challenge with an ageable NTE inhibitor. Much of the early work on protection against OPIDN has been summarized in an excellent review by Carrington (1989).

The seminal report on protection showed that phenyl benzylcarbamate inhibited hen brain NTE in vitro and in vivo (Johnson and Lauwerys, 1969). However, the carbamoylated enzyme undergoes relatively rapid spontaneous reactivation, with a half-life in vivo between 6 and 10 hours postdosing. This reactivation was fortuitous, as it permitted determination of the dose-response and time course of protection. Thus, pretreatment of hens with phenyl benzylcarbamate conferred complete protection against the clinical and histopathological manifestations of OPIDN from a challenge dose of the delayed neurotoxicant, DFP, as long as the level of NTE inhibition produced by the carbamate was >30% at the time when the DFP challenge was given. Positive controls pretreated only with the DMSO vehicle developed severe OPIDN from DFP administration. Parallel experiments with [^{32}P]-DFP showed that the carbamate pretreatment did not affect entry of DFP into the brain.

A follow-up study showed that protection against OPIDN from DFP was also conferred by pretreatment with phenyl *N*-benzyl-*N*-methyl carbamate, phenyl *n*-butylcarbamate, phenylmethanesulfonyl fluoride (PMSF), and *n*-butanesulfonyl fluoride (Johnson, 1970). Again, dose-response and time course experiments demonstrated that protection required at least 30% inhibition of NTE at the time that the challenge dose of DFP was administered, whereas dosing with DFP alone required >70% inhibition of NTE to produce OPIDN. Prolonged inhibition of NTE at a level of 80% for 2 weeks with PMSF or 85% inhibition with *n*-butanesulfonyl fluoride did not produce OPIDN. Finally, protection by PMSF against OPIDN from DFP could be blocked by pretreatment with phenyl benzylcarbamate.

Next, Johnson found that phosphinates (e.g., 4-nitrophenyl di-*n*-butylphosphinate; 4-nitrophenyl di-*n*-pentylphosphinate; and 2,2-dichlorovinyl di-*n*-pentylphosphinate) with high structural similarity to neuropathic phosphates and phosphonates inhibited NTE but were not neuropathic (Johnson, 1974). Upon testing these compounds for their ability to protect against OPIDN induced by DFP, he found complete protection that was afforded for 4–6 days, which matched the time for brain NTE inhibition to fall to 30%, thus again suggesting a threshold of about 70% inhibition and aging of brain NTE in order to precipitate OPIDN.

Subsequent experiments in other laboratories confirmed Johnson's observations of protection against OPIDN. These included a study of the time course of protection of PMSF against OPIDN induced by TOCP in hens (Carrington and Abou-Donia, 1983), as well as PMSF protection against an intra-arterially produced unilateral DFP-induced OPIDN in cats (Baker et al., 1980; Drakontides and Baker, 1983) and hens

(Caroldi et al., 1984). Moreover, an important conclusion from the unilateral experiments was that OPIDN could be produced as a direct effect of OP compounds on axons.

Whereas rats are resistant to clinical signs of OPIDN, they exhibit histopathological findings of axonal degeneration following administration of neuropathic OP compounds. Accordingly, Veronesi and Padilla (1985) were able to show protection by PMSF against spinal cord lesions produced by mipafox in rats. Likewise, protection was conferred by the thiocarbamate, molinate, against OPIDN induced by di-*n*-butyl dichlorvos (DBDCV) in rats as well as in hens (Moretto et al., 2001).

Further work in Johnson's lab extended protective agents in hens to inhibitory but nonaging isomers of the insecticide, EPN (protective against OPIDN from the opposite EPN isomer that both inhibited NTE and aged) (Johnson and Read, 1987). Partial protection was observed from pretreatment of hens with the nerve agent soman against challenge with DFP (Johnson et al., 1988a). Soman-inhibited hen brain NTE was found to age negligibly. Owing to the extremely high acute cholinergic toxicity of soman and NTE data on only a single animal from each group, it was not possible to assess accurately the level of nonaged NTE in soman-predosed hens.

Finally, Johnson had found that the phosphinates he had identified earlier as nonneuropathic NTE inhibitors that were protective against neuropathic OP compounds also had shortcomings for NTE/OPIDN studies; in addition to being difficult to synthesize, these phosphinates exhibited high acute cholinergic toxicity and chemical instability. Ultimately, he discovered that phenyl di-*n*-pentylphosphinate was easy to synthesize and that it displayed high potency against NTE, high chemical stability, and low potency against AChE. Moreover, this new phosphinate conferred complete protection against clinical and histopathological signs of OPIDN induced by DFP challenge at either 2 or 4 days after the prophylactic treatment (Johnson and Read, 1993; Johnson et al., 1988b).

7.4 Potentiation or promotion of OPIDN

In 1990, C.N. Pope and S. Padilla demonstrated that nonaging NTE inhibitors protected against OPIDN only if given prior to dosing with a neuropathic (aging) inhibitor—when the order of dosing was reversed, the nonaging NTE inhibitor potentiated the neuropathic effect of the aging inhibitor (Pope and Padilla, 1990). This surprising phenomenon was studied further by M. Lotti and colleagues, who changed the name in their publications from potentiation to promotion. This terminology was based on the absence of clinical signs of OPIDN following dosing with an aging inhibitor of NTE (i.e., an initiator) vs. the appearance of clinical signs of OPIDN following dosing with an initiator and subsequent dosing with a nonaging inhibitor of NTE (i.e., a promoter) (Lotti and Moretto, 1999).

Potentiation had actually been noticed as early as 1985 in an experiment designed to demonstrate protection afforded by PMSF against mipafox-induced spinal cord lesions in rats (Veronesi and Padilla, 1985). As noted above, although rats and mice are highly resistant to the clinical manifestations of OPIDN, they express NTE/PNPLA6 in their nervous systems (UniProt, 2019b,c) and develop axonal lesions that are particularly evident in the dorsal columns of the spinal cord in response to treatment with neuropathic OP compounds (Lapadula et al., 1985; Padilla and Veronesi, 1985; Veronesi, 1984; Veronesi and Padilla, 1985).

PMSF protected rats against axonopathy when given 4 h prior to mipafox, but protection was absent when PMSF was given 14 days before mipafox. In contrast, when mipafox was given 4 h before PMSF, there was an exacerbation of the axonopathy. However, the authors noted that the lesions in the exacerbated case differed somewhat in appearance and distribution from those produced by mipafox alone (Veronesi and Padilla, 1985). In any event, this experiment added support to Johnson's theory that OPIDN requires both inhibition and aging of NTE, and it was the first report of the potentiation of OPIDN by a nonaging inhibitor of NTE.

The specific basis for potentiation or promotion of OPIDN has not been elucidated, but in view of the fact that potentiators or promoters delay recovery from nerve crush (Moretto et al., 1993), a plausible general explanation involves the notion of interference with mechanisms of repair of axonal injury (Lotti, 2002). In addition, although the first reports of potentiation/promotion indicated that the effective compounds were NTE inhibitors, subsequent work by Lotti and colleagues strongly suggested that the target for promotion/potentiation is not NTE (Lotti, 2002; Lotti and Moretto, 1993; Moretto et al., 1994, 2001).

7.5 NTE-based biomarkers and biosensors

The discovery and characterization of NTE activity in circulating lymphocytes and platelets (Bertoncin et al., 1985; Bleecker et al., 1983; Dudek and Richardson, 1982; Maroni and Bleecker, 1986) made it possible to use blood samples for biomonitoring exposures to neuropathic OP compounds (Lotti et al., 1983; Schwab and Richardson, 1986). With the advent of electrochemical biosensors, NTE activity could be determined in whole blood (Makhaeva et al., 2003; Sigolaeva et al., 2001), thus obviating the need to isolate leukocyte fractions prior to analysis. Moreover, biosensor assays permitted real-time detection of NTE activity and could be extended for the simultaneous assay of several esterases (Kohli et al., 2007; Sigolaeva et al., 2010). Because different OP compounds have varying inhibitory potencies against different esterases, this differential inhibition creates an "esterase profile" for a given inhibitor that serves to predict the type of toxicity or therapeutic effect that it will exert (Makhaeva et al., 2013, 2016a,b).

8. Cellular and molecular biology of NTE

8.1 NTE: The next generation

During the decades of research on NTE since its discovery, most of the studies were focused on the inhibition and aging of its esterase function, the association of these events with OPIDN, and the development of biomarkers and biosensors of exposure to neuropathic OP compounds (Makhaeva et al., 2016a,b; Richardson et al., 2013). Only relatively recently have investigations expanded to include noncatalytic regions of the protein, factors that may regulate its hydrolase function or turnover, interactions with other proteins, NTE knockouts or silencing, and NTE mutations associated with disease (Huang et al., 2016; Hufnagel et al., 2015; Sogorb et al., 2016; Synofzik and Schule, 2017; Synofzik et al., 2014, 2015). These new developments have of course resulted from engaging a new generation of researchers (along with some from earlier days) to continue the exploration of NTE by applying the modern tools of molecular biology, bioinformatics, and computational molecular modeling. Many of the advances in understanding NTE have stemmed from the discovery of SWS ("Swiss Cheese") the *Drosophila* orthologue of human NTE/PNPLA6, so named because of the vacuolated appearance of fruit fly brains arising from certain mutations in the *sws* gene encoding the SWS protein (Kretzschmar et al., 1997; Sunderhaus et al., 2019a,b).

8.2 NTE protein sequences and domains

8.2.1 Canonical and isoform sequences

In the early 1990s, during M.K. Johnson's phased retirement and the move of the MRC Toxicology Unit from Carshalton to the University of Leicester, important new developments in NTE research were realized, including the affinity purification, cloning, and sequencing of the human protein (Glynn, 2006; Glynn et al., 1994, 1999; Lush et al., 1998).

Screening of a human brain cDNA library yielded the first reported sequence for NTE (clone D16) comprising 1327 amino acids, with the catalytic center (defined by Ser966) located within a hydrophobic region (Lush et al., 1998). This sequence is now categorized by UniProt as isoform-2, with isoform-4 regarded as the canonical NTE sequence out of 4 isoforms arising from alternative splicing that have been identified thus far (UniProt, 2019a). The canonical isoform-4 comprises 1375 residues with Ser1014 as the active-site serine residue. Note that the National Center for Biotechnology Information (NCBI) refers to this sequence as isoform-a, accession/version number NP_001159583.1 (NCBI, 2019).

Throughout the remainder of this chapter, except where otherwise indicated, the numbering of the canonical isoform-4 sequence will be used. It should also be noted that despite our knowledge of the sequence and domains of NTE, its three-dimensional structure has still not been determined experimentally.

8.2.2 Four types of domains

Subsequent analysis of the primary amino acid sequence of NTE suggested it to be a multi-domain protein with three distinct regions (Fig. 13): (1) an N-terminal transmembrane (TM) domain; (2) a series of three tandem cyclic nucleotide binding (CNB) homology domains (CNB1, CNB2, and CNB3); and (3) a C-terminal catalytic domain, the patatin NTE

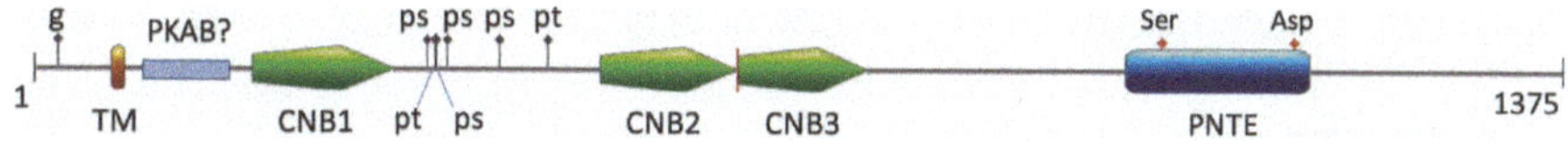

Fig. 13 Domains of the NTE/PNPLA6 protein. The canonical protein sequence-a (NCBI, 2019) or isoform-4 (UniProt, 2019a) comprises 1375 amino acid residues. The established three types of domains and their approximate residue number boundaries include the following: a transmembrane domain (TM, 60–80); three putative cyclic nucleotide binding domains (CNB1, 195–322), (CNB2, 511–633), and (CNB3, 629–749); and a catalytic domain (patatin-homology NTE; PNTE, 981–1147). By analogy to the *Drosophila* homologue, SWS (Wentzell et al., 2014), a potential fourth type of domain (PKA binding, PKAB) is proposed to reside between the TM domain and CNB1. Some specific sites of interest include g (*N*-glycosylation site, Asn20); ps and pt. (phosphorylation sites on serine and threonine, respectively; Ser354, Thr361, Ser362, Ser372, Ser420, and Thr464); and the active site residues in the catalytic domain (Ser1014 and Asp1134) (UniProt, 2019a).

(PNTE) homology domain (Glynn, 2013; Wijeyesakere et al., 2007). More recently, in the *Drosophila* NTE orthologue, SWS, a fourth domain (4) has been described that can bind a catalytic subunit (C3) of protein kinase A (PKA) (Bettencourt da Cruz et al., 2008; Sunderhaus et al., 2019a,b); here, we refer to this region as PKAB. In the following subsections, we will briefly discuss the current state of knowledge of these domains as well as the subcellular localization of NTE, highlighting areas where new research is needed to address gaps in knowledge and understanding. Note that protein domains are often not precisely demarcated. The residues assigned a given domain will sometimes overlap with an adjoining domain, and residue assignments to domains may differ somewhat between data sources, e.g., UniProt and NCBI.

8.2.3 Transmembrane (TM) domain

The N-terminal TM domain of NTE anchors the protein to the cytoplasmic face of the endoplasmic reticulum (ER). Consistent with the nature of mammalian transmembrane domains, its structure is predicted to be an α-helix spanning residues 9–31 in NTE isoform-2. In the case of isoform-4, the TM region consists of residues 60–80, with residues 1–59 being localized within the ER lumen and the remaining residues (81–1375) on the cytoplasmic side (UniProt, 2019a). Interestingly, analysis of the primary sequence of NTE isoform-4 suggests the presence of an *N*-linked glycosylation site at residue Asn20 that is absent in isoform-2 (denoted by "g" in Fig. 13). Virtually nothing is known about a possible role for this potential *N*-linked glycosylation site; however, in other ER-folded proteins such a site serves to direct the nascent polypeptide to the calnexin/calreticulin chaperone cycle (Lamriben et al., 2016; Wijeyesakere et al., 2013). Thus, the role of these glycan-binding ER chaperones in binding or promoting the folding of a subset of NTE molecules represents an intriguing area for further investigation.

8.2.4 Cyclic nucleotide binding (CNB) homology domains

The CNB domains of NTE encompass the following residues by isoform-2 numbering: 163–262 (CNB1), 480–573 (CNB2), and 597–689 (CNB3) (UniProt, 2019a; Wijeyesakere et al., 2007). By the canonical isoform-4 numbering, these domains are contained within the following residues (Fig. 13): 195–322 (CNB1), 511–633 (CNB2), and 629–749 (CNB3) (UniProt, 2019a).

Cyclic nucleotides such as cyclic AMP (cAMP) and cyclic GMP (cGMP) play crucial roles as second messengers that transduce signals in a variety of cellular signaling systems throughout life forms ranging from bacteria to humans (Gancedo, 2013). Thus, the presence of a tandem series of putative CNB domains within the N-terminal locale of NTE suggests that this region may play a role in regulating its hydrolase activity.

When the current domain organization of NTE was first published (Wijeyesakere et al., 2007), very little was known about the structure or function of the three CNB domains. Indeed, in the previous decade, there had been only a single study hinting that these domains might be vestigial remnants incapable of engaging cyclic nucleotides (Dremier et al., 2003). Nevertheless, we have been unwilling to accept that an apparent CNB region could be devoid of purpose, and the existence of three such domains suggests the possibility that NTE requires an exquisite tunability of control. Accordingly, using homology models, we have carried out preliminary molecular docking simulations and found that cyclic nucleotides (cAMP and cGMP) can be docked to the putative binding pockets of all three CNB domains with favorable free energies of binding (S.J.W. and R.J.R., unpublished data).

Recent elegant experiments in *Drosophila* have shown that certain mutations in SWS, the fruit-fly orthologue of NTE, produce neurodegenerative disease and disruption of phospholipid homeostasis that can be rescued by expressing wild-type NTE in the flies (Sunderhaus et al., 2019b). However, expression of NTE that contained disease-causing mutations in either the catalytic domain or the CNB domains suppressed neurological deficits but did not restore lipid homeostasis. These findings provided support for two important conclusions: (1) disruption of the catalytic function of NTE is not sufficient to produce neurodegeneration; and (2) the CNB domains participate in the regulation of NTE catalytic activity.

8.2.5 Catalytic (patatin-homology) domain (PNTE)

By far, the best-characterized aspect of NTE is its hydrolase activity, which resides in the catalytic domain (residues 981–1147) (UniProt, 2019a). While the structure of this region has not been determined experimentally, bioinformatics analyses of the primary sequence of NTE show that the catalytic Ser1014 nucleophile (isoform-4 numbering) is located within a region that has some homology to the plant protein, patatin. Accordingly, one of the names ascribed to this region is the patatin-homology domain of NTE (PNTE) (Fig. 13). Based on the initial published structure for patatin isoform-17 (pat17) (PDB ID 1OXW) (Rydel et al., 2003), the first structural homology model for PNTE was presented (Wijeyesakere et al., 2007). In this paper, it was proposed that PNTE consists of a modified α/β hydrolase fold with a central parallel β-sheet flanked by seven α-helices. Subsequently, a somewhat higher resolution structure of pat17 was published (PDB 4PK9) along with its aged DFP complex (PDB 4PKA) and nonaged complex with methyl arachidonylfluorophosphonate (MAFP) (PDB 4PKB) (Wijeyesakere et al., 2014).

Previously, based on findings from site-directed mutagenesis, the catalytic machinery of NTE had been thought to consist of a novel Ser966-Asp1086-Asp960 catalytic triad (isoform-2 numbering) (Atkins and Glynn, 2000) that was distinct from the canonical Ser-His-Asp/Glu triads found in other α/β hydrolases. However, the current view is that akin to patatin and $cPLA_2$, the catalytic center of NTE consists of a Ser1014-Asp1134 catalytic dyad (isoform-4 numbering) (Fig. 13), with the Asp1134 residue serving as both the general base and general acid during the catalytic process.

Thus, the proposed hydrolysis mechanism for NTE involves Asp1134 serving as a general base at the initial stage of the catalytic process to activate the Ser1014 nucleophile by abstracting a proton. This allows the Ser1014 hydroxyl oxygen to perform a nucleophilic attack on the substrate. The ensuing tetrahedral intermediate, whose negative charge is expected to be stabilized by the oxyanion hole of NTE (formed by Gly986, Gly987, Ala988 and Arg989), would collapse to yield an acylated enzyme intermediate, with Asp1134 donating a proton to the leaving group, thereby serving as a general acid. Subsequent engagement and activation of a water molecule would allow the formation of a second tetrahedral intermediate transition state, resulting in the release of the hydrolyzed substrate and regeneration of the enzyme (Rydel et al., 2003; Wijeyesakere et al., 2007).

In the case of interactions between NTE and neuropathic OP compounds, a similar mechanism is proposed to be responsible for the formation of the organophosphylated enzyme (Wijeyesakere et al., 2007). However, an additional path (aging) is available for OP compounds containing linkages such as P–O–R or P–NH–R that involves loss of a side chain via an S_N2-mechanism (or possibly loss of a proton from a singly substituted nitrogen atom in a phosphoramidate such as mipafox). This reaction results in a permanently inhibited enzyme and, as Johnson postulated as early as 1974 (Johnson, 1974), is thought to be the initiating factor in the pathogenesis of OPIDN (Ehrich and Jortner, 2001; Kropp et al., 2004; Richardson et al., 2013; Wijeyesakere and Richardson, 2010).

The catalytic serine residue also resides in the center of a serine lipase motif, GXSXG, as found, for example in enzymes of the PLA2 superfamily of phospholipases (Ma and Turk, 2001). The NTE sequence contains an additional GXSXG motif in residues 1256–1260; this region appears to be disordered, and it is likely that this second serine lipase motif is a false positive (Via et al., 2007); nevertheless, it would be of interest to determine if the serine residue in this locus were subject to organophosphylation.

8.2.6 Protein kinase A binding (PKAB) domain

Located between the TM domain and the first CNB domain, the PKAB domain in the *Drosophila* NTE orthologue, SWS, has been shown to bind PKA-C3, thereby localizing the kinase to the ER and inhibiting its activity. The PKAB domain also appears to regulate the esterase/phospholipase activity of SWS and to protect against the neurodegenerative effect of PKA-C3 hyperactivity (Bettencourt da Cruz et al., 2008).

However, in a *Drosophila* OPIDN model, TOCP produced 80% inhibition of SWS catalytic activity, resulting in neurodegeneration and motor deficits. Increasing SWS levels by overexpression was not protective, whereas decreasing SWS levels was protective against the neuropathic effects of TOCP, suggesting a gain of toxic function in OP-inhibited SWS. TOCP treatment resulted in a significant decrease in PKA activity. Using yeast two-hybrid assays, mouse NTE was shown to bind to *Drosophila* PKA-C3, and flies overexpressing PKA-C3 were protected against the behavioral deficits produced by TOCP (Wentzell et al., 2014).

Based on the *Drosophila* studies and the mouse NTE binding experiments, similar interactions between human NTE and PKA are expected, but preliminary investigations have not yet confirmed this interaction (J.K.F., unpublished observations). We have designated the provisional site of this domain, as "PKAB?" in Fig. 13; but currently the boundaries of this hypothetical region are unknown.

8.2.7 Posttranslational modifications (PTMs)

In addition to the predicted *N*-linked glycosylation site at Asn20 within the TM domain indicated above, there are phosphorylation sites at Ser354, Thr361, Ser362, Ser372, Ser420, and Thr464 (Isoform-4 numbering) (UniProt, 2019a) (Fig. 13). Such sites presumably would have regulatory functions (Hirabayashi et al., 2004; Walaas and Greengard, 1991), but these have not yet been characterized. However, it is of interest that exposure of HEK293 cells to chlorpyrifos oxon resulted in organophospylation of Tyr residues in 73 proteins, and that many of these residues were phosphorylation sites for protein kinases (Onder et al., 2018). Thus, it is possible that NTE could be organophospylated at sites other than the catalytic Ser residue, and that such chemical modifications of the protein could disrupt normal phosphorylation/dephosphorylation mediated by protein kinases and phosphoprotein phosphatases.

8.3 Subcellular localization of NTE

Given the presence of an N-terminal TM domain, NTE is, unsurprisingly, characterized as a membrane-bound phospholipase. Indeed, its association with the ER membrane had been evident given the early subcellular fractionation studies of hen brain homogenates showing the enzyme to be concentrated within the microsomal fraction (Richardson et al., 1979). These results were further confirmed by subsequent findings demonstrating the need for NTE to be associated with the membrane, which is facilitated by the N-terminal TM domain (Li et al., 2003). Furthermore, limited proteolysis of NTE-containing microsomes revealed the protein to be exposed to the cytoplasm (Li et al., 2003). These findings were further supported by subsequent colocalization studies in primary cultures of mouse hippocampal neurons (Akassoglou et al., 2004), bioinformatics analyses of the primary sequence of the N-terminal region of NTE (Wijeyesakere et al., 2007), and fluorescence microscopy studies of GFP-tagged NTE expressed in HEK 293T cells (Chang et al., 2009). In contrast to the ER localization of NTE in brain, a significant fraction of the NTE activity in peripheral nerve is found in the soluble faction (Vilanova et al., 1993, 1999). Currently, the role of this soluble form of apparent NTE activity is unclear, and it is not known whether it arises as a distinct isoform or if it is released from the ER and transported down the axon as a soluble protein.

8.4 Expression of NTE in human tissues

Early work indicating the presence of NTE in various species and tissues based on enzyme activity has more recently been augmented using next-generation sequencing of RNA transcripts in tissues (RNA-Seq). These results have confirmed that NTE/PNPLA6 is expressed in a wide variety of human tissues. In addition to nervous system and lymphatic tissues, expression of NTE is particular high in kidney, lung, and testis (NCBI, 2019).

8.5 NTE knockouts, silencing, and mutations

8.5.1 Conventional knockouts

Conventional knockout of NTE in mice proved to be lethal after embryonic day 8, but disruption of the *Nte* (*PNPLA6*) gene in a single allele permitted survival with NTE activity levels reduced to about 60% of normal and no apparent structural abnormalities in the nervous system (Winrow et al., 2003). Lethality due to complete inactivation of the *Nte* gene in mice was subsequently attributed to placental failure and impaired vasculogenesis, and again, heterozygous $Nte^{+/-}$ mice displayed normal organ development, despite possessing NTE activities of only 50% of normal levels (Moser et al., 2004). In connection with the role of NTE in placental function, pre-eclampsia in humans has been associated with decreased placental NTE expression (Zhong et al., 2018).

8.5.2 Silencing

Downregulation of NTE gene expression by RNA interference in D3 mouse embryonic stem cells was found to alter multiple pathways involved in development of respiratory, neural, and vascular systems, but such effects did not occur when NTE activity was inhibited by the neuropathic OP compound, mipafox (Sogorb et al., 2016).

A zebrafish model of PNPLA6 insufficiency has been created using a morpholino oligonucleotide knockdown technique in embryos that resulted in developmental abnormalities and motor neuron defects. The study also showed that signaling via the bone morphogenetic protein (BMP) was involved, as phosphorylation of Smad1/5/8 was markedly elevated in the knockdown fish. The phenotype could be rescued by human wild-type *PNPLA6* mRNA but not by mRNA containing active-site mutations that inactivated the catalytic activity of PNPLA6 (Song et al., 2013). The BMP pathway has also been implicated in the development of hereditary spastic paraplegia (HSP) (Zhao and Hedera, 2013).

8.5.3 Conditional knockouts

Conditional knockout (cKO) of NTE in CNS neurons of mice was carried out using the *nestin-cre/loxP* system, which deleted neuronal NTE as of embryonic day 11 (Akassoglou et al., 2004). Brain NTE activity in cKO mice was decreased by about 90% compared to wild type control mice. Brain morphology in cKO mice appeared normal at 2 weeks of age, but at 6 weeks, these mice developed ER disruption and vacuolation of neurons in the hippocampus and thalamus; cerebellar Purkinje cells were also affected. Functionally, cKO mice exhibited significant decreases in rotarod staying times.

A second cKO study in mice was undertaken in order to investigate the role of NTE deficiency in spinal cord axonopathy, neuronal PtdCho homeostasis, and neuronal secretion (Read et al., 2009). The same *nestin-cre/loxP* protocol used by Akassoglou et al. (2004) was followed; therefore, NTE deficiency was induced at embryonic day 11. Axonal degeneration was found in cKO mice aged 18–19 days in the gracile nucleus of the medulla—the site of termination of long ascending tracts in the dorsal columns of the spinal cord. Spinal tract lesions in ascending and descending tracts consisting of swollen axons progressed in severity and number in cKO mice over a period of many months. Hindlimb dysfunction also progressed over the same period. Brain PtdCho levels were elevated by nearly 20% at 1–3 months and remained at this level when assessed at 6–12 months. The rate of secretion of two proteins, reelin and the secreted fragment of the amyloid precursor protein (sAPPa) was impaired by about 20% in cultured cerebellar granule cells from cKO mice, indicating moderate compromise of the constitutive secretory pathway (Read et al., 2009).

In order to compare the cKO results with OPIDN, normal C57BL/6J mice were dosed orally for 5 days with 500 mg/kg of the neuropathic OP compound, MOCP in arachis oil (Read et al., 2009). As noted previously, mice do not develop clinical signs of OPIDN. However, they possess neural NTE that is inhibited by acute dosing with neuropathic OP compounds (Makhaeva et al., 2014, 2016a,b), and they develop spinal cord lesions following chronic dosing with TOCP (Lapadula et al., 1985). Axonal degeneration was found in the gracile nucleus at 1 week after starting MOCP dosing. The type and number of lesions was similar to what was seen in one-month-old cKO mice. However, the lesions in the MOCP-dosed mice did not increase in number or severity over time, and clinical signs of hindlimb dysfunction did not appear. Brain PtdCho levels were modestly elevated by nearly 20% 1 week after the start of MOCP dosing and then returned to baseline during weeks 2–7 rather than remaining elevated as was the case with the cKO of NTE. Taken together, these results were interpreted to indicate that inactivation of NTE by either cKO or sustained inhibition by MOCP resulted in axonal degeneration that was likely due to a deficit in adult axonal maintenance rather than an effect on neurological development (Read et al., 2009).

8.5.4 NTE disease-causing mutations in humans

The first human disease-causing mutations in NTE were reported in 2008 (Rainier et al., 2008). These mutations occurred in the catalytic domain, producing an entity named NTE motor neuron disease (NTE-MND) characterized by progressive spastic paraplegia from childhood leading to muscle wasting in the lower legs and intrinsic muscles of the hands (Rainier et al., 2011). Because this condition is a subset of the hereditary spastic paraplegias (HSPs), the condition is also referred to as SPG39, SPG39/NTE-MND, or SPG39/NTE[PNPLA6]-MND. Constructs of the NTE catalytic domain containing these disease mutations were expressed in human fibroblasts and found to have altered enzymatic properties compared to wild-type controls (Hein et al., 2010a). Constructs from asymptomatic subjects who were heterozygous for a mutation predicted to lack 235 residues from the catalytic domain had only about 40% of control NTE activity (Hein et al., 2010b). This finding was consistent with results from other studies of $Nte^{+/-}$ mice that had 50–60% of control NTE activity with apparently normal nervous system development (Akassoglou et al., 2004; Moser et al., 2004).

Since the first reports of disease-causing mutations in human NTE/PNPLA6, numerous additional cases have been published, indicating that these mutations produce a broad spectrum of neurodegenerative conditions, including Boucher-Neuhauser and Gordon Holmes syndromes (Synofzik et al., 2014; Topaloglu et al., 2014) as well as Oliver-McFarlane and Laurence-Moon syndromes (Hufnagel et al., 2015). These conditions present clinically with a wide range of deficits, including hypogonadism, ataxia, and retinal photoreceptor degeneration with concomitant childhood blindness (Kmoch et al., 2015). Most of the mutations reported thus far have been located in the catalytic domain, but some have occurred in the CNB region of NTE.

When wild-type human NTE was expressed in *sws* null *Drosophila* mutants, both the movement and neurodegenerative phenotypes were suppressed; in addition, elevations of phospholipids were ameliorated to some extent. However, when human NTE containing disease-causing mutations was expressed in *sws* null *Drosophila*, movement and neurodegenerative phenotypes were suppressed to some degree, but none of the human mutants had any effect on lipid levels in the flies. The human NTE mutants included some with mutations in the catalytic domain and others with mutations in the region containing CNB domains. The results were interpreted to indicate that CNB regions could modulate NTE catalytic activity and that the mutant forms of human NTE retained some degree of their biological function, suggesting that disruption of the catalytic function of SWS was not the only factor in disease pathogenesis (Sunderhaus et al., 2019b).

9. Reconciling apparent conflicts between toxicological and genetic data

It has been over 50 years since the discovery of NTE and nearly that long since publication of the theory that both inhibition and aging of >70% of NTE in the nervous system are required in order to produce OPIDN (Johnson, 1969b, 1974). SAR/QSAR data accumulated over this period have largely supported the idea that OPIDN results from a combination of a loss of physiological function (NTE inhibition) coupled with a gain of toxic function (formation of aged NTE with its negatively charged organophosphyl group covalently attached to the active-site serine residue) (Makhaeva et al., 2014; Richardson et al., 2013).

On the other hand, relatively recent molecular biology studies tend to suggest that neurodegenerative and other deleterious effects are produced by a loss of physiological function of NTE alone rather than a combined loss of function and gain of toxic function. These studies included those employing conventional (Winrow et al., 2003) and conditional (Akassoglou et al., 2004; Read et al., 2009) knockouts of the *PNPLA6* gene, knockdown of PNPLA6 in zebrafish, and neurogenetics findings in human subjects with disease-causing NTE mutations (Topaloglu et al., 2014). In particular, these investigations indicate that sustained absence of NTE activity and/or other functions of the protein is/are sufficient to produce neurodegenerative effects.

However, the toxicological data and at least some of the genetics results are not necessarily mutually exclusive. In the case of disease-causing NTE mutations, there are varying degrees of inactivation of catalytic activity, but there is also the presence of mutant forms of NTE that might parallel the situation with inhibited and aged NTE in OPIDN.

Given that the catalytic function of NTE is thought to be concerted deacylation of PtdCho to yield GroPCho +2 FFA, systemic inhibition of NTE activity by neuropathic OP compounds would be expected to result in increases in brain PtdCho levels. However, only a modest and transient increase was produced by repeated dosing of MOCP in mice (Read et al., 2009). Similarly, there was no demonstrable change in brain PtdCho concentrations following a single large dose of TOCP or PMSF in mice (Hou et al., 2009) or hens (Hou et al., 2008), although it is possible that the inhibition levels and/or the timing of phospholipid assays prevented detection of a transient and relatively small increase in PtdCho levels. Indeed, a follow-up study using enhanced lipidomics techniques found significant elevations of PtdCho and LysoPtdCho levels in hen spinal cord microsomal fractions 2 days after a single oral dose of 750 mg/kg of TOCP, and these changes, except for elevated PtdCho, were prevented by pretreatment with 60 mg/kg PMSF 24 h before dosing with TOCP. Interestingly, the largest increase in the lipid profile was seen with sphingomyelin, and there was a decrease in phosphatidylinositol (PtdIns) (Zhu et al., 2016). Substantial and persistent disruptions of lipid homeostasis were found in sciatic nerve of hens given a single dose of 750 mg/kg TOCP—PtdCho levels were significantly elevated as much as ~44% on days 2, 7, and 14 and returned to control levels by day 21 (Xu et al., 2018). Whereas the axonal lesions in OPIDN have been described as Wallerian degeneration (Bouldin and Cavanagh, 1979a), early studies of lipid changes during Wallerian degeneration showed that synthesis of PtdCho was elevated in rat sciatic nerve starting 3 days after a crush injury (Natarajan et al., 1982). Thus, it is possible that increases in PtdCho concentrations during OPIDN or Wallerian degeneration reflect decreased NTE activity, increased PtdCho synthesis, or both.

It is much more difficult to reconcile the cKO NTE results with the toxicological findings. Although there have been a number of caveats raised about conditional knockouts in general and the *nestin-cre/loxP* system in particular (Harno et al., 2013; Liang et al., 2012; Yang et al., 2009), it seems unlikely that the cKO NTE mice could have had a partially disrupted *PNPLA6* gene giving rise to truncated NTE proteins that escaped detection from the polyclonal antibody directed against the catalytic domain region of the enzyme (Akassoglou et al., 2004; Read et al., 2009). Therefore, assuming the cKO experiments resulted in a complete disruption of the *PNPLA6* gene, there would have been no expression of truncated or otherwise abnormal forms of NTE protein to serve as the counterpart to inhibited/aged NTE in OPIDN. However, given that the *nestin-cre/loxP* system initiates the cKO of NTE starting at embryonic day 11, the possibility remains that the resulting neurological abnormalities reflect neurodevelopmental sequelae from the absence of the NTE protein and its enzymatic activity. Although the neuropathological changes in the spinal cords of cKO mice had a similar appearance to those in the MOCP-dosed mice

(Read et al., 2009), axonal degeneration arising from a wide range of causes can have a similar appearance and distribution of lesions (Coleman, 2005).

Likewise, the knockdown experiments in zebrafish indicate that suppression of a critical amount of PNPLA6 is sufficient to cause motor neuron defects and other abnormalities, but here again, the knockdown was done during embryonic development (Song et al., 2013), so that the observed effects were the result of neurodevelopmental phenomena rather than an effect on adult axonal maintenance.

Moreover, when adult zebrafish brain NTE was inhibited over 70% following dosing with CBDP or DFP, no behavioral deficits or spinal cord histopathological lesions were detected. In addition, a thorough lipidomic analysis detected only a small increase in brain lysophospholipids 48 hours after DFP but not following CBDP, and no changes in brain PtdCho resulted from treatment with either inhibitor. The authors concluded that the zebrafish was not a good model of human OPIDN (Faria et al., 2018).

Johnson had noted that all neuropathic OP compounds had one thing in common—the ability to age after inhibiting NTE. Accordingly, he reasoned that the aged protein served as a signal to initiate axonal degeneration. Loss of NTE catalytic activity seemed irrelevant for the induction of lesions, because nonaging compounds inhibited NTE without producing OPIDN. Furthermore, nonaging inhibitors of NTE were not inert, because pretreating animals with nonaging inhibitors conferred protection against OPIDN from subsequent dosing with aging inhibitors, as long as the inhibition level from the pretreatment was ~30% or more (Carrington, 1989; Johnson, 1970, 1974; Johnson and Read, 1993; Moretto et al., 2001).

However, it is also true that aging compounds necessarily result in longer-term inhibition than at least some nonaging inhibitors (e.g., carbamates) because aging inhibitors yield a highly stable organophosphylated NTE that cannot undergo spontaneous reactivation. Thus, recovery of NTE activity following inhibition by ageable inhibitors depends solely upon resynthesis of the enzyme rather than a combination of resynthesis and reactivation. Nevertheless, prolonged inhibition of NTE has also been produced by various dosing regimens with nonaging inhibitors, such as repeated dosing with carbamates, single or repeated dosing with phosphinates, and single or repeated dosing with sulfonyl fluorides, such as PMSF. NTE inhibited by phosphinates can be reactivated in vitro by KF or oximes, but NTE inhibited by PMSF apparently cannot be reactivated either spontaneously or by powerful nucleophiles. Yet, despite prolonged inhibition by these nonaging inhibitors of NTE, no axonal degeneration is produced (Meredith and Johnson, 1988; Richardson et al., 2020).

A possible resolution of the apparent paradox concerning nonaging NTE inhibitors is to propose that they are nonspecific and could theoretically inhibit other serine hydrolases whose inactivation would serve to negate the loss of NTE activity (Wijeyesakere and Richardson, 2010). After all, the human genome encodes >200 serine hydrolases, and at least half of these have not been functionally characterized (Lenfant et al., 2016; Simon and Cravatt, 2010). However, it would seem that the simpler explanation for the nonpathogenicity and protection afforded by nonaging NTE inhibitors is that these compounds lack the distinctive common feature of aging inhibitors—the ability to place a negatively charged phosphyl group in the active site of NTE (Johnson, 1990). This is especially so, considering that some of the nonaging phosphinate inhibitors exhibit high potency and selectivity for NTE. Thus, from a neurotoxicological perspective, we are left with two theoretical possibilities: (1) OPIDN results from the presence of aged NTE; or (2) OPIDN results from a combination of NTE inhibition plus the presence of the aged protein. As a practical matter, in order to produce aged NTE, it is necessary first to inhibit the enzyme with an ageable inhibitor; therefore, the toxicological data ultimately lead to the conclusion that OPIDN requires both inhibition and aging of NTE.

Referring back to the neurogenetics findings summarized above, it would appear that disease-causing NTE mutations could produce neurodegeneration via loss of NTE activity and/or through the presence of abnormal forms of NTE protein. On the other hand, knockdown or cKO of NTE most likely produces neurodegeneration from loss of NTE protein and concomitant absence of its enzymatic activity and/or deficiency of unknown functions of its other domains. In any event, it is important to bear in mind that thus far, the neurogenetics findings do not distinguish between developmental effects versus effects on maintenance of a healthy nervous system in the adult.

10. Suggested future research on elucidating the role of NTE in OPIDN

In consideration of the toxicological and neurogenetics findings reviewed above, we propose two hypothetical models to help guide future research on elucidating the role of NTE in OPIDN: (1) phospholipid homeostasis disruption (PHD); and (2) OP-induced Wallerian degeneration (OPIWAD). These are not necessarily mutually exclusive models.

Indeed, they could very well be viewed as sequential, whereby disruption of axonal transport in the PHD model could serve as a trigger for axonal degeneration via the OPIWAD model.

10.1 Phospholipid homeostasis disruption (PHD)

Fig. 14 is a depiction of a hypothetical phospholipid homeostasis disruption (PHD) model, showing some of the partial pathways of phospholipid metabolism in a neuron. This model represents a synthesis of the "lipid hypothesis" (Wijeyesakere and Richardson, 2010) together with an extension to mammalian systems of a model based on studies in yeast (Glynn, 2005, 2013). In this model, the final stage of PtdCho synthesis takes place in the Golgi, and the product is retrogradely transported to the ER by a PtdCho/PtdIns transport protein (PITP), a mammalian homologue of the yeast transport protein, sec14p (Wyckoff et al., 2010). Despite being named for their originally discovered function of shuttling PtdIns between membrane compartments, PITPs can also accommodate PtdCho as cargo (Carvou et al., 2010; Grabon et al., 2015).

In Fig. 14, *Pathway (1)*, we hypothesize that some fraction of the PtdCho transported to the ER in complex with a PITP is degraded by NTE, liberating PITP in the process (Glynn, 2005, 2013). If this function of NTE were deficient, then an increased fraction of the available PITP pool would remain associated with PtdCho rather than becoming available for transporting PtdIns to the Golgi. The resulting deficiency of PtdIns and concomitantly phosphorylated PtdIns in the Golgi would compromise the constitutive secretory pathway and axonal transport, thus precipitating axonal degeneration (Bartlett et al., 2002; Lorenzo et al., 2014). In this connection, it is noteworthy that disruptions in axonal transport have been shown to precede clinical signs of OPIDN in hens treated with neuropathic OP compounds (Gupta et al., 1997; Moretto et al., 1987; Song et al., 2012). In addition, NTE has been shown to undergo fast axonal transport in hen sciatic nerve (Carrington and Abou-Donia, 1985).

In contrast, *Pathway (2)* in Fig. 14 involves direct deacylation of free PtdCho by NTE. Compromising this function of NTE would not affect PITP levels, so that PtdIns could function normally to support the constitutive secretory pathway and axonal transport. Thus, a neuropathic OP compound would be expected preferentially to inhibit *Pathway (1)*, whereas a nonneuropathic protective NTE inhibitor would be expected preferentially to inhibit *Pathway (2)*.

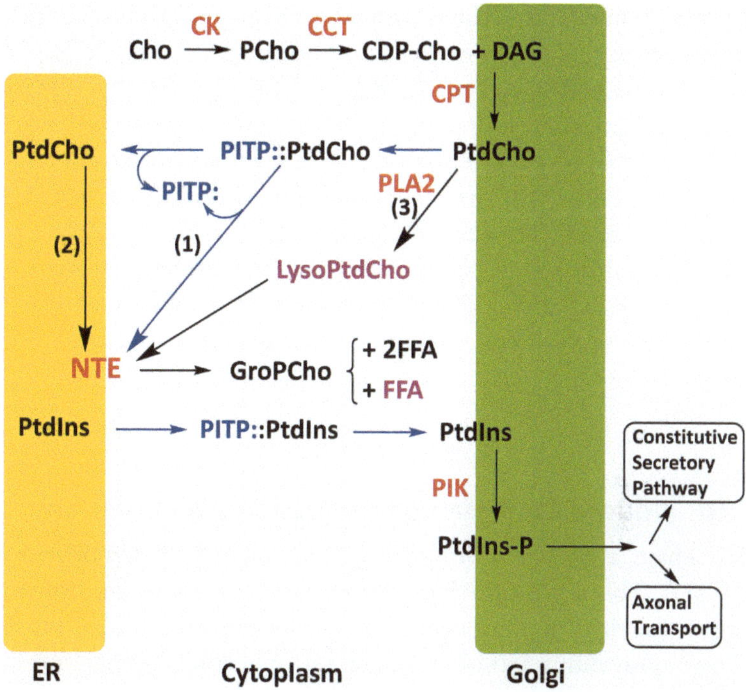

Fig. 14 Phospholipid homeostasis disruption (PHD) model of the role of NTE in OPIDN. Partial pathways of phospholipid metabolism in a neuron are shown. Choline (Cho) taken up by the cell is phosphorylated by choline kinase (CK) to phosphocholine (PCho), which is then converted to cytidine diphosphate-choline (CDP-Cho) via cytidine triphosphate phosphocholine cytidyltransferase (CCT) catalysis. The final step carried out in the Golgi involves addition of a diacylglycerol (DAG) to CDP-Cho catalyzed by choline phosphotransferase (CPT) to form phosphatidylcholine (PtdCho). The PtdCho product forms a complex with a phosphatidylcholine/phosphatidylinositol transport protein (PITP), which conveys PtdCho to the endoplasmic reticulum (ER), releasing uncomplexed PITP back into the cytoplasm. In *Pathway (1)*, some fraction of the PtdCho being transported to the ER is deacylated by NTE to glycerophosphocholine (GroPCho) and two free fatty acids (FFA), releasing free PITP in the process. *Pathway (2)* involves deacylation of PtdCho in the ER to GroPCho +2 FFA. *Pathway (3)* depicts hydrolysis of PtdCho by a phospholipase A2 (PLA2) to lysophosphatidylcholine (LysoPtdCho), which is then further degraded by NTE to GroPCho+FFA. Phosphatidylinositol (PtdIns) in the ER is transported via a PITP to the Golgi, where it is phosphorylated by phosphatidylinositol kinases (PIK) to various phosphorylated PtdIns species (PtdIns-P), which help regulate the constitutive secretory pathway and axonal transport. *Colors*: blue, PITPs and their pathways; green, Golgi; magenta, LysoPtdCho and the FFA formed from its hydrolysis; red, enzymes; yellow, ER. See text for explanation of how NTE inactivation in this model could precipitate OPIDN (Carvou et al., 2010; Glynn, 2005, 2013; Marcucci et al., 2010; Ong et al., 2015).

Pathway (3) in Fig. 14 involves one or more non-NTE phospholipase A2 (PLA2) enzymes hydrolyzing PtdCho directly to LysoPtdCho, which would then be expected to be rapidly degraded by NTE to yield GroPCho and an FFA. However, if NTE were preferentially inhibited, LysoPtdCho could increase to injurious levels, leading to axonal degeneration (De et al., 2003). On the other hand, preferential inhibition of PLA2 activity would block LysoPtdCho formation, thus conferring protection against NTE inhibition (Wijeyesakere and Richardson, 2010). Indeed, inhibitors of various brain PLA2 enzymes are considered an important neuropharmacological class of agents for the treatment of a multiplicity of neurological disorders (Dennis et al., 2011; Farooqui et al., 2006; Ong et al., 2015).

10.2 OP-induced Wallerian degeneration (OPIWAD)

Fig. 15 depicts a simplified model of OP-induced Wallerian degeneration (OPIWAD). This model provides a conceptual framework for understanding potential actions of neuropathic and nonneuropathic NTE inhibitors in initiating or preventing axonal degeneration. Wallerian degeneration and its relationship to OPIDN are discussed below.

In his insightful histopathological observations of OPIDN histopathology produced by TOCP in hens, Cavanagh (1954) described the spatial-temporal evolution of lesions in the spinal cord and peripheral nerves as a process of primary axonal degeneration followed by demyelination that closely resembled Wallerian degeneration induced by cutting or crushing spinal cord tracts or peripheral nerves. He also noted that the lesions were similar to those seen in thiamine deficiency, which results in a slowing of energy metabolism that leads to degeneration of distal axons—the nerve cell components farthest from the cell body. In a follow-up study of OPIDN produced by DFP in cats, Bouldin and Cavanagh (1979a) described the lesions as a "chemical transection" of axons that precipitated Wallerian degeneration.

Originally described by Waller in 1850, Wallerian degeneration was considered for the next 139 years to be a passive degenerative process—a consequence of separating the distal axon from vital substances supplied to it by the cell body (Waller, 1850). Then, in 1989, investigators in Oxford University made the astounding discovery of a strain of mouse whose distal axons remained intact for relatively long periods following transection of a peripheral nerve (Lunn et al., 1989). Ultimately, the cause of this

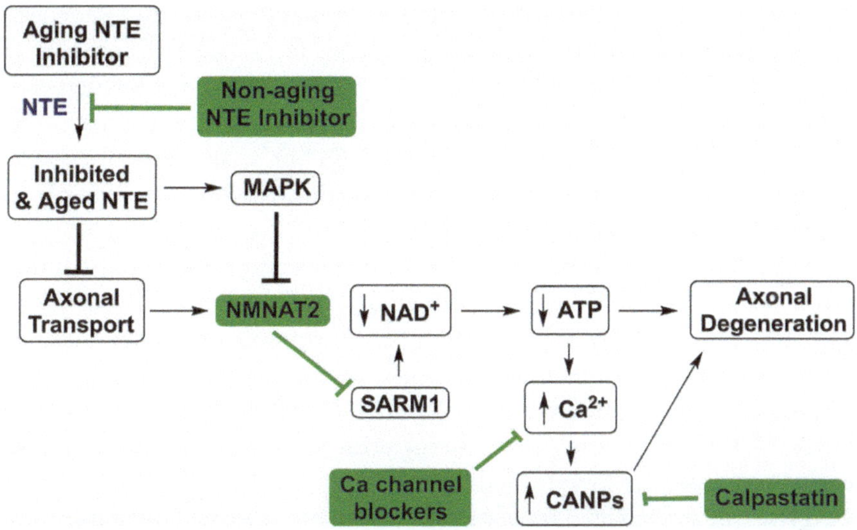

Fig. 15 Simplified model of OP-induced Wallerian degeneration (OPIWAD). Exposure to a neuropathic (aging) OP NTE inhibitor produces inhibited and aged NTE, which triggers Wallerian degeneration via inhibition of axonal transport and/or stimulation of mitogen-activated protein kinase (MAPK) and/or associated kinase cascade pathways. Nicotinamide mononucleotide adenylyltransferase 2 (NMNAT2) is a prosurvival factor continuously supplied to the axon via axonal transport; it blocks sterile alpha and toll-interleukin receptor (TIR) motif containing 1 (SARM1) from depleting nicotinamide adenine dinucleotide (NAD^+) levels. Activation of kinase cascades lowers NMNAT2 levels enabling SARM1 to deplete NAD^+, leading to ATP depletion. The resulting energy crisis increases Ca^{2+} concentrations from external and internal (ER and mitochondrial) sources, thereby activating calcium-activated neutral proteases (CANPs; calpains), which degrade axonal proteins resulting in axonal degeneration. Nonaging NTE inhibitors block formation of aged NTE, conferring protection against OPIDN by blocking initiation. In the execution phase of Wallerian degeneration, axonopathy can be ameliorated by Ca^{2+}-channel blockers and/or CANP inhibitors such as calpastatin. Components of the model that block axonopathy are shown in green; NTE is colored blue (Conforti et al., 2014; Rosell and Neukomm, 2019; Walker et al., 2017).

axonal resiliency was found to be a mutation, aptly named Wallerian degeneration slow (Wld^S). The discovery was serendipitous, as the mutation arose spontaneously and permitted detection coincidentally only after becoming homozygous. The revelation of Wld^S fundamentally changed our concept of Wallerian degeneration from a passive process to an active program of axonal self-destruction—a kind of local apoptosis confined to the axon (Gerdts et al., 2016).

Ultimately, it was established that Wld^S protects axons from degeneration in a gain of function manner by coding for a stable fusion protein that substitutes for the labile cytosolic nicotinamide mononucleotide adenyltransferase 2 (NMNAT2). This is an enzyme that catalyzes the formation of nicotinamide adenine dinucleotide (NAD^+), an essential cofactor for numerous enzymatic pathways, including those involved in redox reactions of energy metabolism and the production of the "energy currency" of the cell, adenosine triphosphate (ATP) (Coleman and Freeman, 2010). In a normal axon, NMNAT2 is degraded rapidly, but it is constantly resupplied from the cell body by axonal transport. Disruption of axonal transport by injury, disease, or certain neurotoxic agents interrupts the supply of NMNAT2, which quickly leads to NAD^+ depletion, followed closely by loss of ATP. The resulting local energy crisis permits intra-axonal calcium levels to rise sharply, thereby activating calcium-activated neutral proteases (CANPs; calpains), which mediate destruction of the axon. Given its central role as a prosurvival factor that helps maintain axons in a healthy state (Gilley and Coleman, 2010), NMNAT2 is tightly regulated by multiple pathways and processes including cAMP signaling, protein phosphorylation cascades, palmitoylation, and ubiquitin ligases (Geden and Deshmukh, 2016; Rosell and Neukomm, 2019).

Complementary studies have identified prodegenerative factors and pathways, including sterile alpha and Toll/interleukin-1 receptor motif-containing 1 (SARM1) and the p38-mitogen-activated protein kinase (MAPK) cascade. Loss of function mutations in SARM1 or blockades of the MAPK pathway confer protection against axonal degeneration induced by various triggers, including physical or chemical agents. Conversely, activation of SARM1 and/or the MAPK pathway triggers axonal degeneration, possibly by decreasing the stability of NMNAT2, thus leading to depletion of NAD^+ and ATP (Loring and Thompson, 2020).

Wallerian degeneration appears to be a biphasic process consisting of an initiation phase of early events contributing to NAD^+ depletion followed after a latent period by an execution phase (Chang et al., 2016; Coleman and Freeman, 2010). The precise events that mark the transition from one stage to another are still being investigated, but it appears that the execution interval entails increased intra-axonal Ca^{2+} concentrations and activation of CANPs that contribute to the dissolution of the axon (Conforti et al., 2014; Song and Xie, 2012). However, calcium entry and/or release of calcium from internal storage sites such as the ER and mitochondria could also occur during the initiation phase, even in toxic neuropathies that do not involve physical trauma to axons. For example, at 24 hours after a neuropathic dose of TOCP in hens, CANP activity was significantly elevated in sciatic nerve, and administration of the calcium-channel blocker, nimodipine, within 24 hours after TOCP was effective at normalizing CANP activity measured at 24 hours and 28 days post-TOCP. This treatment was partially effective at ameliorating clinical signs of OPIDN assessed 28 days post-TOCP (Emerick et al., 2010).

Additional pathways and targets that modulate initiation and/or execution of Wallerian degeneration are being continually discovered. Among these are as-yet unidentified serine proteases whose inhibition delays neurite degeneration in primary neuronal cultures (Conforti et al., 2014). Not surprisingly, mitochondria have been implicated in Wallerian degeneration (Barrientos et al., 2011), and studies are ongoing to determine whether mitochondrial perturbations correspond to cause or effect as well as to initiation, execution, or both phases of axonal degeneration (Conforti et al., 2014). However, recent work indicates that inhibition of mitochondrial respiration is a downstream event in axonal degeneration triggered by activation of SARM1 via c-Jun N-terminal kinase (JNK)-mediated phosphorylation (Murata et al., 2018).

Of particular relevance to OPIDN is the fact that many of the regulatory pathways in Wallerian degeneration involve a balance between the actions of protein kinases and phosphoprotein phosphatases to phosphorylate and dephosphorylate proteins (Ali et al., 2017; Araki and Wakatsuki, 2019; Murata et al., 2018). These processes can operate on both sides of the opposing forces of prosurvival and prodegenerative pathways. For example, it is conceivable that organophosphylation and aging of NTE could serve as a signal to initiate Wallerian degeneration, as proposed previously, but not yet rigorously tested (Richardson, 1984). Furthermore, chemical modification of NTE by nonaging adduction might confer protection against axonopathy by maintaining NMNAT2 levels or blocking activation of SARM1 or MAPK pathways (Walker et al., 2017). Apropos of these concepts, studies of organophosphylation of proteins by in vitro or in vivo exposures to OP insecticides or nerve agents have of course found OP adducts on esteratic Ser residues (Black and Read, 2013; Graham et al., 2016). In addition, these investigations have detected OP adducts on nonesteratic Thy, Tyr, and Lys residues of multiple proteins, including α- and β-tubulin and ubiquitin (John et al., 2010, 2018; Onder et al., 2018; Schmidt et al., 2014; Schopfer and Lockridge, 2018). Moreover, many of the OP-adducted residues are phosphorylation sites for protein kinases.

Other studies have shown that OP compounds can disrupt kinase-mediated phosphorylation in *Drosophila*, rat H9c2 cells, and mouse brain, resulting in hypo- or hyperphosphorylation of specific regulatory sites (Batista et al., 2016; Felemban et al., 2015; Zhu et al., 2010). In addition, OP compounds reacting directly with Lys residues in proteins can form isopeptide crosslinks with Glu or Asp residues, and tubulin modified in this manner forms high molecular weight intermolecular aggregates (Schmidt et al., 2014; Schopfer and Lockridge, 2018). Taken together, these findings clearly demonstrate the possibility that exposure to OP compounds could disrupt physiological or pathogenic processes that are regulated by protein phosphorylation and dephosphorylation events normally mediated by protein kinases and phosphoprotein phosphatases. Such events are now known to include the induction or prevention of Wallerian degeneration. As noted previously, NTE has at least six nonesteratic phosphorylation sites (UniProt, 2019a), and these could be candidates for disruption by organophosphylation.

11. Conclusion

Despite the considerable advances that have been made in understanding the physiological and pathogenic roles of NTE since its discovery, there remains a major obstacle for making further progress—the lack of a 3D structure of the protein. The fact that NTE is a large membrane-bound and multi-domain molecule hinders structural determination by X-ray crystallography. Nevertheless, solving its individual domains can be undertaken using X-ray and NMR techniques (Puthenveetil and Vinogradova, 2019), and determining how its domains interact with each other can be approached by combining site-directed fluorescence, chemical crosslinking, isotope exchange, and mass spectrometry (Hodge et al., 2019; Raghuraman et al., 2019; Xiang et al., 2020). In addition, cryogenic electron microscopy has recently emerged as an ideal technique for solving the structures of "difficult" proteins, and its resolution is now approaching that of X-ray methods (Nwanochie and Uversky, 2019; Schur, 2019). In the meantime, our work on the structure and dynamics of NTE is in progress using a coordinated suite of computational molecular modeling approaches, including iterative threading, global exploration with loop perturbation, and coarse-grained molecular dynamics simulations (Liu et al., 2019; Singh and Lee, 2019). Subsequently, in silico mutagenesis and covalent docking will be used to study alterations in structure, flexibility, and stability of NTE induced by mutations and chemical modification with aging vs. nonaging inhibitors (Doering et al., 2018; Scarpino et al., 2018).

Our current computational work on the NTE catalytic domain has shown that a patatin-like protein, PlpD, is a better template than patatin itself for creating a homology model (RJR, unpublished observations). PlpD is a virulence factor with phospholipase A1 activity secreted by *Pseudomonas aeroginosa,* a bacterium commonly associated with opportunistic and hospital-acquired infections (da Mata Madeira et al., 2016; Foulkes et al., 2019). Similarly, replication of coronaviruses and flaviviruses have been shown to involve manipulation of host phospholipid homeostasis via activation of phospholipase A2 enzymes, and inhibition of these activities compromises viral replication (Liebscher et al., 2018; Müller et al., 2018). These findings have suggested an entirely new direction for future work on NTE that should help inform studies of important human pathogens. Thirty years ago, M.K. Johnson published a "Contemporary Issues in Toxicology" article titled, "Organophosphates and Delayed Neuropathy—Is NTE Alive and Well?" This was a report on the status of our understanding of NTE and the mechanism of OPIDN along with an outline of how to use this knowledge for the intelligent and economic assessment of the relative neurotoxicological risks of exposures to OP compounds. He ended the article with words of encouragement that also served as a call to action for further work leading to ever-greater understanding (Johnson, 1990). It is fitting to requote his words here:

"The saying, 'If you do not use your lamp in the dark, whose fault is it that you stumble?' has been attributed to Gilbert of Newminster (ca. 1200–1250), and Winston S. Churchill in 1941 said, 'Give us the tools and we'll finish the job.' We have the 'lamp'. We have the 'tools'."

References

Abou-Donia, M.B., 1981. Organophosphorus ester-induced delayed neurotoxicity. Annu. Rev. Pharmacol. Toxicol. 21, 511–548.

Akassoglou, K., Malester, B., Xu, J., Tessarollo, L., Rosenbluth, J., Chao, M.V., 2004. Brain-specific deletion of neuropathy target esterase/swisscheese results in neurodegeneration. Proc. Natl. Acad. Sci. U. S. A. 101, 5075–5080.

Aksnes, D.W., Langfeldt, L., Wouters, P., 2019. Citations, citation indicators, and research quality: an overview of basic concepts and theories. SAGE Open 2019, 1–17.

Albers, J.W., 2005. The neurological approach. In: Berent, S., Albers, J.W. (Eds.), Neurobehavioral Toxicology: Neurological and Neuropsychological Perspectives. In: Foundations and Methods, vol. I. Taylor & Francis, New York, pp. 135–188.

Aldridge, W.N., 1964. A method for the characterization of two similar B-esterases present in the chicken nervous system. Biochem. J. 93, 619–623.

Aldridge, W.N., Barnes, J.M., 1966a. Further observations on the neurotoxicity of organophosphorus compounds. Biochem. Pharmacol. 15, 541–548.
Aldridge, W.N., Barnes, J.M., 1966b. Esterases and neurotoxicity of some organophosphorus compounds. Biochem. Pharmacol. 15, 549–554.
Aldridge, W.N., Reiner, E., 1972. Enzyme Inhibitors as Substrates: Interactions of Esterases With Esters of Organophosphorus and Carbamic Acids. North-Holland, Amsterdam. 328pp.
Aldridge, W.N., Barnes, J.M., Johnson, M.K., 1969. Studies on delayed neurotoxicity produced by some organophosphorus compounds. Ann. N.Y. Acad. Sci. 160, 314–322.
Ali, Y.O., Bradley, G., Lu, H.-C., 2017. Screening with an NMNAT2-MSD platform identifies small molecules that modulate NMNAT2 levels in cortical neurons. Sci. Rep. 7, 43846.
Araki, T., Wakatsuki, S., 2019. Regulation of neuronal/axonal degeneration by ZNRF1 ubiquitin ligase. Neurosci. Res. 139, 21–25.
Atkins, J., Glynn, P., 2000. Membrane association of and critical residues in the catalytic domain of human neuropathy target esterase. J. Biol. Chem. 275, 24477–24483.
Atkins, J., Luthjens, L.H., Hom, M.L., Glynn, P., 2002. Monomers of the catalytic domain of human neuropathy target esterase are active in the presence of phospholipid. Biochem. J. 361, 119–123.
Baker, T., Lowndes, H.E., Johnson, M.K., Sandborg, I.C., 1980. The effects of phenylmethanesulfonyl fluoride on delayed organophosphorus neuropathy. Arch. Toxicol. 46, 305–311.
Barnes, J.M., 1975. Assessing hazards from prolonged and repeated exposure to low doses of toxic substances. Br. Med. Bull. 31, 196–200.
Barrientos, S.A., Martinez, N.W., Yoo, S., Jara, J.S., Zamorano, S., Hetz, C., Twiss, J.L., Alvarez, J., Court, F.A., 2011. Axonal degeneration is mediated by the mitochondrial permeability transition pore. J. Neurosci. 31, 966–978.
Bartlett, S.E., Reynolds, A.J., Weible II, M., Hendry, I.A., 2002. Phosphatidylinositol kinase enzymes regulate the retrograde axonal transport of NT-3 and NT-4 in sympathetic and sensory neurons. J. Neurosci. Res. 68, 169–175.
Batista, J.E.S., Sousa, L.R., Martins, I.K., Rodrigues, N.R., Posser, T., Franco, J.L., 2016. Data on the phosphorylation of p38MAPK and JNK induced by chlorpyrifos in *Drosophila melanogaster*. Data Brief 8, 32–34.
Berg, J.M., Tymoczko, J.L., Stryer, L., 2012. Biochemistry, seventh ed. W.J. Freeman, New York. 1054 pp.
Bertoncin, D., Russolo, A., Caroldi, S., Lotti, M., 1985. Neuropathy target esterase in human lymphocytes. Arch. Environ. Health 40, 139–144.
Bettencourt da Cruz, A., Wentzell, J., Kretzschmar, D., 2008. Swiss cheese, a protein involved in progressive neurodegeneration, acts as a noncanonical regulatory subunit for PKA-C3. J. Neurosci. 28, 10885–10892.
Bevis, W.M., 1936. Residuals of "Jake Leg Paralysis" Med. Bull. Veterans' Admin. 13, 157–160.
Bidstrup, P.L., Bonnell, J.A., Beckett, A.G., 1953. Paralysis following poisoning by a new organic phosphorus insecticide (mipafox): report on two cases. Br. Med. J. 1, 1068–1072.
Black, R.M., Read, R.W., 2013. Biological markers of exposure to organophosphorus nerve agents. Arch. Toxicol. 87, 421–437.
Bleecker, M.L., Maroni, M., Sepanski, M., 1983. A biologic marker for organophosphate exposure: neurotoxic esterase activity in human lymphocytes and platelets. Dev. Toxicol. Environ. Sci. 11, 507–512.
Bouldin, T.W., Cavanagh, J.B., 1979a. Organophosphorus neuropathy. I. A teased-fiber study of the spatio-temporal spread of axonal degeneration. Am. J. Pathol. 94, 241–252.

Bouldin, T.W., Cavanagh, J.B., 1979b. Organophosphorus neuropathy. II. A fine-structural study of the early stages of axonal degeneration. Am. J. Pathol. 94, 253–270.

Brooks, J., Erickson, T.B., Kayden, S., Ruiz, R., Wilkinson, S., Burkle Jr., F.M., 2018. Responding to chemical weapons violations in Syria: legal, health, and humanitarian recommendations. Confl. Health 12, 12.

Burke, R.E., O'Malley, K., 2013. Axon degeneration in Parkinson's disease. Exp. Neurol. 246, 72–83.

Carbonell, P., Carlsson, L., Faulon, J.-L., 2013. Stereo signature molecular descriptor. J. Chem. Inf. Model. 53, 887–897.

Carletti, E., Schopfer, L.M., Colletier, J.-P., Froment, M.-T., Nachon, F., Weik, M., Lockridge, O., Masson, P., 2011. Reaction of cresyl saligenin phosphate, the organophosphorus agent implicated in aerotoxic syndrome, with human cholinesterases: mechanistic studies employing kinetics, mass spectrometry, and X-ray structure analysis. Chem. Res. Toxicol. 24, 797–808.

Carletti, E., Colletier, J.-P., Schopfer, L.M., Santoni, G., Masson, P., Lockridge, O., Nachon, F., Weik, M., 2013. Inhibition pathways of the potent organophosphate CBDP with cholinesterases revealed by X-ray crystallographic snapshots and mass spectrometry. Chem. Res. Toxicol. 26, 280–289.

Caroldi, S., Lotti, M., Masutti, A., 1984. Intra-arterial injection of diisopropylfluorophosphate or phenylmethanesulphonyl fluoride produces unilateral neuropathy or protection, respectively, in hens. Biochem. Pharmacol. 33, 3213–3217.

Carrington, C.D., 1989. Prophylaxis and the mechanism for the initiation of organophosphorus compound-induced delayed neurotoxicity. Arch. Toxicol. 63, 165–172.

Carrington, C.D., Abou-Donia, M.B., 1983. The time course of protection from delayed neurotoxicity induced by tri-*o*-cresyl phosphate and *O,O*-diisopropyl phosphorofluoridate by phenyl methyl sulfonyl fluoride in chickens. Toxicol. Lett. 18, 251–256.

Carrington, C.D., Abou-Donia, M.B., 1984. The correlation between the recovery rate of neurotoxic esterase activity and sensitivity to organophosphorus-induced delayed neurotoxicity. Toxicol. Appl. Pharmacol. 75, 350–357.

Carrington, C.D., Abou-Donia, M.B., 1985. Axoplasmic transport and turnaround of neurotoxic esterase in hen sciatic nerve. J. Neurochem. 44, 616–621.

Carvou, N., Holic, R., Li, M., Futter, C., Skippen, A., Cockcroft, S., 2010. Phosphatidylinositol- and phosphatidylcholine-transfer activity of PITPβ is essential for COPI-mediated retrograde transport from the Golgi to the endoplasmic reticulum. J. Cell Sci. 123, 1262–1273.

Cavanagh, J.B., 1954. The toxic effects of tri-ortho-cresyl phosphate on the nervous system: an experimental study in hens. J. Neurol. Neurosurg. Psychiatry 17, 163–172.

Cavanagh, J.B., Holland, P., 1961. Localization of cholinesterases in the chicken nervous system and the problem of the selective neurotoxicity of organophosphorus compounds. Brit. J. Pharmacol. 16, 218–230.

CDC, 2019a. Biomonitoring Summary. Organophosphorus Insecticides: Dialkyl Phosphate Metabolites. https://www.cdc.gov/biomonitoring/OP-DPM_BiomonitoringSummary.html. Accessed 10 January 2020.

CDC, 2019b. Interim CDC Recommendations for Zika Vector Control in the Continental United States. http://www.cdc.gov/zika/public-health-partners/vector-control-us.html. Accessed 26 November 2019.

Chang, P.A., Long, D.X., Wu, Y.J., Sun, Q., Song, F.Z., 2009. Identification and characterization of chicken neuropathy target esterase. Gene 435, 45–52.

Chang, B., Quan, Q., Lu, S., Wang, Y., Peng, J., 2016. Molecular mechanisms in the initiation phase of Wallerian degeneration. Eur. J. Neurosci. 44, 2040–2048.

Cherenkov, P.A., 1934. Visible luminescence of pure liquids under the influence of γ-radiation. Dokl. Akad. Nauk SSSR 2, 451–454. [Russian].
Cherniak, M.G., 1988. Toxicological screening for organophosphorus-induced delayed neurotoxicity: complications in toxicity testing. Neurotoxicology 9, 249–271.
Clothier, B., Johnson, M.K., 1979. Rapid aging of neurotoxic esterase after inhibition by di-isopropyl phosphorofluoridate. Biochem. J. 177, 549–558.
Clothier, B., Johnson, M.K., 1980. Reactivation and aging of neurotoxic esterase inhibited by a variety of organophosphorus esters. Biochem. J. 185, 739–747.
Coleman, M., 2005. Axon degeneration mechanisms: commonality amid diversity. Nat. Rev. Neurosci. 6, 889–898.
Coleman, M.P., Freeman, M.R., 2010. Wallerian degeneration, Wld^S, and Nmnat. Annu. Rev. Neurosci. 33, 245–257.
Conforti, L., Gilley, J., Coleman, M.P., 2014. Wallerian degeneration: an emerging axon death pathway linking injury and disease. Nat. Rev. Neurosci. 15, 394–409.
Correll, L., Ehrich, M., 1991. A microassay for neurotoxic esterase determinations. Fundam. Appl. Toxicol. 16, 110–116.
Czajka, M., Matysiak-Kucharek, M., Jodlowska-Jedrych, B., Sawicki, K., Fal, B., Drop, B., Kruszewski, M., Kapka-Skrzypczak, L., 2019. Organophosphorus pesticides can influence the development of obesity and type 2 diabetes with concomitant metabolic changes. Environ. Res. 178, 108685.
da Mata Madeira, P.V., Zouhir, S., Basso, P., Neves, D., Laubier, A., Salacha, R., Bleves, S., Faudry, E., Contreras-Martel, C., Dessen, A., 2016. Structural basis of lipid targeting and destruction by the type V secretion system of *Pseudomonas aeruginosa*. J. Mol. Biol. 428, 1790–1803.
Davis, C.S., Richardson, R.J., 1980. Organophosphorus compounds. In: Spencer, P.S., Schaumburg, H.H. (Eds.), Experimental and Clinical Neurotoxicology. Williams & Wilkins, Baltimore, pp. 527–544.
Davis, C.S., Richardson, R.J., 1987. Neurotoxic esterase: characterization of the solubilized enzyme and the conditions for its solubilization from chicken brain microsomal membranes with ionic, zwitterionic, or nonionic detergents. Biochem. Pharmacol. 36, 1393–1399.
Davis, C.S., Johnson, M.K., Richardson, R.J., 1985. Organophosphorus compounds. In: O'Donoghue, J.L. (Ed.), Neurotoxicology of Industrial and Commercial Chemicals. vol. II. CRC Press, Boca Raton, pp. 1–23.
Davison, A.N., 1953. Some observations on the cholinesterases of the central nervous system after the administration of organo-phosphorus compounds. Brit. J. Pharmacol. 8, 212–216.
De, S., Trigueros, M.A., Kalyvas, A., David, S., 2003. Phospholipase A_2 plays an important role in myelin breakdown and phagocytosis during Wallerian degeneration. Mol. Cell. Neurosci. 24, 753–765.
Dennis, E.A., Cao, J., Hsu, Y.-H., Magrioti, V., Kokotos, G., 2011. Phospholipase A_2 enzymes: physical structure, biological function, disease implication, chemical inhibition, and therapeutic intervention. Chem. Rev. 111, 6130–6185.
Doering, J.A., Lee, S., Kristiansen, K., Evenseth, L., Barron, M.G., Sylte, I., LaLone, C.A., 2018. *In silico* site-directed mutagenesis informs species-specific predictions of chemical susceptibility derived from the sequence alignment to predict across species susceptibility (SeqAPASS) tool. Toxicol. Sci. 166, 131–145.
Drakontides, A.B., Baker, T., 1983. An electrophysiologic and ultrastructural study of the phenylmethanesulfonyl fluoride protection against a delayed organophosphorus neuropathy. Toxicol. Appl. Pharmacol. 70, 411–422.
Dremier, S., Kopperud, R., Doskeland, S.O., Dumont, J.E., Maenhaut, C., 2003. Search for new cyclic AMP-binding proteins. FEBS Lett. 546, 103–107.

Dudek, B.R., Richardson, R.J., 1982. Evidence for the existence of neurotoxic esterase in neural and lymphatic tissue of the adult hen. Biochem. Pharmacol. 31, 1117–1121.

Ehrich, M., Jortner, B., 2001. Organophosphorus-induced delayed neuropathy. In: Krieger, E. (Ed.), Handbook of Pesticide Toxicology, second ed. Academic Press, San Diego, pp. 987–1012.

Emerick, G.L., Peccinini, R.G., De Oliveira, G.H., 2010. Organophosphorus-induced delayed neuropathy: a simple and efficient therapeutic strategy. Toxicol. Lett. 192, 238–244.

Epstein, T.M., Samanta, U., Kirby, S.D., Cerasoli, D.M., Bahnson, B.J., 2009. Crystal structures of brain group-VIII phospholipase A2 in nonaged complexes with the organophosphorus nerve agents soman and sarin. Biochemistry 48, 3425–3435.

Eto, M., 1974. *Organophosphorus Pesticides: Organic and Biological Chemistry*. CRC Press, Cleveland. 387pp.

Eto, M., 1997. Functions of phosphorus moiety in agrochemical molecules. Biosci. Biotech. Biochem. 61, 1–11.

Faria, M., Fuertes, I., Prats, E., Abad, J.L., Padros, F., Gomez-Canela, C., Casas, J., Estevez, J., Vilanova, E., Pina, B., Raldua, D., 2018. Analysis of the neurotoxic effects of neuropathic organophosphorus compounds in adult zebrafish. Sci. Rep. 8, 4844.

Farooqui, A.A., Ong, W.-Y., Horrocks, L.A., 2006. Inhibitors of brain phospholipase A_2 activity: their neuropharmacological effects and therapeutic importance for the treatment of neurological disorders. Pharmacol. Rev. 58, 591–620.

Felemban, S.G., Garner, A.C., Smida, F.A., Boocock, D.J., Hargreaves, A.J., Dickenson, J.M., 2015. Phenyl saligenin phosphate induced caspase-3 and c-jun N-terminal kinase activation in cardiomyocyte-like cells. Chem. Res. Toxicol. 28, 2179–2191.

Figueiredo, T.H., Apland, J.P., Braga, M.F.M., Marini, A.M., 2018. Acute and long-term consequences of exposure to organophosphate nerve agents in humans. Epilepsia 59 (S2), 92–99.

Fischer, L.R., Culver, D.G., Tennant, P., Davis, A.A., Wang, M., Castellano-Sanchez, A., Khan, J., Polak, M.A., Glass, J.D., 2004. Amyotrophic lateral sclerosis is a distal axonopathy: evidence in mice and man. Exp. Neurol. 185, 232–240.

Foulkes, D.M., McLean, K., Haneef, A.S., Fernig, D.G., Winstanley, C., Berry, N., Kaye, S.B., 2019. Microorganisms 7, 707.

Franjesivic, A.J., Sillart, S.B., Beck, J.M., Vyas, S., Callam, C.S., Hadad, C.M., 2019. Resurrection and reactivation of acetylcholinesterase and butyrylcholinesterase. Chem. A Eur. J. 25, 5337–5371.

Gallazzini, M., Ferraris, J.D., Kunin, M., Morris, R.G., Burg, M.B., 2006. Neuropathy target esterase catalyzes osmoprotective renal synthesis of glycerophosphocholine in response to high NaCl. Proc. Natl. Acad. Sci. U. S. A. 103, 15260–15265.

Gancedo, J.M., 2013. Biological roles of cAMP: variations on a theme in the different kingdoms of life. Biol. Rev. Camb. Philos. Soc. 88, 645–668.

Gavrilescu, N., Peters, R.A., 1931. CLII. Biochemical lesions in vitamin B deficiency. Biochem. J. 25, 1397–1409.

Geden, M.J., Deshmukh, M., 2016. Axon degeneration: context defines distinct pathways. Curr. Opin. Neurobiol. 39, 108–115.

Gerdts, J., Summers, D.W., Milbrandt, J., Diantonio, A., 2016. Axon self-destruction: new links among SARM1, MAPKs, and NAD^+ metabolism. Neuron 89, 449–460.

Gilley, J., Coleman, M.P., 2010. Endogenous Nmnat2 is an essential survival factor for maintenance of healthy axons. PLoS Biol. 8 e1000300.

Glynn, P., 2000. Neural development and neurodegeneration: two faces of neuropathy target esterase. Prog. Neurobiol. 61, 61–74.

Glynn, P., 2003. NTE: one target protein for different toxic syndromes with distinct mechanisms? Bioessays 25, 742–745.
Glynn, P., 2005. Neuropathy target esterase and phospholipid deacylation. Biochim. Biophys. Acta 1736, 87–93.
Glynn, P., 2006. A mechanism for organophosphate-induced delayed neuropathy. Toxicol. Lett. 162, 94–97.
Glynn, P., 2007. Axonal degeneration and neuropathy target esterase. Arh. Hig. Rada Toksikol. 58, 355–358.
Glynn, P., 2013. Neuronal phospholipid deacylation is essential for axonal and synaptic integrity. Biochim. Biophys. Acta 1831, 633–641.
Glynn, P., Read, D.J., Guo, R., Wylie, S., Johnson, M.K., 1994. Synthesis and characterization of a biotinylated organophosphorus ester for detection and affinity purification of a brain serine esterase: neuropathy target esterase. Biochem. J. 301, 551–556.
Glynn, P., Read, D.J., Lush, M.J., Atkins, J., 1999. Molecular cloning of neuropathy target esterase (NTE). Chem. Biol. Interact. 119–120, 513–517.
Goethe J.W. von, *Zur Farbenlehre*. J.G., 1810. Cotta'schen Buchhandlung, Tübingen, preface, xxv.
Gottlieb, S., Marsh, P.B., 1946. Quantitative determination of phenolic fungicides. Ind. Eng. Chem. Anal. Ed. 18, 16–19.
Grabon, A., Lhan, D., Bankaitis, V.A., 2015. Phosphatidylinositol transfer proteins and instructive regulation of lipid kinase biology. Biochim. Biophys. Acta 1851, 724–735.
Graham, L.A., Johnson, D., Carter, M.D., Stout, E.G., Erol, H.A., Isenberg, S.L., Mathews, T.P., Thomas, J.D., Johnson, R.C., 2016. A high-throughput UHPLC-MS/MS method for the quantification of five aged butyrylcholinesterase biomarkers from human exposure to organophosphorus nerve agents. Biomed. Chromatogr. 31 e3830.
GraphPad, 2016a. Confidence Intervals of the EC50. GraphPad Prism Curve Fitting Guide. http://www.graphpad.com/guides/prism/7/curve-fitting/index.htm?reg_why_prism_fits_the_logec50_rat.htm. Accessed 15 November 2018.
GraphPad, 2016b. Advice: Avoid Scatchard, Lineweaver-Burk and Similar Transforms. GraphPad Prism Curve Fitting Guide. https://www.graphpad.com/guides/prism/7/curve-fitting/avoidscatchard_lineweaver_burkeandsimilartransforms.htm?toc=0&printWindow. Accessed 15 November 2018.
Greiner, A.J., Richardson, R.J., Worden, R.M., Ofoli, R.Y., 2010. Influence of lysophospholipid hydrolysis by the catalytic domain of neuropathy target esterase on the fluidity of bilayer lipid membranes. Biochim. Biophys. Acta 1798, 1533–1539.
Gupta, R.P., Abdel-Rahman, A., Wilmarth, K.W., Abou-Donia, M.B., 1997. Alteration in neurofilament axonal transport in the sciatic nerve of the diisopropyl phosphorofluoridate (DFP)-treated hen. Biochem. Pharmacol. 53, 1799–1806.
Hageman, G., Pal, T.M., Nihom, J., Mackenzie Ross, S.J., van den Berg, M., 2019. Aerotoxic syndrome, discussion of possible diagnostic criteria. Clin. Toxicol. 1–3. https://doi.org/10.1080/15563650.2019.1649419. 2019 Aug 7.
Harno, E., Cottrell, E.C., White, A., 2013. Metabolic pitfalls of CNS Cre-based technology. Cell Metab. 18, 21–28.
Hein, N.D., Stuckey, J.A., Rainier, S.R., Fink, J.K., Richardson, R.J., 2010a. Constructs of human neuropathy target esterase catalytic domain containing mutations related to motor neuron disease have altered enzymatic properties. Toxicol. Lett. 196, 67–73.
Hein, N.D., Rainier, S., Richardson, R.J., Fink, J.K., 2010b. Motor neuron disease due to neuropathy target esterase mutation: enzyme analysis of fibroblasts from human subjects yields insights into pathogenesis. Toxicol. Lett. 199, 1–5.
Henschler, D., 1958. Die Trikresylphosphate-Vergiftung: Experimentelle Klärung von Problemen der Ätiologie und Pathogenese. Klin. Wochenschr. 36, 663–674.

Hirabayashi, T., Murayama, T., Shimizu, T., 2004. Regulatory mechanism and physiological role of cytosolic phospholipase A_2. Biol. Pharm. Bull. 27, 168–1173.

Hodge, E.A., Benhaim, M.A., Lee, K.K., 2019. Bridging protein structure, dynamics, and function using hydrogen/deuterium-exchange mass spectrometry. Protein Sci. 2019, 1–13.

Hou, W.Y., Long, D.X., Wang, H.P., Wang, Q., Wu, Y.J., 2008. The homeostasis of phosphatidylcholine and lysophosphatidylcholine was not disrupted during tri-*o*-cresyl phosphate-induced delayed neurotoxicity in hens. Toxicology 252, 56–63.

Hou, W.Y., Long, D.X., Wu, Y.J., 2009. The homeostasis of phosphatidylcholine and lysophosphatidylcholine in nervous tissues of mice was not disrupted after administration of tri-*o*-cresyl phosphate. Toxicol. Sci. 109, 276–285.

Howland, R.D., Lowndes, H.E., Baker, T., Richardson, R.J., 1980. DFP mononeuropathy: evidence for a peripheral site of initiation. Brain Res. 184, 248–251.

Huang, F.F., Chang, P.A., Sun, L.X., Qin, W.Z., Han, L.P., Chen, R., 2016. The destruction box is involved in the degradation of the NTE family proteins by the proteasome. Mol. Biol. Rep. 43, 1285–1292.

Hufnagel, R.B., Arno, G., Hein, N.D., Hersheson, J., Prasad, M., Anderson, Y., Krueger, L.A., Gregory, L.C., Stoetzel, C., Jaworek, T.J., Hull, S., Li, A., Plagnol, V., Willen, C.M., Morgan, T.M., Prows, C.A., Hegde, R.S., Riazuddin, S., Grabowski, G.A., Richardson, R.J., Dieterich, K., Huang, T., Revesz, T., Martinez-Barbera, J.P., Sisk, R.A., Jefferies, C., Houlden, H., Dattani, M.T., Fink, J.K., Dollfus, H., Moore, A.T., Ahmed, Z.M., 2015. Neuropathy target esterase impairments cause Oliver-McFarlane and Laurence-Moon syndromes. J. Med. Genet. 52, 85–94.

Hutton, J.T., Rock, R., Vidal, O.J., Puente, A.E., 2011. Neuropathy. In: Noggle, C., Dean, R., Horton, A.M. (Eds.), Encyclopedia of Neuropsychological Disorders. Springer Publishing Company, New York, pp. 531–534.

IUPAC, 1978. Nomenclature of phosphorus-containing compounds of biochemical importance. Biochem. J. 171, 1–19.

Jamal, G.A., 1995. Long term neurotoxic effects of chemical warfare organophosphate compounds (sarin). Adverse Drug React. Toxicol. Rev. 14, 83–84.

Jianmongkol, S., Berkman, C.E., Thompson, C.M., Richardson, R.J., 1996. Relative potencies of the four stereoisomers of isomalathion for inhibition of hen brain acetylcholinesterase and neurotoxic esterase *in vitro*. Toxicol. Appl. Pharmacol. 139, 342–348.

John, H., Breyer, F., Thumfart, J.O., Höchstetter, H., Thiermann, H., 2010. Matrix-assisted laser desorption/ionization time-of-flight mass spectrometry (MALDI-TOF MS) for detection and identification of albumin phosphylation by organophosphorus pesticides and G- and V-type nerve agents. Anal. Bioanal. Chem. 398, 2677–2691.

John, H., Van Der Schans, M.J., Koller, M., Spruit, H.E.T., Worek, F., Thiermann, H., Noort, D., 2018. Fatal sarin poisoning in Syria 2013: forensic verification within an international laboratory network. Forensic Toxicol. 36, 61–71.

Johnson, M.K., 1968. An enzyme in hen brain hydrolyzing phenyl phenylacetate: a possible connection with the delayed neurotoxic effect of some organophosphorus compounds. Biochem. J. 110, 13P. [Abstract].

Johnson, M.K., 1969a. A phosphorylation site in brain and the delayed neurotoxic effect of some organophosphorus compounds. Biochem. J. 111, 487–495.

Johnson, M.K., 1969b. The delayed neurotoxic effect of some organophosphorus compounds: identification of the phosphorylation site as an esterase. Biochem. J. 114, 711–717.

Johnson, M.K., 1969c. Delayed neurotoxic action of some organophosphorus compounds. Br. Med. Bull. 25, 231–235.

Johnson, M.K., 1969d. Counting of Cherenkov radiation from ^{32}P in nonaqueous media. Anal. Biochem. 29, 348–350.

Johnson, M.K., 1970. Organophosphorus and other inhibitors of brain 'neurotoxic esterase' and the development of delayed neurotoxicity in hens. Biochem. J. 120, 523–531.

Johnson, M.K., 1973. Brain "neurotoxic esterase" Hoppe Seylers Z. Physiol. Chem. 354, 6–7. [Abstract].

Johnson, M.K., 1974. The primary biochemical lesion leading to the delayed neurotoxic effects of some organophosphorus esters. J. Neurochem. 23, 785–789.

Johnson, M.K., 1975a. Structure-activity relationships for substrates and inhibitors of hen brain neurotoxic esterase. Biochem. Pharmacol. 24, 797–805.

Johnson, M.K., 1975b. The delayed neuropathy caused by some organophosphorus esters: mechanism and challenge. CRC Crit. Rev. Toxicol. 3, 289–316.

Johnson, M.K., 1975c. Organophosphorus esters causing delayed neurotoxic effects: mechanism of action and structure/activity studies. Arch. Toxicol. 34, 259–288.

Johnson, M.K., 1977. Improved assay of neurotoxic esterase for screening organophosphates for delayed neurotoxicity potential. Arch. Toxicol. 37, 113–115.

Johnson, M.K., 1982. The target for initiation of delayed neurotoxicity by organophosphorus esters: biochemical studies and toxicological applications. In: Hodgson, E., Bend, J.R., Philpot, R.M. (Eds.), Reviews in Biochemical Toxicology. vol. 4. Elsevier, Amsterdam, pp. 141–212.

Johnson, M.K., 1988. Sensitivity and selectivity of compounds interacting with neuropathy target esterase: further structure-activity studies. Biochem. Pharmacol. 37, 4095–4104.

Johnson, M.K., 1990. Organophosphates and delayed neurotoxicity—is NTE alive and well? Toxicol. Appl. Pharmacol. 102, 385–399.

Johnson, M.K., Lauwerys, R., 1969. Protection by some carbamates against the delayed neurotoxic effects of di-isopropyl phosphorofluoridate. Nature 222, 1066–1067.

Johnson, M.K., Read, D.J., 1987. The influence of chirality on the delayed neuropathic potential of some organophosphorus esters: neuropathic and prophylactic effects of stereoisomeric esters of ethyl phenylphosphonic acid (EPN oxon and EPN) correlate with quantitites of aged and unaged neuropathy target esterase *in vivo*. Toxicol. Appl. Pharmacol. 90, 103–115.

Johnson, M.K., Read, D.J., 1993. Prophylaxis against and promotion of organophosphate-induced delayed neuropathy by phenyl di-*n*-pentylphosphinate. Chem. Biol. Interact. 87, 449–455.

Johnson, M.K., Richardson, R.J., 1983. Biochemical endpoints: neurotoxic esterase assay. Neurotoxicology 4, 311–320.

Johnson, M.K., Read, D.J., Benschop, H.P., 1985. Interaction of the four stereoisomers of soman (pinacolyl methylphosphonofluoridate) with acetylcholinesterase and neuropathy target esterase of hen brain. Biochem. Pharmacol. 34, 1945–1951.

Johnson, M.K., Willems, J.L., De Bisschop, H.C., Read, D.J., Benschop, H.P., 1988a. High doses of soman protect against organophosphorus-induced delayed polyneuropathy but tabun does not. Toxicol. Appl. Pharmacol. 92, 34–41.

Johnson, M.K., Bird, I., Meredith, C., 1988b. Phenyl di-*n*-pentylphosphinate: a convenient reactivatable inhibitor for studies on neuropathy target esterase (NTE) and protection against organophosphate-induced delayed polyneuropathy. Chem. Biol. Interact. 87, 449–455.

Jokanović, M., Kosanović, M., Brkić, D., Vukomanović, P., 2011. Organophosphate induced delayed polyneuropathy in man: an overview. Clin. Neurol. Neurosurg. 113, 7–10.

Jokanović, M., Segrt, Z., Stojilijković, M.P., Ristić, D., Kovać, B., 2019. Organophosphate-induced delayed polyneuropathy in man: clinical presentation, mechanisms and treatment. Am. J. Rare Dis. Diagn. Ther. 2, 8–11.

Jones, P.F., Johnson, K.E., 1973. Estimation of phenols by the 4-aminoantipyrine method: identification of the colored reaction products by proton magnetic resonance spectroscopy. Can. J. Chem. 51, 2860–2868.

Karam, C., Dyck, P.J., 2015. Toxic neuropathies. Semin. Neurol. 35, 448–457.

Karczmar, A.G., 2007. Anticholinesterases and war gases. In: Karczmar, A.G. (Ed.), Exploring the Vertebrate Central Cholinergic Nervous System. Springer, New York, pp. 237–310.

Kayyali, U.S., Moore, T.B., Randall, J.C., Richardson, R.J., 1991. Neurotoxic esterase (NTE) assay: optimized conditions based on detergent-induced shifts in the phenol/4-aminoantipyrine chromophore spectrum. J. Anal. Toxicol. 15, 86–89.

Kienesberger, P.C., Oberer, M., Lass, A., Zechner, R., 2009. Mammalian patatin domain containing proteins: a family with diverse lipolytic activities involved in multiple biological functions. J. Lipid Res. 50 (Suppl), S63–S68.

Kmoch, S., Majewski, J., Ramamurthy, V., Cao, S., Fahiminiya, S., Ren, H., Macdonald, I.M., Lopez, I., Sun, V., Keser, V., Khan, A., Stranecky, V., Hartmannova, H., Pristoupilova, A., Hodanova, K., Piherova, L., Kuchar, L., Baxova, A., Chen, R., Barsottini, O.G., Pyle, A., Griffin, H., Splitt, M., Sallum, J., Tolmie, J.L., Sampson, J.R., Chinnery, P., Care4rare, C., Banin, E., Sharon, D., Dutta, S., Grebler, R., Helfrich-Foerster, C., Pedroso, J.L., Kretzschmar, D., Cayouette, M., Koenekoop, R.K., 2015. Mutations in PNPLA6 are linked to photoreceptor degeneration and various forms of childhood blindness. Nat. Commun. 6, 5614.

Kohli, N., Srivastava, D., Sun, J., Richardson, R.J., Lee, I., Worden, R.M., 2007. Nanostructured biosensor for measuring neuropathy target esterase activity. Anal. Chem. 79, 5196–5203.

Kretzschmar, D., Hasan, G., Sharma, S., Heisenberg, M., Benzer, S., 1997. The *Swiss Cheese* mutant causes glial hyperwrapping and brain degeneration in *Drosophila*. J. Neurosci. 17, 7425–7432.

Kropp, T.J., Richardson, R.J., 2003. Relative inhibitory potencies of chlorpyrifos oxon, chlorpyrifos methyl oxon, and mipafox for acetylcholinesterase versus neuropathy target esterase. J. Toxicol. Environ. Health A 66, 1145–1157.

Kropp, T.J., Glynn, P., Richardson, R.J., 2004. The mipafox-inhibited catalytic domain of human neuropathy target esterase ages by reversible proton loss. Biochemistry 43, 3716–3722.

Lamriben, L., Graham, J.B., Adams, B.M., Hebert, D.N., 2016. N-glycan-based ER molecular chaperone and protein quality control system: the calnexin binding cycle. Traffic 17, 308–326.

Lapadula, D.M., Patton, S.E., Campbell, G.A., Abou-Donia, M.B., 1985. Characterization of delayed neurotoxicity in the mouse following chronic oral administration of tri-*o*-cresyl phosphate. Toxicol. Appl. Pharmacol. 79, 83–90.

Lenfant, N., Bourne, Y., Marchot, P., Chatonnet, A., 2016. Relationships of human α/β hydrolase fold proteins and other organophosphate-interacting proteins. Chem. Biol. Interact. 259 (Pt. B), 343–351.

Li, Y., Dinsdale, D., Glynn, P., 2003. Protein domains, catalytic activity, and subcellular distribution of neuropathy target esterase in mammalian cells. J. Biol. Chem. 278, 8820–8825.

Liang, H., Hippenmeyer, S., Ghashghaei, H.T., 2012. A nestin-cre transgenic mouse is insufficient for recombination in early embryonic neural progenitors. Biol. Open 1, 1200–1203.

Liang, T., Xue, F., Hang, W., Wen, B., Zhang, Q., Chen, J., Liu, X., Chen, J., 2019. Neuron-specific apolipoprotein E4 (1-272) fragment induces tau hyperphosphorylation and axonopathy via triggering endoplasmic reticulum stress. J. Alzheimers Dis. 71, 597–611.

Liebscher, S., Ambrose, R.L., Aktepe, T.E., Mikulasova, A., Prier, J.E., Gillespie, L.K., Lopez-Denman, A.J., Rupasinghe, T.W.T., Tull, D., McConville, M.J., Mackenzie, J.M., 2018. Phospholipase A2 activity during the replication cycle of the flavivirus West Nile virus. PLoS Pathog. 14, e1007029.

Liu, J., Zhou, X.-G., Zhang, Y., Zhang, G.-J., 2019. CGLFold: a contact-assisted de novo protein structure prediction using global exploration and loop perturbation sampling algorithm. Bioinformatics btz943. (in press).

Lock, E., Smith, A., Glynn, P., 2019. Obituary: Dr. Martin Keith Johnson, 1930–2018. Neurotoxicology 71, 168–169.

Lorenzo, D.N., Badea, A., Davis, J., Hosteller, J., He, J., Zhong, G., Zhuang, X., Bennett, V., 2014. A PIK3C3-ankyrin-B-dynactin pathway promotes axonal growth and multiorganelle transport. J. Cell Biol. 207, 735–752.

Loring, H.S., Thompson, P.R., 2020. Emergence of SARM1 as a potential therapeutic target for Wallerian-type diseases. Cell Chem. Biol. 27, 1–13.

Lotti, M., 2002. Promotion of organophosphate induced delayed polyneuropathy by certain esterase inhibitors. Toxicology 181–182, 245–248.

Lotti, M., Johnson, M.K., 1978. Neurotoxicity of organophosphorus pesticides: predictions can be based on in vitro studies with hen and human enzymes. Arch. Toxicol. 41, 215–221.

Lotti, M., Johnson, M.K., 1980. Repeated small doses of a neurotoxic organophosphate: monitoring of neurotoxic esterase in brain and spinal cord. Arch. Toxicol. 45, 263–271.

Lotti, M., Moretto, A., 1993. The search for the physiological functions of NTE: is NTE a receptor? Chem. Biol. Interact. 87, 407–416.

Lotti, M., Moretto, A., 1999. Promotion of organophosphate induced delayed polyneuropathy by certain esterase inhibitors. Chem. Biol. Interact. 119–120, 519–524.

Lotti, M., Moretto, A., 2005. Organophosphate-induced delayed polyneuropathy. Toxicol. Rev. 24, 37–49.

Lotti, M., Becker, C.E., Aminoff, M.J., Woodrow, J.E., Seiber, J.N.J., Talcott, R.E., Richardson, R.J., 1983. Occupational exposure to the cotton defoliants DEF and merphos: a rational approach to monitoring organophosphorous-induced delayed neurotoxicity. J. Occup. Med. 25, 517–522.

Lowry, O.H., Rosebrough, N.J., Farr, A.L., Randall, R.J., 1951. Protein measurement with the Folin phenol reagent. J. Biol. Chem. 193, 265–275.

Lunn, E.R., Perry, V.H., Brown, M.C., Rosen, H., Gordon, S., 1989. Absence of Wallerian degeneration does not hinder regeneration in peripheral nerve. Eur. J. Neurosci. 1, 27–33.

Lush, M.J., Li, Y., Read, D.J., Willis, A.C., Glynn, P., 1998. Neuropathy target esterase and a homologous *Drosophila* neurodegeneration-associated mutant protein contain a novel domain conserved from bacteria to man. Biochem. J. 332, 1–4.

Ma, Z., Turk, J., 2001. The molecular biology of the group VIA Ca^{2+}-independent phospholipase A2. Prog. Nucleic Acid Res. Mol. Biol. 67, 1–33.

Main, A.R., 1980. Cholinesterase inhibitors. In: Hodgson, E., Guthrie, F.E. (Eds.), Introduction to Biochemical Toxicology. Elsevier, New York, pp. 193–223.

Makhaeva, G.F., Sigolaeva, L.V., Zhuravleva, L.V., Eremenko, A.V., Kurochkin, I.N., Malygin, V.V., Richardson, R.J., 2003. Biosensor detection of neuropathy target esterase in whole blood as a biomarker of exposure to neuropathic organophosphorus compounds. J. Toxicol. Environ. Health A 66, 599–610.

Makhaeva, G.F., Malygin, V.V., Strakhova, N.N., Sigolaeva, L.V., Sokolovskaya, L.G., Eremenko, A.V., Kurochkin, I.N., Richardson, R.J., 2007. Biosensor assay of neuropathy target esterase in whole blood as a new approach to OPIDN risk assessment: review of progress. Hum. Exp. Toxicol. 26, 273–282.

Makhaeva, G.F., Radchenko, E.V., Palyulin, V.A., Rudakova, E.V., Aksinenko, A.Y., Sokolov, V.B., Zefirov, N.S., Richardson, R.J., 2013. Organophosphorus compound esterase profiles as predictors of therapeutic and toxic effects. Chem. Biol. Interact. 203, 231–237.

Makhaeva, G.F., Rudakova, E.V., Hein, N.D., Serebryakova, O.G., Kovaleva, N.V., Boltneva, N.P., Fink, J.K., Richardson, R.J., 2014. Further studies toward a mouse model for biochemical assessment of neuropathic potential of organophosphorus compounds. J. Appl. Toxicol. 34, 1426–1435.

Makhaeva, G.F., Rudakova, E.V., Serebryakova, O.G., Aksinenko, A.Y., Lushchekina, S.V., Bachurin, S.O., Richardson, R.J., 2016a. Esterase profiles of organophosphorus compounds in vitro predict their behavior in vivo. Chem. Biol. Interact. 259, 332–342.

Makhaeva, G.F., Rudakova, E.V., Sigolaeva, L.V., Kurochkin, I.N., Richardson, R.J., 2016b. Neuropathy target esterase in mouse whole blood as a biomarker of exposure to neuropathic organophosphorus compounds. J. Appl. Toxicol. 36, 1468–1475.

Marcucci, H., Paoletti, L., Jackowski, S., Banchio, C., 2010. Phosphatidylcholine biosynthesis during neuronal differentiation and its role in cell fate determination. J. Biol. Chem. 285, 25382–25393.

Maroni, M., Bleecker, M.L., 1986. Neuropathy target esterase in human lymphocytes and platelets. J. Appl. Toxicol. 6, 1–7.

Masoud, A., Sandhir, R., 2012. Increased oxidative stress is associated with the development of organophosphate-induced delayed neuropathy. Hum. Exp. Toxicol. 31, 1214–1227.

Masson, P., Nachon, F., 2017. Cholinesterase reactivators and bioscavengers for pre- and post-exposure treatments of organophosphorus poisoning. J. Neurochem. 142 (Suppl. 2), 26–40.

Meredith, C., Johnson, M.K., 1988. Neuropathy target esterase: rates of turnover *in vivo* following covalent inhibition with phenyl di-*n*-pentylphosphinate. J. Neurochem. 51, 1097–1101.

Michalovicz, L.T., Locker, A.R., Kelly, K.A., Miller, J.V., Barnes, Z., Fletcher, M.A., Miller, D.B., Klimas, N.G., Morris, M., Lasley, S.M., O'Callaghan, J.P., 2019. Corticosterone and pyridostigmine/DEET exposure attenuate peripheral cytokine expression: supporting a dominant role for neuroinflammation in a mouse model of Gulf War illness. Neurotoxicology 70, 26–32.

Moloney, E.B., De Winter, F., Verhaagen, J., 2014. ALS as a distal axonopathy: molecular mechanisms affecting neuromuscular junction stability in the presymptomatic stages of the disease. Front. Neurosci. 8, 252.

Monroy-Noyola, A., Trujillo, B., Yescas, P., Martinez-Salazar, F., Garcia-Jimenez, S., Rios, C., Vilanova, E., 2015. Stereospecific hydrolysis of a phosphoramidate used as an OPIDP model by human sera with PON1 192 alloforms. Arch. Toxicol. 89, 1801–1809.

Moretto, A., Lotti, M., Sabri, M.I., Spencer, P.S., 1987. Progressive deficit of retrograde axonal transport is associated with the pathogenesis of di-*n*-butyl dichlorvos axonopathy. J. Neurochem. 49, 1515–1522.

Moretto, A., Capodicasa, E., Peraica, M., Lotti, M., 1991. Age sensitivity to organophosphate-induced delayed polyneuropathy: biochemical and toxicological studies in developing chicks. Biochem. Pharmacol. 41, 1497–1504.

Moretto, A., Capodicasa, E., Peraica, M., Lotti, M., 1993. Phenylmethanesulfonyl fluoride delays the recovery from crush of peripheral nerves in hens. Chem. Biol. Interact. 87, 457–462.

Moretto, A., Bertolazzi, M., Lotti, M., 1994. The phosphorothioic acid *O*-(2-chloro-2,3,3-triflurorocyclobutyl) *O*-ethyl *S*-propyl ester exacerbates organophosphate polyneuropathy without inhibition of neuropathy target esterase. Toxicol. Appl. Pharmacol. 129, 133–137.

Moretto, A., Gardiman, G., Panfilo, S., Colle, M.-A., Lock, E.A., Lotti, M., 2001. Effects of S-ethyl hexahydro-1H-azepine-a-carbothioate (Molinate) on di-n-butyl dichlorovinyl phosphate (DBDCVP) neuropathy. Toxicol. Sci. 62, 274–279.

Morgan, J.P., 1982. The Jamaica ginger paralysis. JAMA 248, 1864–1867.

Morgan, J.P., Tulloss, T.C., 1976. The Jake Walk Blues: a toxicologic tragedy mirrored in American popular music. Ann. Intern. Med. 85, 804–808.

Mortensen, S.R., Brimijoin, S., Hooper, M.J., Padilla, S., 1998. Comparison of the in vitro sensitivity of rat acetylcholinesterase to chlorpyrifos-oxon: what do tissue IC50 values represent? Toxicol. Appl. Pharmacol. 148, 46–49.

Moser, M., Li, Y., Vaupel, K., Kretzschmar, D., Kluge, R., Glynn, P., Buettner, R., 2004. Placental failure and impaired vasculogenesis result in embryonic lethality for neuropathy target esterase-deficient mice. Mol. Cell. Biol. 24, 1667–1679.

Müller, C., Hardt, M., Schwudke, D., Neuman, B.W., Pleschka, S., Ziebuhr, J., 2018. Inhibition of cytosolic phospholipase $A_2\alpha$ impairs an early step of coronavirus replication in cell culture. J. Virol. 92, e01463–e01517.

Murata, H., Khine, C.C., Nishikawa, A., Yamamoto, K.-I., Kinoshita, R., Sakaguchi, M., 2018. C-Jun N-terminal kinase (JNK)-mediated phosphorylation of SARM1 regulates NAD^+ cleavage activity to inhibit mitochondrial respiration. J. Biol. Chem. 293, 18933–18943.

Natarajan, V., Yao, J.K., Dyck, P.J., Schmid, H.H.O., 1982. Early stimulation of phosphatidylcholine biosynthesis during Wallerian degeneration of rat sciatic nerve. J. Neurochem. 38, 1419–1428.

Navre, M., 2019. Why using pIC50 instead of IC50 will change your life. https://www.collaborativedrug.com/why-using-pic50-instead-of-ic50-will-change-your-life/. Accessed 7 February 2020.

NCBI, 2019. Neuropathy Target Esterase Isoform-A [Homo sapiens]. https://www.ncbi.nlm.nih.gov/protein/NP_001159583.1. Accessed 19 December 2019.

Neumann, R., Peter, H.H., 1987. Insecticidal organophosphates: nature made them first. Experientia 43, 1235–1237.

Nozaki, H., Aikawa, N., Shinozawa, Y., Hori, S., Fujisima, S., Takuma, K., Sagoh, M., 1995. Sarin poisoning in Tokyo subway. Lancet 345, 980–981.

Nwanochie, E., Uversky, V.N., 2019. Structure determination by single-particle cryo-electron microscopy: only the sky (and intrinsic disorder) is the limit. Int. J. Mol. Sci. 20, 4186.

Okamura, T., Takasu, N., Ishimatsu, S., Miyanoki, S., Mitsuhashi, A., Kumada, K., Tanaka, K., Hinohara, S., 1996. Report on 640 victims of the Tokyo subway sarin attack. Ann. Emerg. Med. 28, 129–135.

Onder, S., Schopfer, L.M., Tacal, O., Blake, T.A., Johnson, R.C., Lockridge, O., 2018. Mass spectral detection of diethoxyphospho-tyrosine adducts on proteins from HEK293 cells using monoclonal antibody depY for enrichment. Chem. Res. Toxicol. 31, 520–530.

Ong, W.-Y., Farooqui, T., Kokotos, G., Farooqui, A.A., 2015. Synthetic and natural inhibitors of phospholipases A_2: their importance for understanding and treatment of neurological disorders. ACS Chem. Neurosci. 6, 814–831.

Ordentlich, A., Barak, D., Kronman, C., Benschop, H.P., De Jong, L.P.A., Ariel, N., Barak, R., Segall, Y., Velan, B., Shafferman, A., 1999. Exploring the active center of human acetylcholinesterase with stereomers of an organophosphorus inhibitor with two chiral centers. Biochemistry 38, 3055–3066.

Ospina, M., Jayatilaka, N.K., Wong, L.Y., Restrepo, P., Calafat, A.M., 2018. Exposure to organophosphate flame retardant chemicals in the U.S. General population: data from the 2013-2014 national health and nutrition examination survey. Environ. Int. 110, 32–41.

Padilla, S., Veronesi, B., 1985. The relationship between neurological damage and neurotoxic esterase inhibition in rats acutely exposed to tri-*ortho*-cresyl phosphate. Toxicol. Appl. Pharmacol. 78, 78–87.

Page, W.F., 2003. Long-term health effects of exposure to sarin and other anticholinesterase chemical warfare agents. Mil. Med. 168, 239–245.

Pelzer, E.A., Melzer, C., Schonberger, A., Hess, M., Timmermann, L., Eggers, C., Tittgemeyer, M., 2019. Axonal degeneration in Parkinson's disease—basal ganglia circuitry and D2 receptor availability. Neuroimage Clin. 23, 101906.

Peters, R.A., 1963. Biochemical Lesions and Lethal Synthesis. International Series of Monographs on Pure and Applied Biology, vol. 18. Pergamon Press, Oxford, p. 321pp.

Peters, R.A., 1969. The biochemical lesion and its historical development. Br. Med. Bull. 25, 223–226.

Peters, R.A., Stocken, L.A., Thompson, R.H.S., 1945. British anti-lewisite (BAL). Nature 156, 616–619.

Petroianu, G., 2012. Natural phosphor ester cholinesterase inhibitors. Mil. Med. Sci. Lett. 81, 82–83.

Pope, C.N., Padilla, S., 1989. Modulation of neurotoxic esterase activity *in vitro* by phospholipids. Toxicol. Appl. Pharmacol. 97, 272–278.

Pope, C.N., Padilla, S., 1990. Potentiation of organophosphorus-induced delayed neurotoxicity by phenylmethylsulfonyl fluoride. J. Toxicol. Environ. Health 31, 261–273.

Poulsen, E., Aldridge, W.N., 1964. Studies on esterases in the chicken central nervous system. Biochem. J. 90, 182–189.

Puthenveetil, R., Vinogradova, O., 2019. Solution NMR: a powerful tool for structural and functional studies of membrane proteins in reconstituted environments. J. Biol. Chem. 294, 15914–15931.

Quinn, D.M., Topczewski, J., Yasapala, N., Lodge, A., 2017. Why is aged acetylcholinesterase so difficult to reactivate? Molecules 22. e1464.

Raghuraman, H., Chatterjee, S., Das, A., 2019. Site-directed fluorescence approaches for dynamic structural biology membrane peptides and proteins. Front. Mol. Biosci. 6, 96.

Rainier, S., Bui, M., Mark, E., Thomas, D., Tokarz, D., Ming, L., Delaney, C., Richardson, R.J., Albers, J.W., Matsunami, N., Stevens, J., Coon, H., Leppert, M., Fink, J.K., 2008. Neuropathy target esterase gene mutations cause motor neuron disease. Am. J. Hum. Genet. 82, 780–785.

Rainier, S., Albers, J.W., Dyck, P.J., Eldevik, O.P., Wilcock, S., Richardson, R.J., Fink, J.K., 2011. Motor neuron disease due to neuropathy target esterase gene mutation: clinical features of the index families. Muscle Nerve 43, 19–25.

Read, D.J., Li, Y., Chao, M.V., Cavanagh, J.B., Glynn, P., 2009. Neuropathy target esterase is required for adult vertebrate axon maintenance. J. Neurosci. 29, 11594–11600.

Read, D.J., Li, Y., Chao, M.V., Cavanagh, J.B., Glynn, P., 2010. Organophosphates induce distal axonal damage, but not brain oedema, by inactivating neuropathy target esterase. Toxicol. Appl. Pharmacol. 245, 108–115.

Reinen, J., Nematollahi, L., Fidder, A., Vermeulen, N.P., Noort, D., Commandeur, J.N., 2015. Characterization of human cytochrome P450s involved in the bioactivation of tri-*ortho*-cresyl phosphate (ToCP). Chem. Res. Toxicol. 28, 711–721.

Reiss, R., Chang, E.T., Richardson, R.J., Goodman, M., 2015. A review of epidemiologic studies of low-level exposures to organophosphorus insecticides in non-occupational populations. Crit. Rev. Toxicol. 45, 531–641.

Reuters, 2016. Pesticides a Controversial Weapon Against Zika. http://www.reuters.tv/v/MOI/2016/09/02/pesticides-a-controversial-weapon-against-zika. Accessed 15 December 2019.

Richardson, R.J., 1984. Neurotoxic esterase: Normal and pathogenic roles. In: Narahashi, T. (Ed.), Cellular and Molecular Neurotoxicology. Raven Press, New York, pp. 285–294.

Richardson, R.J., 1992. Interactions of organophosphorus compounds with neurotoxic esterase. In: Chambers, J.E., Levi, P.E. (Eds.), Organophosphates: Chemistry, Fate, and Effects. Academic Press, San Diego, pp. 299–323.

Richardson, R.J., 2010. Anticholinesterase insecticides. In: McQueen, C.A. (Ed.), Comprehensive Toxicology. vol. 13. Academic Press, Oxford, pp. 433–444.

Richardson, R.J., Makhaeva, G.F., 2014. Organophosphorus compounds. In: Wexler, P. (Ed.), Encyclopedia of Toxicology, third ed. vol. 3. Elsevier, Ltd., Oxford, pp. 714–719

Richardson, R.J., Davis, C.S., Johnson, M.K., 1979. Subcellular distribution of marker enzymes and of neurotoxic esterase in adult hen brain. J. Neurochem. 32, 607–615.

Richardson, R.J., Hein, N.D., Wijeyesakere, S.J., Fink, J.K., Makhaeva, G.F., 2013. Neuropathy target esterase (NTE): overview and future. Chem. Biol. Interact. 203, 238–244.

Richardson, R.J., Fink, J.K., Worden, R.M., Wijeyesakere, S.J., Makhaeva, G.F., 2020. Neuropathy target esterase as a biomarker and biosensor of delayed neuropathic agents. In: Gupta, R.C. (Ed.), Handbook of Toxicology of Chemical Warfare Agents, third ed. Elsevier, San Diego. (in press).

Ritchie, R.J., Prvan, T., 1996. Current statistical methods for estimating the K_m and V_{max} of Michaelis-Menten kinetics. Biochem. Educ. 24, 196–206.

Rosell, A.L., Neukomm, L.J., 2019. Axon death signalling in Wallerian degeneration among species and in disease. Open Biol. 9, 190118.

Rydel, T.J., Williams, J.M., Krieger, E., Moshiri, F., Stallings, W.C., Brown, S.M., Pershing, J.C., Purcell, J.P., Alibhai, M.F., 2003. The crystal structure, mutagenesis, and activity studies reveal that patatin is a lipid acyl hydrolase with a Ser-Asp catalytic dyad. Biochemistry 10, 6696–6708.

Sanson, B., Nachon, F., Colletier, J.P., Froment, M.T., Toker, L., Greenblatt, H.M., Sussman, J.L., Ashani, Y., Masson, P., Silman, I., Weik, M., 2009. Crystallographic snapshots of nonaged and aged conjugates of soman with acetylcholinesterase, and of a ternary complex of the aged conjugate with pralidoxime. J. Med. Chem. 52, 7593–7603.

Santayana, G., 1905. *The Life of Reason*. Scribner's Sons, New York, p. 284.

Scarpino, A., Ferenczy, G.G., Keseru, G.M., 2018. Comparative evaluation of covalent docking tools. J. Chem. Inf. Model. 58, 1441–1458.

Schenone, M., Dancik, V., Wagner, B.K., Clemons, P.A., 2013. Target identification and mechanism of action in chemical biology and drug discovery. Nat. Chem. Biol. 9, 232–240.

Schmidt, C., Breyer, F., Blum, M.-M., Thiermann, H., Worek, F., John, H., 2014. V-type nerve agents phosphonylate ubiquitin at biologically relevant lysine residues and induce intramolecular cyclization by an isopeptide bond. Anal. Bioanal. Chem. 406, 5171–5185.

Schopfer, L.M., Lockridge, O., 2018. Chlorpyrifos oxon promotes tubulin aggregation via isopeptide cross-linking between diethoxyphospho-Lys and Glu or Asp: implications for neurotoxicity. J. Biol. Chem. 293, 13566–13577.

Schur, F.K., 2019. Toward high-resolution in situ structural biology with cryo-electron tomography and subtomogram averaging. Curr. Opin. Struct. Biol. 58, 1–9.

Schwab, B.W., Richardson, R.J., 1986. Lymphocyte and brain neurotoxic esterase: dose and time dependence of inhibition in the hen examined with three organophosphorus esters. Toxicol. Appl. Pharmacol. 83, 1–9.

Senanayake, N., 1981. Tri-cresyl phosphate neuropathy in Sri Lanka: a clinical and neurophysiological study with a three year follow up. J. Neurol. Neurosurg. Psychiatry 44, 775–780.

Shafferman, A., Ordentlich, A., Barak, D., Stein, D., Ariel, N., Velan, B., 1996. Aging of phosphylated human acetylcholinesterase: catalytic processes mediated by aromatic and polar residues in the active centre. Biochem. J. 318, 833–840.

Shakespeare, W., 1623. *The Tempest*, Act 2, Scene 1.

Sigolaeva, L.V., Makower, A., Eremenko, A.V., Makhaeva, G.F., Malygin, V.V., Kurochkin, I.N., Scheller, F.W., 2001. Bioelectrochemical analysis of neuropathy target esterase activity in blood. Anal. Biochem. 290, 1–9.

Sigolaeva, L., Makhaeva, G., Rudakova, E., Boltneva, N., Porus, M., Dubacheva, G., Eremenko, A., Kurochkin, I., Richardson, R.J., 2010. Biosensor analysis of blood esterases for organophosphorus compounds exposure assessment: approaches to simultaneous determination of several esterases. Chem. Biol. Interact. 187, 312–317.

Simon, G.M., Cravatt, B.F., 2010. Activity-based proteomics of enzyme superfamilies: serine hydrolases as a case study. J. Biol. Chem. 285, 11051–11055.

Singh, N., Lee, W., 2019. Recent advances in coarse-grained models for biomolecules and their applications. Int. J. Mol. Sci. 20, 3774.

Smith, M.I., Lillie, R.D., 1931. The histopathology of tri-*ortho*-cresyl phosphate poisoning. The etiology of the so-called ginger paralysis (third report). Arch. Neurol. Psychiatry 26, 976–992.

Smith, H.V., Spalding, J.M., 1959. Outbreak of paralysis in Morocco due to ortho-cresyl phosphate poisoning. Lancet 2, 1019–1021.

Smith, M.I., Elvove, E., Valaer, P.J., Frazier, W.H., Mallory, G.E., 1930a. Pharmacological and chemical studies of the cause of the so-called ginger paralysis: a preliminary report. Public Health Rep. 45, 1703–1716.

Smith, M.I., Elvove, E., Frazier, W.H., 1930b. The pharmacological action of certain phenol esters, with special reference to the etiology of so-called ginger paralysis. Public Health Rep. 45, 2509–2607.

Smith, M.I., Engel, E.W., Stohlman, E.F., 1932. Further studies on the pharmacology of certain phenol esters with special reference to the relation of chemical constitution and physiologic action. Natl. Inst. Health Bull. 160, 1–53.

Sogorb, M.A., Pamies, D., Estevan, C., Estevez, J., Vilanova, E., 2016. Roles of NTE protein and encoding gene in development and neurodevelopmental toxicity. Chem. Biol. Interact. 259, 352–357.

Song, F., Xie, K., 2012. Calcium-dependent neutral cysteine protease and organophosphate-induced delayed neuropathy. Chem. Biol. Interact. 200, 114–118.

Song, F., Han, X., Zeng, T., Zhang, C., Zou, C., Xie, K., 2012. Changes in beclin-1 and micro-calpain expression in tri-ortho-cresyl phosphate-induced delayed neuropathy. Toxicol. Lett. 210, 276–284.

Song, Y., Wang, M., Mao, F., Shao, M., Zhao, B., Song, Z., Shao, C., Gong, Y., 2013. Knockdown of PNPLA6 protein results in motor neuron defects in zebrafish. Dis. Model. Mech. 6, 404–413.

Spencer, P.S., Schaumburg, H.H., 1977. Ultrastructural studies of the dying-back process. IV. Differential vulnerability of PNS and CNS fibers in experimental central-peripheral distal axonopathies. J. Neuropathol. Exp. Neurol. 36, 300–320.

Stokin, G.B., Lillo, C., Falzone, T.L., Brusch, R.G., Rockenstein, E., Mount, S.L., Raman, R., Davies, P., Masliah, E., Williams, D.S., Goldstein, L.S.B., 2005. Axonopathy and transport deficits early in the pathogenesis of Alzheimer's disease. Science 307, 1282–1288.

Strelow, J.M., 2017. A perspective on the kinetics of covalent and irreversible inhibition. SLAS Discov. 22, 3–20.

Sujkowski, A., Rainier, S., Fink, J.K., Wessells, R.J., 2015. Delayed induction of human NTE (PNPLA6) rescues neurodegeneration and mobility defects of Drosophila swiss cheese (sws) mutants. PLoS ONE 10, e0145356.

Sunderhaus, E.R., Law, A.D., Kretzschmar, D., 2019a. ER responses play a key role in Swiss-cheese/neuropathy target esterase-associated neurodegeneration. Neurobiol. Dis. 130, 104520.

Sunderhaus, E.R., Law, A.D., Kretzschmar, D., 2019b. Disease-associated PNPLA6 mutations maintain partial functions when analyzed in *Drosophila*. Front. Neurosci. 13, 1207.

Suzuki, T., Morita, H., Ono, K., Maekawa, K., Nagai, R., Yazaki, Y., 1995. Sarin poisoning in Tokyo subway. Lancet 345, 980.

Synofzik, M., Schule, R., 2017. Overcoming the divide between ataxias and spastic paraplegias: shared phenotypes, genes, and pathways. Mov. Disord. 32, 332–345.

Synofzik, M., Gonzalez, M.A., Lourenco, C.M., Coutelier, M., Haack, T.B., Rebelo, A., Hannequin, D., Strom, T.M., Prokisch, H., Kernstock, C., Durr, A., Schols, L., Lima-Martinez, M.M., Farooq, A., Schule, R., Stevanin, G., Marques Jr., W., Zuchner, S., 2014. PNPLA6 mutations cause Boucher-Neuhauser and Gordon Holmes syndromes as part of a broad neurodegenerative spectrum. Brain 137, 69–77.

Synofzik, M., Kernstock, C., Haack, T.B., Schols, L., 2015. Ataxia meets chorioretinal dystrophy and hypogonadism: Boucher-Neuhauser syndrome due to PNPLA6 mutations. J. Neurol. Neurosurg. Psychiatry 86, 580–581.

Tagliaferro, P., Kareva, T., Oo, T.F., Yarygina, O., Kholodilov, N., Burke, R.E., 2015. An early axonopathy in a hLRRK2(R1441G) transgenic model of Parkinson disease. Neurobiol. Dis. 82, 359–371.

Talevi, A., 2015. Multi-target pharmacology: possibilities and limitations of the "skeleton key approach" from a medicinal chemist perspective. Front. Pharmacol. 6, 205.

Thompson, C.M., Richardson, R.J., 2004. Anticholinesterase insecticides. In: Marrs, T.C., Ballantyne, B. (Eds.), Pesticide Toxicology and International Regulation (Current Toxicology Series). John Wiley & Sons, Ltd., Chichester, pp. 89–127

Topaloglu, A.K., Lomniczi, A., Kretzschmar, D., Dissen, G.A., Kotan, L.D., McArdle, C.A., Koc, A.F., Hamel, B.C., Guclu, M., Papatya, E.D., Eren, E., Mengen, E., Gurbuz, F., Cook, M., Castellano, J.M., Kekil, M.B., Mungan, N.O., Yuksel, B., Ojeda, S.R., 2014. Loss-of-function mutations in PNPLA6 encoding neuropathy target esterase underlie pubertal failure and neurological deficits in Gordon Holmes syndrome. J. Clin. Endocrinol. Metab. 99, E2067–E2075.

UniProt, 2019a. UniProtKB—Q8IY17 (PLPL6_HUMAN). http://www.uniprot.org/uniprot/Q8IY17. Accessed 26 November 2019.

UniProt, 2019b. UniProtKB—M0R715 (M0R715_RAT). https://www.uniprot.org/uniprot/M0R715, 2019b. Accessed 17 December 2019.

UniProt, 2019c. UniProtKB—Q3TRM4 (PLPL6_MOUSE). https://www.uniprot.org/uniprot/Q3TRM4. Accessed 17 December 2019.

USEPA, 2019. Success in Mosquito Control: An Integrated Approach. https://www.epa.gov/mosquitocontrol/success-mosquito-control-integrated-approach. Accessed 26 November 2019.

Van Tienhoven, M., Atkins, J., Li, Y., Glynn, P., 2002. Human neuropathy target esterase catalyzes hydrolysis of membrane lipids. J. Biol. Chem. 277, 20942–20948.

Varadaraju, C., Tamilselvan, G., Enoch, I.M.V.M., Selvakumar, M., 2018. Phenol sensing studies by 4-aminoantipyrine method—a review. Org. Med. Chem. Int. J. 5, 555657.

Venkova, K., Christov, A., Kamaluddin, Z., Kobalka, P., Siddiqui, S., Hensley, K., 2014. Semaphorin 3A signaling through neurophilin-1 is an early trigger for distal axonopathy in the SOD1G93A mouse model of amyotrophic lateral sclerosis. J. Neuropathol. Exp. Neurol. 73, 702–713.

Veronesi, B., 1984. A rodent model of organophosphorus-induced delayed neuropathy: distribution of central (spinal cord) and peripheral nerve damage. Neuropathol. Appl. Neurobiol. 10, 357–368.

Veronesi, B., Padilla, S., 1985. Phenylmethylsulfonyl fluoride protects rats from mipafox-induced delayed neuropathy. Toxicol. Appl. Pharmacol. 81, 258–264.

Veronesi, B., Padilla, S., Blackmon, K., Pope, C., 1991. Murine susceptibility to organophosphorus-induced delayed neuropathy (OPIDN). Toxicol. Appl. Pharmacol. 107, 311–324.

Via, A., Gherardini, P.F., Ferraro, E., Ausiello, G., Tomba, G.S., Helmer-Citterich, M., 2007. False occurrences of functional motifs in protein sequences highlight evolutionary constraints. BMC Bioinformatics 8, 68.

Vilanova, E., Barril, J., Carrera, V., 1993. Biochemical properties and possible toxicological significance of various forms of NTE. Chem. Biol. Interact. 87, 369–381.

Vilanova, E., Escudero, M.A., Barril, J., 1999. NTE soluble isoforms: new perspectives for targets of neuropathy inducers and promoters. Chem. Biol. Interact. 119–120, 525–540.

Walaas, S.I., Greengard, P., 1991. Protein phosphorylation and neuronal function. Pharmacol. Rev. 43, 299–349.

Walker, L.J., Summers, D.W., Sasaki, Y., Brace, E.J., Milbrandt, J., DiAntonio, A., 2017. MAPK signaling promotes axonal degeneration by speeding the turnover of the axonal maintenance factor NMNAT2. eLife 6, e22540.

Waller, A., 1850. Experiments on the section of the glossopharyngeal and hypoglossal nerves of the frog, and observations of the alterations produced thereby in the structure of their primitive fibres. Philos. Trans. R. Soc. Lond. A 140, 423–429.

Watson, J.C., Dyck, P.J.B., 2015. Peripheral neuropathy: a practical approach to diagnosis and symptom management. Mayo Clin. Proc. 90, 940–951.

Wentzell, J.S., Cassar, M., Kretzschmar, D., 2014. Organophosphate-induced changes in the PKA regulatory function of Swiss cheese/NTE lead to behavioral deficits and neurodegeneration. PLoS ONE 9, e87526.

WHO (Ed.), 2011. Vector-Borne Disease Control: A 10-Year Assessment (2000–2009), fifth ed. WHO, Geneva. 41pp.

WHO, 2016. Vector Control Operations Framework for Zika Virus. WHO, Geneva. 10pp.

Wijeyesakere, S.J., Richardson, R.J., 2010. Neuropathy target esterase. In: Krieger, R. (Ed.), Hayes' Handbook of Pesticide Toxicology, third ed. Elsevier/Academic Press, San Diego, pp. 1435–1478.

Wijeyesakere, S.J., Richardson, R.J., Stuckey, J.A., 2007. Modeling the tertiary structure of the patatin domain of neuropathy target esterase. Protein J. 26, 165–172.

Wijeyesakere, S.J., Rizvi, S.M., Raghavan, M., 2013. Glycan-dependent and –independent interactions contribute to cellular substrate recruitment by calreticulin. J. Biol. Chem. 288, 35104–35116.

Wijeyesakere, S.J., Richardson, R.J., Stuckey, J.A., 2014. Crystal structure of patatin-17 in complex with aged and non-aged organophosphorus compounds. PLoS One 9, e108245.

Winrow, C.J., Hemming, M.L., Allen, D.M., Quistad, G.B., Casida, J.E., Barlow, C., 2003. Loss of neuropathy target esterase in mice links organophosphate exposure to hyperactivity. Nat. Genet. 33, 477–485.

Witschi, H., 2002. Profiles in toxicology: John Morrison Barnes, 1913-1975. Toxicol. Sci. 68, 267–269.

Witschi, H., Lock, E., 2001. Profiles in toxicology: Carshalton in the 1960s. Toxicol. Sci. 62, 4–5.

Worek, F., Wille, T., Koller, M., Thiermann, H., 2016. Toxicology of organophosphorus compounds in view of an increasing terrorist threat. Arch. Toxicol. 90, 2131–2145.

Wu, S.-Y., Casida, J.E., 1994. Neuropathy target esterase inhibitors: enantiomeric separation and stereospecificity of 2-substituted-4*H*-1,3,2-benzodioxaphosphorin 2-oxides. Chem. Res. Toxicol. 7, 77–91.

Wyckoff, G.J., Solidar, A., Yoder, M.D., 2010. Phosphatidylinositol transfer proteins: sequence motifs in structural and evolutionary analyses. J. Biomed. Sci. Eng. 3, 65–77.

Xiang, Y., Shen, Z., Shi, Y., 2020. Chemical cross-linking and mass spectrometric analysis of the endogenous yeast exosome complexes. Methods Mol. Biol. 2062, 383–400.

Xu, H.-Y., Wang, P., Sun, Y.-J., Xu, M.-Y., Zhu, L., Wu, Y.-J., 2018. Activation of neuregulin 1/ErbB signaling is involved in the development of TOCP-induced delayed neuropathy. Front. Mol. Neurosci. 11, 129.

Yang, S.H., Bergo, M.O., Farber, E., Qiao, X., Fong, L.G., Young, S.G., 2009. Caution! Analyze transcripts from conditional knockout alleles. Transgenic Res. 18, 483–489.

Yang, F.-W., Zhao, T.-P., Ren, F.-Z., Pang, G.-F., Li, Y.-X., 2020. Assessment of the endocrine-disrupting effects of diethyl phosphate, a nonspecific metabolite of organophosphorus pesticides, by *in vivo* and *in silico* approaches. Environ. Int. 135, 105383.

Yuan, Q., Zhang, H., Zhang, X., Huang, P., Liu, R., Song, Y.Q., Wu, W., Lin, Z.X., 2019. Association between axonopathy and amyloid plaques in the spinal cord of the transgenic mice of Alzheimer's disease. Neuroscience 409, 152–161.

Zaccheo, O., Dinsdale, D., Meacock, P.A., Glynn, P., 2004. Neuropathy target esterase and its yeast homologue degrade phosphatidylcholine to glycerophosphocholine in living cells. J. Biol. Chem. 279, 24024–24033.

Zhao, J., Hedera, P., 2013. Hereditary spastic paraplegia-causing mutations in atlastin-1 interfere with BMPRII trafficking. Mol. Cell Neurosci. 52, 87–96.

Zhong, T., Chen, J., Ling, Y., Yang, B., Xie, X., Yu, D., Zhang, D., Ouyang, J., Kuang, H., 2018. Down-regulation of neuropathy target esterase in preeclampsia placenta inhibits human trophoblast cell invasion via modulating MMP-9 levels. Cell. Physiol. Biochem. 45, 1013–1022.

Zhu, H., O'Brien, J.J., O'Callaghan, J.P., Miller, D.B., Zhang, Q., Rana, M., Tsui, T., Peng, Y., Tomesch, J., Hendrick, J.P., Wennogle, L.P., Snyder, G.L., 2010. *Nerve agent exposure elicits site-specific changes in protein phosphorylation in mouse brain*. Brain Res. 1342, 11–23.

Zhu, L., Wang, P., Sun, Y.J., Xu, M.Y., Wu, Y.J., 2016. Disturbed phospholipid homeostasis in endoplasmic reticulum initiates tri-*o*-cresyl phosphate-induced delayed neurotoxicity. Sci. Rep. 6, 37574.

Zhuang, Q., Franjesevic, A.J., Corrigan, T.S., Coldren, W.H., Dicken, R., Sillart, S., DeYong, A., Yoshino, N., Smith, J., Fabry, S., Fitzpatrick, K., Blanton, T.G., Joseph, J., Yoder, R.J., McElroy, C.A., Ekici, L.D., Callam, C.S., Hadad, C.M., 2018. Demonstration of *in vitro* resurrection of aged acetylcholinesterase after exposure to organophosphorus chemical nerve agents. J. Med. Chem. 61, 7034–7042.

CHAPTER TWO

Neurotoxicity of organophosphate nerve agents

Ramesh C. Gupta*
Toxicology Department, Breathitt Veterinary Center, Murray State University, Hopkinsville, KY, United States
*Corresponding author: e-mail address: rgupta@murraystate.edu

Contents

1. Introduction

Organophosphate (OP) nerve agents (commonly called nerve agents) were developed prior to, during and after World War II in Germany, the United States, the United Kingdom, Russia, and some other countries. Tabun was the first nerve agent (GA) synthesized in 1936 by the German

Advances in Neurotoxicology, Volume 4
ISSN 2468-7480
https://doi.org/10.1016/bs.ant.2019.11.001

chemist Gerhard Schrader. This led to the discovery of other nerve agents, such as sarin (GB), cyclosarin (GF), soman (GD), VX, Novichok (including Russian VX), and Chinese VX. These are common nerve agents known to academicians, governmental agencies, and the public. Some nerve agents have been used in wars and by dictators, extremists, cult leaders, and terrorist groups as chemical weapons of mass destruction (CWMD). Some of these incidents include the Iraq-Iran conflict (1980s), the Aum Shinrikyo's attacks with sarin in a Tokyo subway and in Matsumoto, Japan (1994–1995) (Okumura et al., 2015; Tu, 2002), and the Syrian government's nerve agent attacks on the Syrian people (2013 and onward) (Dolgin, 2013; John et al., 2018; Rosman et al., 2014). The Organization for the Prohibition of Chemical Weapons (OPCW) and the United Nations have concluded that sarin has been used by the Syrian military on multiple occasions (Guterres, 2017; OPCW, 2018). Recently, nerve agents have been used to target individuals of national and international significance. For example, in March of 2018, North Korea used VX to assassinate the half-brother of the nation's leader Kim Jong Un. Again, in the same year, former Russian spy Sergei Skripal and his daughter Yulia Skripal in Salisbury, and Dawn Sturgess and Charlie Rowley in Amesbury, UK, were poisoned with a nerve agent of the Novichok group (Chai et al., 2018; Nepovimova and Kuca, 2018).

Nerve agents produce toxicity primarily by irreversible inactivation of acetylcholinesterase (AChE) in nervous tissue and skeletal muscles. These nerve agents not only target nervous system and skeletal muscles, but affect other organs and systems involving noncholinergic mechanisms as well. Biomarkers of nerve agent(s) exposure include AChE inactivation and the detection of nerve agents or their metabolite(s) in body tissue/fluid (Barr et al., 2004; Driskell et al., 2002). Biomarkers of effect and susceptibility have also been identified. Onset of toxicity of nerve agents can occur within a few minutes, followed by maximal severity in <10 min, and death can ensue within a few min to a few hours depending on the dose. Therapy for nerve agent poisoning includes atropine sulfate and oximes, but this can be futile unless antidotal treatment is given in time. This chapter provides an overview of OP nerve agents' significance, chemistry, neurotoxicity, mechanisms, biomarkers and treatment.

2. A brief historical perspective

In the early 1930s, Dr. Gerhard Schrader, a noted chemist at the I.G. Farbenindustrie laboratory Leverkusen, Germany, synthesized OP compounds, such as parathion for insecticidal use. Prior to World War II, under the authority of the German Ministry of Defense, his priority shifted from insecticides to toxic chemical warfare agents (CWA). They synthesized diisopropylphosphorofluoridate (DFP), which was not toxic enough to be used as a nerve agent, and too toxic to be used as insecticide. DFP is still used as a surrogate of nerve agents. This led to the development of highly toxic OP nerve agents/gases of the G series: tabun (GA) in 1937, sarin (GB) in 1938, and soman (GD) in 1944. In 1954 VX, which is many times more toxic than compounds of the G series, was synthesized by Professor Ranajit Ghosh at the ICI Company in the UK. During the same time period, another compound of the V series (VR or RVX, R 033), with super toxicity, was developed by the Soviet military as a chemical weapon of mass destruction (CWMD).

Nerve agents have been developed and stockpiled by many countries. Human and animal casualties have occurred on a small and large scale from OP nerve agents due to their mishandling or intentional use in wars, and by dictators and terrorists. In 1950, two separate accidental exposures occurred at Dugway Proving Ground, wherein workers exhibited symptoms of toxicity 3 days after a sarin test. On March 13, 1968, the United States Army at Dugway Proving Ground conducted three tests, one of them involved jet aircraft spraying the nerve agent VX, resulting in the deaths of >6000 sheep in Skull Valley, Utah. Since August 2013, the Syrian government has repeatedly used sarin and other CWMD against rebels in Syria.

Since the terrorist attacks in the United States on September 11, 2001, it is much more likely that terrorists will use nerve agents or other deadly chemicals to kill people or draw attention. For a detailed history and the likely threat from nerve agents, readers are referred to other publications (Bajgar et al., 2015; Johnson et al., 2015; Nepovimova and Kuca, 2018; Worek et al., 2016a,b).

3. Chemistry of OP nerve agents

OP nerve agents of the G-series include tabun (GA), sarin (GB), soman (GD), and cyclosarin (GF), and those of the V series include VX, VR, VG, Novichok, Chinese VX, VE, VM, and others. Soman, sarin, and cyclosarin are phosphonofluoridates, and VX is a phosphonothiolate (Black and Harrison, 1996; Gupta, 2006; Watson et al., 2015). Tabun also has a cyano group. Sarin and VX each have two isomers, while soman has four isomers (C+ P−, C− P−, C+ P+, and C− P+). The nerve agents of the G-series and V-series are all viscous liquids of varying volatility (vapor density relative to air between 4.86 and 6.33) with faint odors ("faintly fruity," "spicy," or odor of "camphor"). Agent VX is an amber-colored liquid with a vapor density of 9.2 and is considered odorless. In general, OP nerve agents are volatile liquids and persist for a short time in the environment, whereas VX is persistent and the least volatile. The vapor pressures of the nerve agents [2.9 mmHg (sarin), 0.4 mmHg (soman), 0.07 mmHg (tabun), and 0.044 mmHg (cyclosarin)] are high, and the lethal vapor risk follows in descending order, with sarin exhibiting the highest lethal vapor risk (Figueiredo et al., 2018; Gupta, 2015). Nerve agent vapors possess little or no olfactory warning properties (Watson et al., 2006, 2015).

In recent years, Novichok has received enormous attention around the world. Novichok, which means "newcomer" in Russian, was a type of nerve agent developed in the Soviet Union for use as a chemical warfare agent in the 1970–1990s. These are binary munitions (such as Novichok-5 and Novicok-7), in which two less toxic precursors have to be mixed to produce the active agent. It is believed that several hundred Novichok agents were synthesized, all with different structures and characteristics. They all are OP compounds and one of the earlier compounds carried the codename VR. This led to the synthesis of A-230, A-232 (binary analog, Novichok-5), and A-234 (binary analog, Novichok-7), depending on the group (methyl, methoxy or ethoxy) attached to P. At this time, very little information is available about the physico-chemical properties of A-agents and the NOVICHOKs (Nepovimova and Kuca, 2018). Some physico-chemical properties and toxicity of nerve agents are listed in Table 1.

4. Route of exposure

Nerve agents are likely to enter the body through dermal, inhalation, conjunctival, and upper gastrointestinal tract (GIT) routes. Because these agents are essentially colorless, odorless, tasteless, and nonirritating to the skin, their entry into the body may not be perceived by the individual until grave signs and symptoms appear (Gupta, 2015; Munro et al., 1994). The volatility of sarin is greater than tabun, yet both agents are of greatest concern for acute inhalation exposures. Compared to other nerve agents, sulfur-containing VX is less volatile and more resistant to detoxification, and thereby more efficient at skin penetration than as an inhalation threat. In a recent in vitro investigation, Dalton et al. (2018) demonstrated that an

Table 1 Chemical and toxicological properties of OP nerve agents.

Nerve agent (CAS no.)	Chemical structure	Chemical name/formula	Mol. wt	Toxicity LD_{50} (mg/kg)
Tabun/GA (77-81-6)	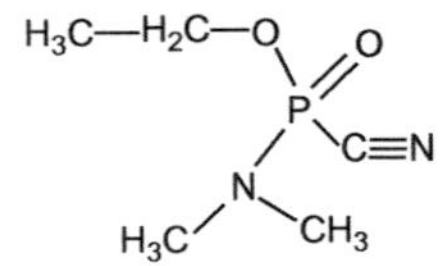	Ethyl dimethylamido cyanophosphate ($C_5H_{11}N_2O_2P$)	162.13	*Dermal*:
				Human = 14–21
				Monkey = 9.3
				Dog = ~45
				Rabbit: 3
				Rat: 12.6
				Subcutaneous:
				Rat = 0.162
				Mouse = 0.25
				Guinea pig = 0.12
				Rabbit = 0.375
				Intravenous:
				Human = 0.014
				Monkey = ~0.05
				Dog = 0.084
				Rabbit = 0.063

Continued

Table 1 Chemical and toxicological properties of OP nerve agents.—Cont'd

Nerve agent (CAS no.)	Chemical structure	Chemical name/formula	Mol. wt	Toxicity LD_{50} (mg/kg)
				Rat = 0.07
				Mouse = 0.15
				Oral:
				Rat = 1.06
				Human = 0.357–0.714
Sarin/GB (107-44-8)	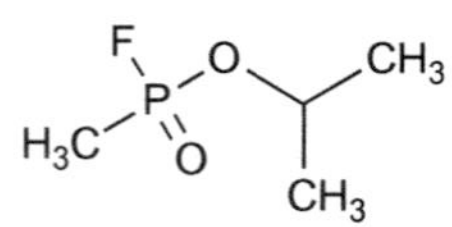	*O*-isopropyl methyl-phosphonofluoridate ($C_4H_{10}FO_2P$)	140.10	*Dermal*:
				Human = 24
				Dog = 10.8
				Rabbit: 4.4
				Rat: 2.5
				Subcutaneous:
				Rat = 0.103–0.108
				Mouse = 0.17
				Guinea pig = 0.03
				Rabbit = 0.03
				Intravenous:
				Human = 0.014

				Monkey = 0.020
				Dog = 0.010
				Rabbit = 0.0147
				Rat = 0.045–0.063
				Mouse = 0.083–0.113
				Oral:
				Rat = 0.10
				Human = 0.071–0.286
Cyclosarin/GF (329-99-7)	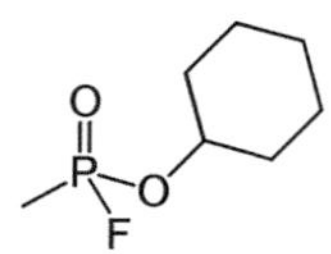	*O*-cyclohexyl methyl-phosphonofluoridate ($C_7H_{14}FO_2P$)	180.2	*Intramuscular:*
				Mouse = 0.171–0.21
				Oral:
				Human = 0.14
Soman/GD (96-64-0)	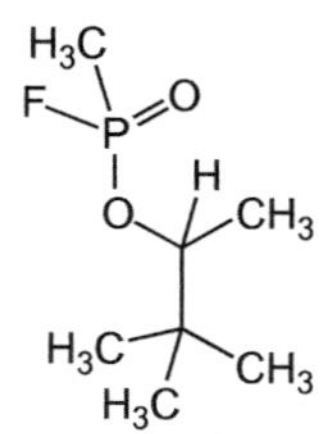	Pinacolyl methyl-phosphonofluoridate ($C_7H_{16}FO_2P$)	182.178	*Dermal:*
				Human = 5.0
				Subcutaneous:
				Rat = 0.07–0.165
				Mouse = 0.156
				Guinea pig = 0.024

Continued

Table 1 Chemical and toxicological properties of OP nerve agents.—Cont'd

Nerve agent (CAS no.)	Chemical structure	Chemical name/formula	Mol. wt	Toxicity LD_{50} (mg/kg)
				Rabbit = 0.02
				Intravenous:
				Human = 0.025
				Rat = 0.0445
				Mouse = 0.035
				Guinea pig = 0.0275
				Oral:
				Human = 0.071–0.286
VX (50782-69-9)	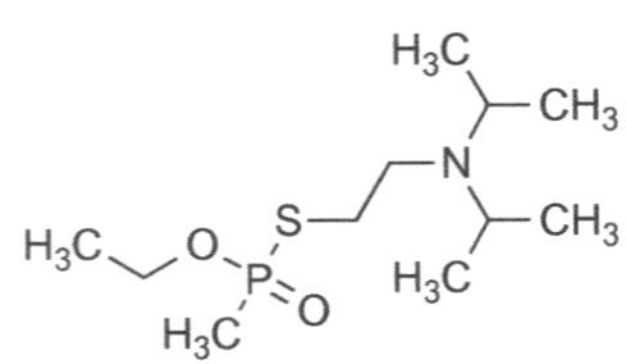 VX	*S*-(2-diisopropyl-aminoethyl) *O*-ethyl methyl-phosphonothiolate ($C_{11}H_{26}NO_2PS$)	267.38	*Dermal:*
				Human = 0.04–0.142
				Monkey = ~0.065
				Dog = 0.054
				Rabbit = 0.025
				Guinea pig = 0.125
				Rat: 0.10
				Subcutaneous:
				Rat = 0.012

				Mouse = 0.022
				Guinea pig = 0.0084
				Rabbit = 0.014–0.066
				Intravenous:
				Human = 0.008
				Monkey = 0.0084
				Dog = 0.0063
				Rabbit = 0.0084
				Rat = 0.007–0.01
				Mouse = 0.02
				Oral:
				Human = 0.043–0.143
RVX (159939-87-4)	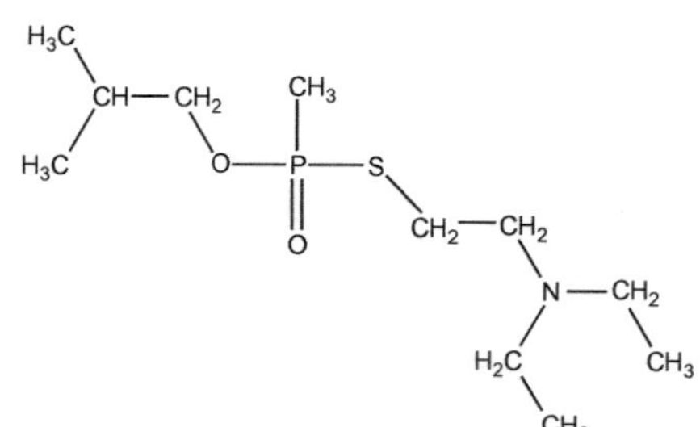	Phosphonothioic acid methyl-, *S*-[2-(diethylamino) ethyl] *O*-(2-methylpropyl) ester ($C_{11}H_{26}SNPO_2$)	276.37	*Dermal:*
				Mice = 0.016
				Rabbit = 0.014
				Cat = 0.01
				Dog = 0.0157
				Human = 0.1–0.01

Continued

Table 1 Chemical and toxicological properties of OP nerve agents.—Cont'd

Nerve agent (CAS no.)	Chemical structure	Chemical name/formula	Mol. wt	Toxicity LD_{50} (mg/kg)
VG	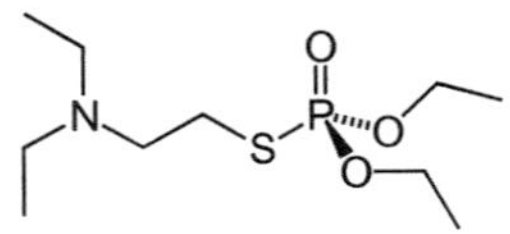	*O*,*O*-diethyl *S*-[2-diethylaminoethyl] phosphorothiolate ($C_{10}H_{24}NO_3PS$)	269.341	–
Chinese VX	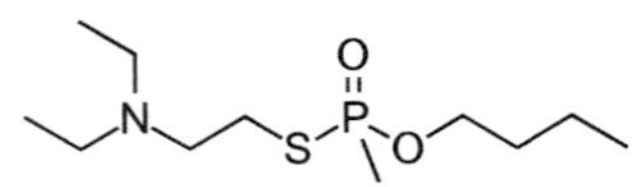	2-(diethylamino)ethyl]-*O*-butyl methylphosphonothioate ($C_{11}H_{26}NO_2PS$)	267.37	–

Note: LD_{50} doses reported for humans are estimated doses.

aqueous solution of VX in contact with the skin resulted in increased penetration rates when compared to neat VX applied to dry skin. By any route of exposure, VX is more toxic than tabun, sarin, cyclosarin or soman in all tested species.

5. Mechanism of action and neurotoxicity

OP nerve agents may exert status epilepticus in three phases involving cholinergic and noncholinergic mechanisms (see Table 2).

5.1 Cholinergic

The cholinergic system consists of several key cholinergic elements: choline, acetylcholine, acetylcholinesterase, and choline acetyltransferase. OP nerve agents inhibit the enzyme acetylcholinesterase (AChE), which hydrolyzes a neurotransmitter acetylcholine (ACh) at nerve synapses and neuromuscular junctions (NMJs). These agents also inhibit butyrylcholinesterase (Gupta et al., 1987, 1988, 1991; Matson et al., 2018), which is known to play a role in the regulation of ACh levels in the CNS (Reid et al., 2013). Toxic effects are generally seen when >50% of the AChE is inhibited, and death typically ensues when >90% of the AChE is inhibited in the brain and diaphragm

Table 2 Proposed consequences of neurochemical events leading to *status epilepticus* and neuropathology following soman exposure.

I. Cholinergic phase (0–5 min)
  1. Soman inhibits brain AChE; ACh levels increase
  2. Elevated ACh triggers seizure activity in susceptible brain regions
  3. Seizure activity rapidly spreads and begins to perturb other neurotransmitter systems. Anticholinergic drugs at low doses can block or terminate seizures during steps 1–3. If seizure is controlled now, no brain damage will occur

II. Transition phase: Mixed cholinergic-noncholinergic (>5 to <40 min)
  4. Catecholamine and amino acid neurotransmitters become progressively more disrupted. Higher anticholinergic doses required. Processes leading to brain damage begin in sensitive areas

III. Noncholinergic phase (≥40 min)
  5. Large changes in catecholamine and amino acid neurotransmitters are evident. Binding of excitatory amino acid receptors significantly increased. Scopolamine no longer effective; drugs with mixed anticholinergic-anti-NMDA activity are effective. Processes leading to brain damage now wide-spread but still can be limited in extent

Adapted from Shih, T.-M., McDonough Jr., J.H., 1997. Neurochemical mechanisms in soman-induced seizures. J. Appl. Toxicol. 17(4), 255–264. (Courtesy of Shih T-M and McDonough Jr. JH).

muscle. A wide brain regional variability in AChE inhibition by different nerve agents is noteworthy (Gupta et al., 1987, 1988, 1991; Jones et al., 2000; McLean et al., 1992). For example, in an acute toxicity, VX (12 μg/kg, sc) caused 97–99% AChE inhibition in cortex, hippocampus, and brainstem, while only 60% in striatum (Gupta et al., 1991). Shih (1982) determined whether brain regional variability of AChE inhibition was reflected in the levels of ACh and choline (Ch). The time course of changes in ACh and Ch in six discrete brain regions of rats following a single injection of soman (0.9 LD_{50}, 120 μg/kg, SC) was determined. Soman produced a maximal ACh elevation in the brain stem at 20 min (34.4%), in cerebellum at 40 min (51.9%), in cortex and striatum at 2 h (320.3% and 35.2%, respectively), and in hippocampus and midbrain at 3 h (94.5% and 56.8%, respectively). ACh levels remained above normal approximately 30 min in the brain stem; 2 h in the midbrain, cerebellum, and striatum; 8 h in the cortex; and 16 h in the hippocampus. Ch levels were elevated in all areas except the striatum. Ch maxima occurred at 10–40 min and returned to control levels approximately 3 h after soman injection. Variability of AChE inhibition among skeletal muscles (slow fiber soleus, fast fiber extensor digitorum longus, and mixed fiber diaphragm) by nerve agents and DFP has also been reported (Gupta et al., 1985, 1987, 1988, 1991; McLean et al., 1992).

Accumulation of ACh at nerve terminals, due to irreversible inactivation of AChE by phosphorylation, leads to overstimulation of ACh receptors, thereby causing clinical signs of hyperactivity (Millard et al., 1999). These chain of events eventually result in muscarinic, nicotinic, and CNS effects. Muscarinic effects, due to increased parasympathetic nerve activity, include miosis, visual disturbances (accommodation disorder and blurred vision), increased salivary and bronchial secretions, rhinorrhea, bronchospasm, bradycardia, and increased gastrointestinal peristaltic activity (abdominal pain, anorexia, nausea, vomiting, and diarrhea), increased sweating and lacrimation (often bloody dacryorrhea). Nicotinic effects, due to hyper stimulation of neuromuscular junction (NMJ) activity, results in fasciculations, convulsions, muscle weakness, and paralysis of muscles (including respiratory muscles). CNS effects, due to elevated ACh, may include anxiety, headaches, excitement, ataxia, somnolence, disorientation, coma, and seizures (Okumura et al., 2015). Acute respiratory insufficiency is the most important cause of immediate death. With sarin, some pharmacological actions can be conflicting such as miosis or mydriasis, and bradycardia or tachycardia.

Individuals dermally exposed to VX may show no symptoms for up to 30 min, but thereafter may rapidly suffer loss of consciousness, convulsions, difficulty breathing, profuse secretions from the nose and mouth, generalized muscle twitching, paralysis and death. Signs and symptoms from vapor exposure generally manifest within seconds to minutes. These typically include miosis or pinpoint pupils, eye pain, runny nose, difficulty breathing and coughing. Exposure to higher doses often results in severe dyspnea, fasciculations, flaccid paralysis, convulsions, loss of consciousness, apnea and death.

Reactivation of inhibited AChE by dephosphorylation is not possible once the nerve agent-AChE complex undergoes "aging," which is thought to be the consequence of the loss of an alkyl or alkoxy group. The aging $t_{1/2}$ for soman, sarin, tabun, and VX were reported to be 1.3 min, 5 h, 46 h, and 48 h, respectively (Sirin et al., 2012; Watson et al., 2006). Although with VX AChE undergoes very slow aging, it will also undergo spontaneous reactivation. The rate of reactivation of red blood cell AChE appears to be about 1% per hour.

In some studies, OP nerve agents and pesticides have been shown to interact directly with ACh receptors and exert cholinergic effects (Bakry et al., 1988; Menking et al., 1990; Tattersall, 1990). Pittel et al. (2006, 2018) demonstrated that VX and echothiophate act as function-specific blockers via a nonclassical path of antagonistic activity, implying the involvement of allosteric/ectopic-binding site in M1 and M3 mAChRs. Aas et al. (1987) reported alterations in muscarinic ACh receptors in rat bronchi and lung tissue after subacute inhalation exposure to soman. Soman and VX have also been reported to directly interact with nicotinic receptors and the ACh receptor-ion channel complex (Chen and Chi, 1986). VX was shown to bind and block postjunctional glutamate receptors, nAChR-ion channels, and mAChRs. Rao et al. (1987) reported that VX caused an increase in ACh release at NMJs in the frog via interaction with the nicotinic ACh receptor-ion channel complex.

5.2 Noncholinergic

It is well established that OP's toxicity involves noncholinergic transmitters, receptors, enzymes, and other molecules (Costa, 2018; Li et al., 2008; Lockridge et al., 2015; Lockridge and Schopfer, 2006; Marrs and Maynard, 2013; Matson et al., 2018; Solberg and Belkin, 1997; Vanova et al., 2018). Lallement et al. (1991a,b) suggested that soman-induced overstimulation of glutamatergic receptors contributed to maintenance of seizure. In another study, VX increased the frequency of GABA- and glutamate-mediated miniature postsynaptic currents- a Ca^{2+}-dependent effect reported to be unrelated to AChE inhibition (Rocha et al., 1999). Chabebo et al. (1999) reported that 0.3–1 nM of sarin reduced the amplitude of GABA-mediated postsynaptic currents but had no effect on the amplitude of glutamatergic-mediated postsynaptic currents. The observed effect was thought to be due to the direct interaction of sarin with muscarinic ACh receptors present on presynaptic GABAergic neurons. Selective reduction in the action potential-dependent release of GABA in the hippocampus may account for sarin-induced seizures. Bakshi et al. (2000) noted that the effects of nerve agents on GABAergic transmission in the CNS may have profound implications for behavioral effects in laboratory animals and humans and may also contribute to the induction of convulsions at higher doses.

The progression of events includes initially high cholinergic activity followed by activation of glutamatergic neurons, inflammation, and neuronal injury (Zaja-Milatovic and Gupta, 2015). The resulting brain damage appears to be associated with oxidative stress and consequent degeneration of pyramidal neurons from the CA1 hippocampal region of the brain. Findings also revealed that seizure-induced early increases in biomarkers of global free radical damage (F2-IsoPs) and the selective peroxidation biomarkers of neuronal membranes (F4-NeuroPs) were accompanied by dendritic degeneration of pyramidal neurons in the CA1 hippocampal area, as evaluated by Golgi impregnation and Neurolucida-assisted morphometry. This investigation explored oxidative injury, different pathways to attenuate biomarkers of oxidative damage associated with anti-AChE exposure, and the extent to which such attenuation is accompanied by rescue from neurodegeneration.

Neurological injury caused by nerve agent-induced seizures is accompanied by an inflammatory reaction involving gliosis, and an induction of inflammatory mediators including prostaglandins, cytokines, cell adhesion proteins, and matrix metalloproteinases (MMPs). Animals exposed to soman at doses producing convulsions exhibited a rapid increase in active astrocytes and the accumulation of glial fibrillary acidic protein (GFAP) (Zimmer et al., 1997).

OP compounds induce activation of glutamatergic neurons, generation of reactive oxygen species (ROS) and reactive nitrogen species (RNS) leading to neurodegeneration. In addition to oxidative damage and interference with adenylyl cyclase cell signaling, chlorpyrifos caused inhibition of DNA synthesis to a greater extent in glioma cell lines (C6 cells) than neuronal cell lines (P12) (Garcia et al., 2001). These effects are independent of cholinergic receptors as the cholinergic antagonist fails to block chlorpyrifos-induced inhibition of DNA synthesis. The findings were also consistent with exposure to another OP compound, diazinon, suggesting that anticholinesterase compounds target glial cells by additional mechanisms of cholinergic toxicity (Walker and Nidiri, 2002).

Involvement of noncholinergic mechanisms in OP toxicity is also supported by evidence that these agents induce activation of glutamatergic neurons. For example, soman-induced seizure increased extracellular glutamate in the pyriform cortex (Wade et al., 1987) and the cornu ammonis (CA) region of the hippocampus (Lallement et al., 1992), followed by the activation of NMDA receptors in the CA1 region. Overstimulation of glutamate receptors causes synaptic and cellular degeneration in the hippocampus (Munirathinam and Bahr, 2004). Moreover, glutamate stimulates ACh release (Anderson et al., 1994) further contributing to excitatory stimulation and propagation of seizures, and thus like a brushfire, it propagates excitotoxic neurodegeneration in vulnerable brain regions. Microdialysis studies revealed an immediate increase in extracellular glutamate concentrations in the septum, pyriform cortex, hippocampal regions and amygdala following soman-triggered seizures (Lallement et al., 1991a,b; Wade et al., 1987). Furthermore, blockage of specific glutamate receptors reduces neuropathogenic responses, including nerve agents' toxicity (Sparenborg et al., 1992).

Death from nerve agent toxicity appears to be due to respiratory failure, which may involve depression of the brain's respiratory center, neuromuscular blockage of the respiratory muscles, bronchial constriction, and increased lung secretions.

5.3 Oxidative stress

Oxidative stress has been implicated as one of the mechanisms in neurotoxicity of OP nerve agents (Gupta et al., 2001; Naughton and Terry, 2018; Pearson and Patel, 2016; Vanova et al., 2018; Zaja-Milatovic and Gupta, 2015; Zaja-Milatovic et al., 2009). In addition to the activation of NMDA receptors, glutamate release also leads to massive Ca^{2+} fluxes into the postsynaptic cells, generation of ROS and RNS, and ensuing neurodegeneration. Following an acute exposure to DFP (1.5 mg/kg, SC), time-course analysis of cerebral biomarkers of oxidative damage revealed significant increases in F_2-isoprostanes (F_2-IsoPs, 142%) and F_4-neuroprostanes (F_4-NeuroPs, 225%) within 30 min. Both F_2-IsoPs and F_4-NeuroPs showed transient increase reaching maximum levels at 1 h with a return to basal levels by 6 h (Zaja-Milatovic et al., 2009). The selective increase in F_4-NeuroPs indicates that neurons are specifically targeted by this mechanism. Thus, DFP produced a transient increase in both F_2-IsoPs and F_4-NeuroPs, which are sensitive and specific in vivo biomarkers of oxidative damage to arachidonic acid (AA) and docosahexaenoic acid (DHA), respectively.

At the time of maximal severity, i.e., 1 h, DFP caused >3-fold increase in citrulline (a marker of RNS) compared to control (139.2 ± 8.4 nmol/g tissue) in cerebral tissue. At the same time, significant depletion of high-energy phosphates (HEPs) such as ATP and phosphocreatine (2.23 ± 0.05 and 8.90 ± 0.33 μmol/g tissue, respectively) occurred. Alterations in HEPs are considered indicators of mitochondrial dysfunction that increases neuronal vulnerability to oxidative and neuronal injuries.

6. Dendritic damage

Dendritic morphology of randomly selected pyramidal neurons from the CA1 hippocampal area in control and DFP-exposed rats (1.25 mg/kg, SC) was evaluated by the Sholl method (Zaja-Milatovic et al., 2009). Sholl analysis represents a quantitative method for morphometric neuronal studies with consecutive-circles (50, 51–100, 101–150 μm from the center of soma) that analysis specifies dendritic geometry. Zaja-Milatovic et al. (2009) demonstrated that DFP treatment (1.25 mg/kg, SC) caused a significant decrease in total dendrite length in proximal (0–50 μm) and intermediate (51–100 μm) Sholl compartments of CA1 pyramidal neurons. DFP treatment also caused a significant decrease in spinal density (number of spines per 100 μm of dendrites) in all three (proximal, intermediate and distal) Sholl compartments of CA1 pyramidal neurons. DFP-induced high intensity seizures and oxidative/nitrosative damage to neuronal membranes at 1 h which coincided with dendritic degeneration of pyramidal neurons in the CA1 hippocampal area. Neurolucida-assisted morphometry showed that DFP-induced brain hyperactivity targeted the dendritic system with profound dendritic regression of hippocampal neurons.

Recently, Figueiredo et al. (2018) summarized the findings of many previous studies, providing evidence that prolonged seizures from OP nerve agents cause the neuropathology (Shih and McDonough, 1997; Shih et al., 2003). Neurodegeneration occurs most frequently in the amygdala, followed by the pyriform cortex, hippocampus, cerebral cortex, thalamus, and caudate/putamen (Aroniadou-Anderjaska et al., 2009; Baze, 1993; Carpentier et al., 1991; Shih et al., 2003).

7. Nerve agents and alterations at the blood–brain barrier

The blood–brain barrier (BBB) is formed and maintained through a dynamic interaction between cerebral and endothelial cells constituting the anatomical basis of the BBB and other neighboring cells, such as astroglia, pericytes, perivascular microglia, and neurons. The cross-talk between these cells endows endothelial cells with a unique BBB phenotype comprising not only the morphological barrier of endothelial tight junctions, but also the enzymatic and metabolic barriers, as well as the uptake and efflux transport systems (Abbott et al., 2010). The BBB is incapable of preventing the exchange of toxins and toxicants from the blood to the brain when its integrity (structure, function, or permeability) is compromised by chemicals or their toxicity (Gupta et al., 2015).

Nerve agents are small lipophilic molecules and they can easily penetrate the BBB by free diffusion in the CNS. Carpentier et al. (1990) found an increased BBB permeability to the EB–albumin complex observed macroscopically in rats intoxicated by 85 μg/kg of soman. Soman produced seizures and reversible BBB opening to the greatest extent after 30–60 min of paroxysmal electroencephalographic (EEG) discharges when signs of cerebral hyperactivity (epileptic EEG pattern and hypoxia) were also at their height. Carpentier et al. (1990) detected the BBB opening in the amygdaloid complex and in some cortical regions (cingulum, entorhinal, and pyriform complex). The thalamus was the most frequently and intensely affected structure, and the hippocampus remained free of exudated immunoreactive plasmatic gamma-immunoglobulin G (IgG). Using [^{3}H]hexamethonium iodide as an indicator of BBB permeability, Jones et al. (2000) observed no significant change in cortex, midbrain, and cerebellum in rats after a single sarin treatment (100 μg/kg, im). The brainstem exhibited a significant decrease in BBB permeability (~58% of control). Various studies suggested that nerve agent (soman or VX)-induced BBB alterations are predominantly related to convulsions and seizures or to a direct cytotoxic action of nerve agents (McDonough et al., 1987). Ashani and Catravas (1981) reported that soman-induced damage to BBB corroborates with a high degree of AChE inhibition in intoxicated rats. In several other studies, OP nerve agents and pesticides are known to cause BBB damage, and their increased access leads to greater AChE inhibition in the brain, while in some other studies BBB damage by nerve agents appears to be independent of AChE inhibition (reviewed in Gupta et al., 2015).

8. Electrophysiological alterations (central vs peripheral) by OP nerve agents

Nonlethal toxic exposure to AChE inhibitors produces prominent motor, behavioral, and autonomic symptoms. The motor symptoms are fasciculations, fibrillations, and body tremors. Fasciculations and fibrillations are due to antidromic neural discharge from excess junctional ACh; tremors are of a central origin (Laskowski and Dettbarn, 1979). In a number of studies, a difference in the pattern of motor symptoms in rats treated with soman or a nerve agent surrogate DFP was observed (Gupta et al., 1985, 1986, 1987). Soman produced complex posturing movements and tremors affecting virtually the entire body, while DFP produced movements that were similar to fasciculations or myokymia. In addition, muscle fiber necrosis was more frequent at symptom-producing doses of DFP than at equivalent doses of soman.

Electrophysiological effects of nerve agents have been reported in many studies. Misulis et al. (1987) recorded electromyographic activity from normal and acutely denervated rat gastrocnemius muscle following administration of soman (100 μg/kg, sc) or DFP (1.5 mg/kg, sc). The motor symptoms induced on the denervated side represented activity originating at the nerve terminal, while those of the contralateral nondenervated muscle represented the sum of peripheral plus central descending activity. The findings indicated that soman and DFP produced different responses. The majority of motor symptoms induced by soman were due to impulses descending from the CNS, and a portion of these symptoms were epileptiform activity. This activity was not generated at spinal levels. In contrast, DFP produced motor symptoms mainly by peripheral action that was dependent on a functioning nerve terminal. Interestingly, at lethal doses, both agents exerted central and peripheral effects. In essence, different patterns of electrical activity were associated with each of the motor symptoms, animals with greater motor symptoms had greater frequencies of motor unit discharge, and muscle fiber necrosis correlates with peripherally generated high-frequency repetitive discharges.

Soman and DFP produced very different amounts and patterns of muscle electrical activity. Soman produced mainly centrally generated motor unit activity with little peripheral muscle fiber discharge. In contrast, DFP produced more peripherally generated activity. There are at least four possible explanations for these differences in electrophysiological effects: (1) nonspecific effects of the inhibitors in muscle or CNS independent of changes in AChE activity, (2) differing affinities of central and peripheral forms of AChE for each inhibitor, (3) varying transport rates across the BBB which may select the degree of central actions of the individual agents, favoring soman, and (4) DFP may be degraded before it reaches the CNS in sufficient concentrations to produce central effects.

9. Nerve agents and the neuromuscular junction

Rats cutely intoxicated with soman (100 μg/kg, sc), sarin (110 μg/kg, sc), tabun (200 μg/kg, sc) or VX (12 μg/kg, sc) developed toxic signs within 5–15 min after soman, sarin, or tabun, and 20–50 min after VX. Signs of maximum severity occurred within 1–2 h, and lasted for 5–7 h. By 24 h, only fine tremors were seen and other signs were subsided, and rats became free of signs after 72 h. Pharmacokinetic studies suggest that skeletal muscles receive a significant amount of nerve agents. Necrosis in slow-twitch fiber soleus, fast-twitch fiber extensor digitorum longus (EDL), and mixed fiber-containing diaphragm and at the neuromuscular junction (NMJ) by soman, sarin, tabun, VX, and DFP have been reported in several studies (Gupta and Dettbarn, 1992; Gupta et al., 1985, 1986, 1987, 1988, 1991).

10. Intermediate syndrome and delayed polyneuropathy

There are only a few studies that have addressed long-term effects of nerve agents in general and VX in particular. Unlike OP pesticides (Costa, 2018; De Bleecker, 2006; Richardson et al., 2015), there is no evidence that nerve agents produce an intermediate syndrome or OP-induced delayed neuropathy (OPIDN). In a number of studies following low dose exposure to nerve agents, only a few delayed effects have been reported. Munro et al. (1994) reported that at supralethal doses, sarin can cause delayed neuropathy in antidote-protected chickens, but there is no evidence that it produces this syndrome in humans at any doses. VX shows no potential for inducing delayed neuropathy in any species.

11. Neuropsychiatric effects and PTSD

Nerve agent-induced brain damage is defined as progressive damage to the brain, resulting from the cholinergic neuronal excitotoxicity and dysfunction caused by OP-induced irreversible AChE inactivation (Chen, 2012). This delayed secondary neuronal damage that occurs mainly in the cholinergic regions of the brain which contain dense accumulations of cholinergic neurons and the majority of cholinergic projections, might be largely responsible for persistent profound neuropsychiatric and neurological impairments (memory, cognitive, mental, emotional, and motor and sensory deficits) in the victims of OP poisoning (Burchfiel and Duffy, 1982; Chen, 2012; Collombet, 2011; Naughton and Terry, 2018; Pearce et al., 1999).

The neuroprotective strategies may help limit or prevent secondary neuronal damage in the early stages of OP poisoning and attenuate the subsequent neuropsychiatric and neurological impairments, thus reducing the long-term disability caused by exposure to OPs (Chen, 2012).

12. Nerve agents and brain cancer

In a 10-year follow-up study, Chen (2012) found a twofold increased risk of brain cancer death among Gulf-War deployed veterans exposed to GB and GF at Khamisiyah, Iraq in March 1991. Furthermore, the risk of brain cancer death, which increased with the time and amount of exposure, is believed to develop over 10–20 years in veterans exposed to sarin (Bullman et al., 2005). There is no reported evidence that VX causes cancer.

13. Biomarkers of OP nerve agent exposure and effect

In a number of studies, OP nerve agents or their metabolites in body fluids and tissues have been suggested to serve as biomarkers of exposure. Driskell et al. (2002) and Barr et al. (2004) developed an isotope dilution GC-MS-MS method for quantitating human urinary metabolites of the OP nerve agents (tabun, sarin, cyclosarin, soman, VX, and RVX). Urine samples were acidified, extracted into ether-acetonitrile, derivatized by methylation with diazomethane, and analyzed by GC-MS-MS. The limits of detection were reported to be <1 μg/L (1 ppb) for all analytes. Recently, developed a GC-MS-MS-based robust method for the detection of six OP nerve agents (tabun, sarin, soman, cyclosarin, VX, and RVX) and the six corresponding breakdown products (ethyl hydrogen dimethyl phosphoramidate (EDPA), isopropyl methylphosphonic acid (IMPA), pinacolyl methylphosphonic acid (PMPA), cyclohexyl methylphosphonic acid (CMPA), ethyl methylphosphonic acid (EMPA), and 2-(methyl)propyl methylphosphonic acid (MMPA)), with a sensitivity of picogram was developed. Ewing and Valenzuela (2018) demonstrated that OP compounds can be detected at the parts-per-trillion (ppt) level using atmospheric flow tube-mass spectrometry (AFT-MS). The Centers for Disease Control's (CDC's) final recommendations for protecting human health from potential adverse effects of exposure to nerve agents in 2008 listed limits ranging from 6×10^{-7} to 1×10^{-6} mg/m^3 (roughly 0.1 ppt$_v$) in the general population (CDC, 2003).

Tsuchihashi et al. (1998) identified two metabolites (ethyl methylphosphonic acid and 2-diisopropylamino-(ethyl) methyl sulfide) of VX in the serum of a Osaka victim exposed to VX, using GC-MS and GC-MS-MS. Miki et al. (1998) developed a GC-MS-negative-ion chemical ionization-based method to determine the main metabolite alkylmethylphosphonic acids (AMPAs; alkyl = Et, i-Pr, and pinacolyl), the main metabolites of the OP nerve agent alkylmethylphosphonic acid (AMPA) in biofluids (urine, serum and saliva) with a detection limits of 60 pg/mL. A hydrolysis product of sarin, *O*-isopropyl methylphosphonic acid (IMPA), can be detected in biological samples using tandem mass spectrometry (MS/MS) in multiple reaction monitoring (MRM) mode, as well as by high-resolution mass spectrometry (HR-MS) without fragmentation following liquid chromatography (LC) separation and electrospray ionization (ESI) in negative mode, ESI (−). Recently, Tan et al. (2017) demonstrated that the gold-plated 3D printed stainless steel electrode outperformed that of the glassy carbon electrode for the analysis of nitroaromatic explosives (2,4-dinitrotoluene and 2,4,6-trinitrotoluene) and an OP pesticide (fenitrothion), displaying higher sensitivities and broader analyte detection range, and this could be used for nerve agents as well. For further details on detection of nerve agents in biological fluids, tissues and environmental samples, readers are referred to other publications (Anzai, 2015; John et al., 2018; Seto, 2015; Van Der Schans, 2015).

Several methods have been developed for quantitation of AChE activity in blood. The monitoring of AChE activity is a reliable biomarker for nerve agent exposure and toxic effects (Jun et al., 2015; Reiner and

Simeon-Rudolf, 2006). Bloch-Shilderman et al. (2018) suggests that ChE activity is the most sensitive biomarker of VX exposure. Systemic toxic effects are seen in approximately 50% of subjects when 75% of RBC AChE is inhibited. Recovery of AChE from inhibition is also used as an indicator of recovery of signs and symptoms of nerve agent toxicity. It needs to be pointed out that in the case of sarin, recovery of RBC AChE activity does not reflect the time course of recovery of AChE activity in tissues. Thus, monitoring of RBC AChE activity is of questionable utility in assessing recovery from nerve agent toxicity.

Compared to RBC AChE, plasma ChE activity is less relevant, because it has not yet been assigned any biological function. Also, plasma ChE activity can be affected by factors such as gender, age, race, oral contraceptives, disease states, circadian patterns, and nutritional status (Munro et al., 1994). Both VX and sarin cause significantly less inhibition of plasma ChE than RBC ChE. Of course, VX causes plasma ChE inhibition significantly less than that caused by sarin. Plasma ChE provides a protective function, complexing with nerve agents and thus reducing their free concentration available to complex with tissue AChE. Recently, Zhao et al. (2018) developed a highly sensitive immunoassay strategy utilizing a nanozyme- and ambient light-based smartphone platform for detection of BuChE as a biomarker of OPs exposure. However, in a previous study, Jones et al. (2000) suggested that plasma BuChE rapidly recovers after sarin or is at least inhibited by VX, and therefore it could not be used as a reliable biomarker of nerve agents' exposure or effect. The Test-Mate OP Kit developed by the EQM Corporation has been chosen by the US Army to monitor potential OP poisoning. The test kit uses a single drop of blood to measure levels of both AChE and BuChE activity. The test is to be used in the field and in clinical labs to monitor levels of OP exposure.

Brain AChE inhibition and the degree of toxicity show a better correlation in that tabun, soman, sarin, and VX injected into rats produced a dose-dependent inhibition of brain AChE, with lethal doses producing >90% inhibition (Gupta et al., 1987, 1988, 1991; Sivam et al., 1984). Recently, Fu et al. (2019) suggested that Y263 site in human serum albumin can be used as a potential biomarker to detect OP nerve agent exposure, using LC-MS/MS method for the verification of new phosphonylated sites.

14. Treatment

Exposure to a nerve agent can be fatal unless therapy is instituted in a timely fashion. Treatment of nerve agent toxicity requires a three-pronged approach: (1) Decontamination is crucial to prevent continued exposure to the patient and to emergency response staff (Capoun and Krykorkova, 2014; Clarkson and Gordon, 2015; Matar et al., 2019), (2) Clinicians should immediately administer atropine sulfate (2–6 mg, IV, every 5–10 min) until muscarinic symptoms (bradycardia, bronchorrhea and bronchospasm) are resolved, and (3) Intravenous administration of an oxime pyridine-2-aldoxime methochloride (2-PAM, 1–2 g) or obidoxime (250 mg). 2-PAM can be repeated every 3–6 h to maintain its plasma concentrations (Wiener and Hoffman, 2004). It needs to be emphasized that oxime should be administered before AChE can be aged (Sharma et al., 2015). Also, no single oxime can reactivate AChE across all OPs; instead, the efficacy of each oxime varies greatly from one OP to the next (Moffett et al., 2018). 2-PAM offers little to no reactivation against tabun, cyclosarin, or soman. Recently, Pashirova et al. (2018) found that a combination of two oximes [2-PAM and 3-HPA (6-(5-(6,7-dimethoxy-3,4-dihydroisoquinolin-2(1H)-yl)pentyl)-3-hydroxy piconaldehyde oxime) more effective than 2-PAM alone in reactivation of AChE and neuroprotection against paraoxon-poisoned rats.

The US Army currently fields the Antidote Treatment-Nerve Agent Auto-Injector (ATNAA) containing 600 mg 2-PAM and 2.1 mg atropine. In experimental studies, various oximes have been evaluated for AChE reactivation against nerve agents (Katalinic et al., 2018; Moffett et al., 2018; Wei et al., 2016; Wilhelm et al., 2018; Worek et al., 2016a). In recent studies, a broad spectrum oxime MMB4 has been found to be more effective and safer than 2-PAM or HLö7 (Moffett et al., 2018; Wilhelm et al., 2018). Additionally, Kassa et al. (2018) found that three newly developed bispyridinium nonoxime compounds (MB408, MB442, and MB444), in combination with standard antidotes (atropine and oxime HI-6 or obidoxime), provided superior protection compared to standard antidotal treatment against acute toxicity of sarin and cyclosarin in mice. In a number of studies, the use of uncharged reactivators has been suggested against OP nerve agent-induced AChE inhibition (Mercey et al., 2011; Renou et al., 2013). The use of novel nonoxime AChE reactivators against OPs (paraoxon, sarin, cyclosarin, and VX), basing on the common structure 4-amino-2-((diethylamino)methyl)phenol, has also been suggested by Horn et al. (2018).

Recent studies suggest that seizure from nerve agents or surrogate DFP poisoning can be prevented and treated with midazolam (Chapman et al., 2015; Reddy and Reddy, 2015) or diazepam (Chai et al., 2018), and central respiratory depression can be antagonized by naloxone (Škrbić et al., 2017).

Reymond et al. (2018) reported that HI-6 dimethanesulfonate (HI-6 DMS) proved superior to pralidoxime methylsulfate (2-PAM) against Russian VX (VR) poisoning in cynomolgus monkeys. As expected with VR, inhibition of blood AChE was observed but without any apparent impact on the clinical recovery of the animals. A single administration of the HI-6 DMS (33 mg/kg) in combination with atropine sulfate (0.25 mg/kg) and avizafone (0.175 mg/kg) provided 50% survival rate in monkeys exposed to 15LD_{50} of VR. Sit et al. (2018) reported that ionizing zwitterionic aldoximes appear to be viable alternatives to existing antidotes for prophylaxis and treatment of individuals exposed to OP nerve agents.

In experimental studies, agents such as pyridostigmine, physostigmine, hyoscine, diphenhydramine, benactyzine, huperzine A, cyclopentyl adenosine, mecamylamine or IV lipid emulsion have been found to be effective as preventative or therapeutic measure (Bird et al., 2002; Chai et al., 2017; Eckert et al., 2007; Eisenkraft and Falk, 2016; Fleisher et al., 1970; Miller et al., 1993; Tuovinen, 2004; Weger and Szinicz, 1981). NMDA receptor antagonist memantine HCl (18 mg/kg, SC), in combination with atropine sulfate (16 mg/kg, SC), was reported to be highly effective against acute toxicity of nerve agents (soman, sarin, tabun and VX) and the OP insecticide methyl parathion in rats (Gupta and Dettbarn, 1992; Gupta and Kadel, 1990; McLean et al., 1992). Prophylactic treatment was more effective than therapeutic treatment. Stojiljković et al. (2019) also reported prophylactic potential of memantine alone or in combination with atropine/HI-6/diazepam against soman poisoning. Memantine provides prophylactic action by protection of AChE and NMDA receptor antagonistic action. Ketamine (another NMDA receptor antagonist) when added to standard therapy (atropine, 2-PAM and midazolam) improved neurobiological outcome in rats intoxicated with sarin (Lewine et al., 2018). In the past, pyridostigmine bromide, in combination with atropine sulfate, has also been used as preventative treatment of nerve agents.

In a number of studies, enzyme-based nerve agent bioscavengers have been found to be very effective in attenuating toxicity (Masson, 2015; Nachon et al., 2013; Sterri and Fonnum, 2015). Lushchekina and Masson (2018) emphasized the catalytic bioscavengers (second-generation bioscavengers) taking cholinesterase inhibition, aging and reactivation into consideration. These biopharmaceuticals are intended to degrade toxic OPs on the skin for decontamination or in the bloodstream for pretreatment and postexposure treatment of OP poisoning. Their catalytic efficiency is expected to be as high as possible ($s_{cat}/K_m > 10^6\,M^{-1}\,min^{-1}$) (Goldsmith and Ashani, 2018; Goldsmith et al., 2017). Mammalian paraoxonases (chimeric paraoxonase-1, PON 1) and phosphotriesterases (PTE) already fulfill this requirement (Bigley et al., 2015; Goldsmith et al., 2017; Rochu et al., 2007). Lushchekina and Masson (2018) suggested that the efficacy of ChE-based catalytic bioscavengers can be improved by: (1) introducing new general base to activate water molecules; (2) increasing accessibility of the phosphorus atom for the water molecule by reducing hydrophobicity around phosphoryl adduct; (3) creating a proton acceptor to facilitate release of the phosphate ion; and (4) creating a reaction pathway favoring reactivation over aging. Lushchekina and Masson (2018) also pointed out that the recent revelation of the chemical structures of highly toxic agents called Novichoks (Costanzi et al., 2018; Hosseini et al., 2016) pose new difficulties that should stimulate research on novel catalytic bioscavengers capable of hydrolyzing P—N bond at high rate and/or oxidizing OP side chains. Hydrolysis by PTEs and self-reactivation of phosphorylated ChEs seems virtually impossible. Therefore, future works need to develop functional associations of enzyme cocktails, encapsulation of enzymes for increasing their stability and life time in the bloodstream, and preventing immune response (Lushchekina and Masson, 2018).

On behalf of the American College of Medical Toxicology, Stolbach et al. (2018) suggested two alternative or contingency countermeasures for nerve agents:

1. The use of expired atropine, diazepam, and pralidoxime auto-injectors and vials if non-expired drugs are unavailable. In March 2016, the FDA issued a memorandum supporting the use of atropine, diazepam, and pralidoxime auto-injectors for up to 2 years beyond the manufacturer's initial expiration date (US FDA, 2016). This memorandum supported the contention that a drug is not necessarily unstable or degraded after the expiration date (Schier et al., 2004). In such a scenario, clinicians can simply administer more medication until the desired clinical effect is observed (ACMT, 2015).
2. Investigation, development, and identification of alternative countermeasures-commonly stocked drugs that are not approved for nerve agent poisoning but are in the same therapeutic class as approved drugs (e.g., glycopyrrolate instead of atropine; and lorazepam instead of diazepam).

15. Concluding remarks and future perspective

OP nerve agents exert their toxicity by multiple mechanisms (cholinergic and noncholinergic). The primary effects of nerve agents appear to be on the CNS and skeletal muscles, but of course other organs are affected as well. There is no evidence of intermediate syndrome, delayed polyneuropathy or tolerance development with nerve agents' toxicity. Treatment rests with blockage of secretory activity and excitotoxicity. Atropine sulfate in combination with an oxime are specific antidotes for the treatment of acute toxicity. Several adjunctive therapies are indicated for additional benefits. Future studies need to look for broad spectrum oximes which can penetrate the blood–brain barrier to be effective against OP nerve agents. Currently, the investigations on novel bioscavengers and catalytic bioscavengers for pretreatment and treatment of OP poisoning are underway.

References

Aas, P., Veiteberg, T.A., Fonnum, F., 1987. Acute and sub-acute inhalation of an organophosphate induced alteration of cholinergic muscarinic receptors. Biochem. Pharmacol. 36, 1261–1266.

Abbott, N.J., Patabendije, A.A., Dolman, D.E., et al., 2010. Structure and function of the blood-brain barrier. Neurobiol. Dis. 37, 13–25.

American College of Medical Toxicology (ACMT), American Academy of Clinical Toxicology, 2015. Antidote shortage in the USA: impact and response. J. Med. Toxicol. 11, 144–146.

Anderson, J.J., Kuo, S., Chase, T.N., 1994. Endogenous excitatory amino acids tonically stimulate striatal acetylcholine release through NMDA but not AMPA receptors. Neurosci. Lett. 176, 264–268.

Anzai, J.-I., 2015. Biosensors for the detection of OP nerve agents. In: Gupta, R.C. (Ed.), Handbook of Toxicology of Chemical Warfare Agents, second ed. Academic Press/ Elsevier, Amsterdam, pp. 925–934.

Aroniadou-Anderjaska, V., Figueiredo, T.H., Apland, J.P., et al., 2009. Primary brain targets of nerve agents: the role of the amygdala in comparison to the hippocampus. Neurotoxicology 30, 772–776.

Ashani, Y., Catravas, G.N., 1981. Seizure-induced changes in the permeability of the blood-brain barrier following administration of anticholinesterase drugs to rats. Biochem. Pharmacol. 30, 2593–2601.

Bajgar, J., Fusek, J., Kassa, J., et al., 2015. Global impact of chemical warfare agents used before and after 1945. In: Gupta, R.C. (Ed.), Handbook of Toxicology of Chemical Warfare Agents, second ed. Academic Press/Elsevier, Amsterdam, pp. 7–15.

Bakry, N.M., el-Rashidi, A.H., Eldefrawi, A.T., et al., 1988. Direct actions of organophosphate anticholinesterases on nicotinic and muscarinic acetylcholine receptors. J. Biochem. Toxicol. 3, 235–259.

Bakshi, K.S., Pang, S.N.J., Snyder, R., 2000. Review of the Army's health risk assessment for oral exposure to six chemical-warfare agents. J. Toxicol. Environ. Health A 59, 281–526.

Barr, J.R., Driskell, W.J., Aston, L.S., et al., 2004. Quantitation of metabolites of the nerve agents sarin, soman, cyclohexylsarin, VX, and Russian VX in human urine using isotope-dilution gas chromatography-tandem mass spectrometry. J. Anal. Toxicol. 28 (5), 372–378.

Baze, W.B., 1993. Soman-induced morphological changes: an overview in the non-human primate. J. Appl. Toxicol. 13, 173–177.

Bigley, A.N., Mabanglo, M.F., Harvey, S.P., et al., 2015. Variants of phosphotriesterases for the enhanced detoxification of the chemical warfare agent VR. Biochemistry 54, 5502–5512.

Bird, S.B., Gaspari, R.J., Lee, W.J., et al., 2002. Diphenhydramine as a protective agent in a rat model of acute, lethal organophosphate poisoning. Acad. Emerg. Med. 9, 1369–1372.

Black, R.M., Harrison, J.M., 1996. The chemistry of organophosphorus chemical warfare agents. In: Hartley, F.R. (Ed.), The Chemistry of Organophosphorus Compounds. In: vol. 4. John Wiley & Sons, Hoboken, NJ, pp. 783–798.

Bloch-Shilderman, E., Rabinovitz, I., Egoz, I., et al., 2018. Determining a threshold sub-acute dose leading to minimal physiological alterations following prolonged exposure to the nerve agent VX in rats. Arch. Toxicol. 92, 873–892.

Bullman, T.A., Mahan, C.M., Kang, H.K., et al., 2005. Mortality in US Army Gulf War veterans exposed to 1991 Khamisiyah chemical munitions destruction. Am. J. Public Health 95, 1382–1388.

Burchfiel, J.L., Duffy, F.H., 1982. Organophosphate neurotoxicity: chronic effects of sarin on the electroencephalogram of monkey and man. Neurobehav. Toxicol. Teratol. 4, 767–778.

Capoun, T., Krykorkova, J., 2014. Comparison of selected methods for individual decontamination of chemical warfare agents. Toxics 2, 307–326.

Carpentier, P., Delamanche, I.S., Bert, M.L., et al., 1990. Seizure-related opening of the blood-brain barrier induced by soman: possible correlation with the acute neuropathology observed in poisoned rats. Neurotoxicology 11, 493–508.

Carpentier, P., Lambrinidis, M., Blandchet, G., 1991. Early dendritic changes in hippocampal pyramidal neurons (field CA1) of rats subjected to acute soman intoxication: a light microscopic study. Brain Res. 541, 293–299.

Centers for Disease Control and Prevention (CDC), 2003. Final recommendations for protecting human health from potential adverse effects of exposure to agents GA, GB and VX. Fed. Regist. 68 (196), 58348–58351.

Chabebo, S.R., Santos, M.D., Albuquerque, E.X., 1999. The organophosphate sarin, at low concentrations, inhibits the evoked release of GABA in rat hippocampal slices. Neurotoxicology 20, 871–882.

Chai, P.R., Boyer, E.W., Al-Nahhas, H., et al., 2017. Toxic chemical weapons of assassination and warfare: nerve agents VX and sarin. Toxicol. Commun. 1, 21–23.

Chai, P.R., Hayes, B.D., Erickson, T.B., et al., 2018. Novichok agents: a historical, current, and toxicological perspective. Toxicol. Commun. 2 (1), 45–48.

Chapman, S., Yaakov, G., Egoz, I., et al., 2015. Sarin-induced brain damage in rats is attenuated by delayed administration of midazolam. Neurotoxicology 49, 132–138.

Chen, Y., 2012. Organophosphate-induced brain damage: mechanisms, neuropsychiatric and neurological consequences, and potential therapeutic strategies. Neurotoxicology 33, 391–400.

Chen, S.M., Chi, M.G., 1986. Direct effect of VX and soman on nicotinic receptors. Acta Pharmacol. Sin. 7, 401–406.

Clarkson, E.D., Gordon, R.K., 2015. Rapid decontamination of chemical warfare agents from the skin. In: Gupta, R.C. (Ed.), Handbook of Toxicology of Chemical Warfare Agents, second ed. Academic Press/Elsevier, Amsterdam, pp. 1127–1139.

Collombet, J.M., 2011. Nerve agent intoxication: recent neuropathophysiological findings and subsequent impact on medical management prospects. Toxicol. Appl. Pharmacol. 255, 229–241.

Costa, L.G., 2018. Organophosphorus compounds at 80: some old and new issues. Toxicol. Sci. 162 (1), 24–35.

Costanzi, S., Machado, J.-H., Mitchell, M., 2018. Nerve agents: what they are, how they work, how to encounter them. ACS Chem. Neurosci. 19 (5), 873–885.

Dalton, C., Graham, S., Jenner, J., 2018. Effect of aqueous dilution on the absorption of the nerve agent VX through skin *in vitro*. Toxicol. In Vitro 53, 121–125.

De Bleecker, J.L., 2006. Intermediate syndrome in organophosphate poisoning. In: Gupta, R.C. (Ed.), Toxicology of Organophosphate and Carbamate Compounds. Academic Press/Elsevier, Amsterdam, pp. 371–380.

Dolgin, E., 2013. Syrian gas attack reinforces need for better anti-sarin drugs. Nat. Med. 19, 1194–1195.

Driskell, W.J., Shih, M., Needham, L.L., et al., 2002. Quantitation of organophosphorus nerve agent metabolites in human urine using isotope dilution gas chromatography-tandem mass spectrometry. J. Anal. Toxicol. 26 (1), 6–10.

Eckert, S., Eyer, P., Worek, F., 2007. Reversible inhibition of acetylcholinesterase by carbamates or huperzine A increases residual activity of the enzyme upon soman challenge. Toxicology 233, 180–186.

Eisenkraft, A., Falk, A., 2016. The possible role of intravenous lipid emulsion in the treatment of chemical warfare agent poisoning. Toxicol. Rep. 3, 202–210.

Ewing, R.G., Valenzuela, B.R., 2018. Selective reagent ions for the direct vapor detection of organophosphorus compounds below parts-per-trillion levels. Anal. Chem. 90, 7583–7590.

Figueiredo, T.H., Apland, J.P., Braga, M.F.M., et al., 2018. Acute and long-term consequences of exposure to organophosphate nerve agents in humans. Epilepsia 59 (S2), 92–99.

Fleisher, J.H., Harris, L.W., Miller, G.R., et al., 1970. Antagonism of sarin poisoning in rats and guinea pigs by atropine, oximes, and mecamylamine. Toxicol. Appl. Pharmacol. 16, 40–47.

Fu, F., Sun, F., Lu, X., et al., 2019. A novel potential biomarker on Y263 site in human serum albumin poisoned by six nerve agents. J. Chromatogr. B 1104, 168–175.

Garcia, S.J., Seidler, F.J., Crumpton, T.L., et al., 2001. Does the developmental neurotoxicity of chlorpyrifos involve glial targets? Macromolecule synthesis, adenyl cyclase signaling, nuclear transcription factors and formation of reactive oxygen in C6 glioma cells. Neurotoxicology 22, 129–154.

Goldsmith, M., Ashani, Y., 2018. Catalytic bioscavengers as countermeasures against organophosphate nerve agents. Chem. Biol. Interact. 292, 50–64.

Goldsmith, M., Aggarwal, N., Ashani, Y., et al., 2017. Overcoming an optimization plateau in the directed evolution of highly efficient nerve agent bioscavengers. Protein Eng. Des. Sel. 30, 333–345.

Gupta, R.C., 2006. Classification and uses of organophosphates and carbamates. In: Gupta, R.C. (Ed.), Toxicology of Organophosphates and Carbamates. Academic Press/Elsevier, Amsterdam, pp. 5–24.

Gupta, R.C. (Ed.), 2015. Handbook of Toxicology of Chemical Warfare Agents, second ed. Academic Press/Elsevier, Amsterdam.

Gupta, R.C., Dettbarn, W.-D., 1992. Potential of memantine, d-tubocurarine, and atropine in preventing acute toxic myopathy induced by organophosphate nerve agents: soman, sarin, tabun, and VX. Neurotoxicology 3, 649–662.

Gupta, R.C., Kadel, W.L., 1990. Methyl parathion acute toxicity: prophylaxis and therapy with memantine and atropine. Arch. Int. Pharmacodyn. Therap. 305, 208–221.

Gupta, R.C., Patterson, G.T., Dettbarn, W.-D., 1985. Mechanisms involved in the development of tolerance to DFP toxicity. Fundam. Appl. Toxicol. 5, S17–S28.

Gupta, R.C., Patterson, G.T., Dettbarn, W.-D., 1986. Mechanisms of toxicity and tolerance to diisopropylphosphorofluoridate at the neuromuscular junction. Toxicol. Appl. Pharmacol. 84, 541–550.

Gupta, R.C., Patterson, G.T., Dettbarn, W.-D., 1987. Biochemical and histochemical alterations following acute soman intoxication in the rat. Toxicol. Appl. Pharmacol. 87, 393–407.

Gupta, R.C., Patterson, G.T., Dettbarn, W.-D., 1988. Acute tabun toxicity: biochemical and histochemical consequences in brain and skeletal muscle of rat. Toxicology 46, 329–342.

Gupta, R.C., Patterson, G.T., Dettbarn, W.-D., 1991. Comparison of cholinergic and neuromuscular toxicity following acute exposure to sarin and VX in rat. Fundam. Appl. Toxicol. 16, 449–458.

Gupta, R.C., Milatovic, D., Dettbarn, W.-D., 2001. Nitric oxide modulates high-energy phosphates in brain regions of rats intoxicated with diisopropylphosphorofluoridate or carbofuran: prevention by *N-tert*-butyl-α-phenylnitrone or vitamin E. Arch. Toxicol. 75, 346–356.

Gupta, R.C., Pitt, J., Zaja-Milatovic, S., 2015. Blood-brain barrier damage and dysfunction by chemical toxicity. In: Handbook of Toxicology of Chemical Warfare Agents, second ed. Academic Press/Elsevier, Amsterdam, pp. 725–739.

Guterres, A., 2017. Seventh report of the Organization for the Prohibition of Chemical Weapons-United nations Joint Investigative Mechanisms (S/2017/904). .

Horn, G., de Koning, M.C., van Grol, M., et al., 2018. Interactions between acetylcholinesterase, toxic organophosphorus compounds and a short series of structurally related non-oxime reactivators: analysis of reactivation and inhibition kinetics *in vitro*. Toxicol. Lett. 299, 218–225.

Hosseini, S.E., Saeidian, H., Amozadeh, A., et al., 2016. Fragmentation pathways and structural characterization of organophosphorus compounds related to the chemical weapons convention by electron ionization and electrospray ionization tandem mass spectrometry. Rapid. Commun. Mass Spectrom. 30, 2585–2593.

John, H., Van der Schans, M.J., Koller, M., et al., 2018. Fatal sarin poisoning in Syria 2013: forensic verification within an international laboratory network. Forensic Toxicol. 36 (1), 61–71.

Johnson, N.H., Larsen, J.C., Meek, E., 2015. Historical perspective of chemical warfare agents. In: Gupta, R.C. (Ed.), Handbook of Toxicology of Chemical Warfare Agents, second ed. Academic Press/Elsevier, Amsterdam, pp. 7–15.

Jones, K.H., Dechkovskaia, A.M., Herrick, E.A., et al., 2000. Subchronic effects following a single sarin exposure on blood-brain and blood-testes barrier permeability, acetylcholinesterase, and acetylcholine receptors in the central nervous system of rat: a dose-response study. J. Toxicol. Environ. Health A 61, 695–707.

Jun, D., Bajgar, J., Kuca, K., Kassa, J., 2015. Monitoring of blood cholinesterase activity in workers exposed to nerve agents. In: Gupta, R.C. (Ed.), Handbook of Toxicology of Chemical Warfare Agents, second ed. Academic Press/Elsevier, Amsterdam, pp. 967–976.

Kassa, J., Timperley, C.M., Bird, M., et al., 2018. Some benefit from non-oximes MB408, MB442 and MB444 in combination with the oximes HI-6 or obidoxime and atropine in antidoting sarin or cyclosarin poisoned mice. Toxicology 408, 95–100.

Katalinic, M., Sinko, G., Hrvat, G., et al., 2018. Oxime-assisted reactivation of tabun-inhibited acetylcholinesterase analyzed by active site mutations. Toxicology 406, 104–113.

Lallement, G., Carpentier, P., Pernot-Marino, I., et al., 1991a. Involvement of the different rat hippocampal glutamatergic receptors in development of seizures induced by soman. Neurotoxicology 4, 655–664.

Lallement, G., Carpentier, P., Collet, A., et al., 1991b. Effects of soman-induced seizures on different extracellular amino acid levels and on glutamate uptake in rat hippocampus. Brain Res. 563, 234–340.

Lallement, G., Denoyer, M., Collet, A., et al., 1992. Changes in hippocampal acetylcholine and glutamate extracellular levels during soman-induced seizures: influence of septal cholinoceptive cells. Neurosci. Lett. 139, 104–107.

Laskowski, M.B., Dettbarn, W.-D., 1979. An electrophysiological analysis of the effects of paraoxon at the neuromuscular junction. J. Pharmacol. Exp. Ther. 210, 269–274.

Lewine, J.D., Weber, W., Gigliotti, A., et al., 2018. Addition of ketamine to standard-of-care countermeasures for acute organophosphate poisoning improves neurobiological outcomes. Neurotoxicology 69, 37–46.

Li, B., Nachon, F., Froment, M.-T., et al., 2008. Binding and hydrolysis of soman by human serum albumin. Chem. Res. Toxicol. 21, 421–431.

Lockridge, O., Schopfer, L.M., 2006. Biomarkers of organophosphate exposure. In: Gupta, R.C. (Ed.), Toxicology of Organophosphate and Carbamate Compounds. Academic Press/Elsevier, Amsterdam, pp. 703–711.

Lockridge, O., Schopfer, L.M., Masson, P., 2015. Biomarkers of exposure to organophosphorus poisons: a new motif for covalent binding to tyrosine in proteins that have no active site serine. In: Gupta, R.C. (Ed.), Handbook of Toxicology of Chemical Warfare Agents, second ed. Academic Press/Elsevier, Amsterdam, pp. 953–965.

Lushchekina, S., Masson, P., 2018. Catalytic bioscavengers against organophosphorus agents: mechanistic issues of self-reactivating cholinesterases. Toxicology 409, 91–102.

Marrs, T.C., Maynard, R.L., 2013. Neurotransmission systems as targets for toxicants: a review. Cell Biol. Toxicol. 29, 381–396.

Masson, P., 2015. Catalytic bioscavengers: the new generation of bioscavenger-based medical countermeasures. In: Gupta, R.C. (Ed.), Handbook of Toxicology of Chemical Warfare Agents, second ed. Academic Press/Elsevier, Amsterdam, pp. 1107–1123.

Matar, H., Price, S.C., Chilcott, R.P., 2019. Further studies of the efficacy of military, commercial and novel skin decontaminants against the chemical warfare agents sulfur mustard, soman and VX. Toxicol. In vitro 54, 263–268.

Matson, L.M., Lee-Stubbs, R.B., Cadieux, C.L., et al., 2018. Assessment of mouse strain differences in baseline esterase activities and toxic response to sarin. Toxicology 410, 10–15.

McDonough Jr., J.H., McLeod Jr., C.G., Nipwoda, M.T., 1987. Direct microinjection of soman and VX into the amygdala produces repetitive limbic convulsions and neuropathology. Brain Res. 435, 123–127.

McLean, M.J., Gupta, R.C., Dettbarn, W.-D., et al., 1992. Prophylactic and therapeutic efficacy of memantine against seizures produced by soman in the rat. Toxicol. Appl. Pharmacol. 112, 95–103.

Menking, D.E., Thompson, R.G., Wolff, V.L., 1990. Irreversible organophosphate effects on nicotinic acetylcholine receptor/ion channel complex Def. Techn. Inf. Cent. Rep. CR DEC-TR-154. .

Mercey, G., Verdelet, T., Saint-André, G., et al., 2011. First efficient uncharged reactivators for the dephosphorylation of poisoned human acetylcholinesterase. Chem. Commun. 475, 5295–5297.

Miki, A., Katagi, M., Tsuchihashi, H., et al., 1998. Determination of alkylmethylphosphonic acids, the main metabolites of organophosphorus nerve agents, in biofluids by gas chromatography-mass spectrometry and liquid-liquid-solid-phase-transfer-catalyzed pentafluorobenzylation. J. Anal. Toxicol. 23 (2), 86–93.

Millard, C.B., Kryger, G., Ordentlich, A., et al., 1999. Crystal structures of aged phosphonylated acetylcholinesterase: nerve agent reaction products at the atomic level. Biochemistry 38, 7032–7039.

Miller, S.A., Blick, D.W., Kerenyi, S.Z., et al., 1993. Efficacy of physostigmine as a pretreatment for organophosphate poisoning. Pharmacol. Biochem. Behav. 44, 343–347.

Misulis, K.E., Clinton, M.E., Dettbarn, W.-D., Gupta, R.C., 1987. Differences in central and peripheral neural actions between soman and diisopropylfluorophosphate, organophosphorus inhibitors of acetylcholinesterase. Toxicol. Appl. Pharmacol. 89, 391–398.

Moffett, M.C., McDonough, J.H., McMonagle, J.D., et al., 2018. Behavioral, pharmacokinetic, and cardiovascular evaluation of candidate oximes in African green monkey. Int. J. Toxicol. 37 (5), 352–363.

Munirathinam, S., Bahr, B.A., 2004. Repeated contact with subtoxic soman leads to synaptic vulnerability in hippocampus. J. Neurosci. Res. 77, 739–746.

Munro, N., Ambrose, K.R., Watson, A.P., 1994. Toxicity of the organophosphate chemical warfare agents GA, GB, and VX: implications for public protection. Environ. Health Perspect. 102, 18–38.

Nachon, F., Brazzolotto, X., Trovaslet, M., Masson, P., 2013. Progress in the development of enzyme-based nerve agent bioscavengers. Chem. Biol. Interact. 206, 536–544.

Naughton, S.X., Terry Jr., A.V., 2018. Neurotoxicity in acute and repeated organophosphate exposure. Toxicology 408, 101–112.

Nepovimova, E., Kuca, K., 2018. Chemical warfare agent NOVICHOK-Mini-review of available data. Food Chem. Toxicol. 121, 343–350.

Okumura, T., Taki, K., Suzuki, K., et al., 2015. The Tokyo subway sarin attack: acute and delayed health effects in survivors. In: Gupta, R.C. (Ed.), Handbook of Toxicology of Chemical Warfare Agents, second ed. Academic Press/Elsevier, Amsterdam, pp. 27–35.

OPCW, 2018. Organization for the Prohibition of Chemical Weapons (OPCW). website (www.opcw.org).

Pashirova, T.N., Braiki, A., Zueva, I.V., et al., 2018. Combination delivery of two oxime-loaded lipid nanoparticles: time-dependent additive action for prolonged rat brain protection. J. Control. Release 290, 102–111.

Pearce, P.C., Crofts, H.S., Muggleton, N.G., et al., 1999. The effects of acutely administered low dose sarin on cognitive behavior and the electroencephalogram in the common marmoset. J. Psychopharmacol. 13, 128–135.

Pearson, J.N., Patel, M., 2016. The role of oxidative stress in organophosphate nerve agent toxicity. Ann. N. Y. Acad. Sci. 1378 (1), 17–24.

Pittel, Z., Barak, D., Segall, Y., 2006. Function-specific blockage of M(1) and M(3) muscarinic acetylcholine receptors by VX and echothiophate. Brain Res. 1085, 102–110.

Pittel, Z., Lazar, S., Gez, R., et al., 2018. Early changes in M2 muscarinic acetylcholine receptors (mAChRs) induced by sarin intoxication may be linked to long lasting neurological effects. Neurotoxicology 65, 248–254.

Rao, K.S., Aracava, Y., Rickett, D.L., et al., 1987. Noncompetitive blockade of the nicotinic acetylcholine receptor-ion channel complex by an irreversible cholinesterase inhibitor. J. Pharmacol. Exp. Ther. 240, 337–344.

Reddy, S.D., Reddy, D.S., 2015. Midazolam as an anticonvulsant antidote for organophosphate intoxication—a pharmacotherapeutic appraisal. Epilepsia 56, 813–821.

Reid, G.A., Chilukuri, N., Darvesh, S., 2013. Butyrylcholinesterase and the cholinergic system. Neuroscience 234, 53–68.

Reiner, E., Simeon-Rudolf, V., 2006. Methods for measuring cholinesterase activities in human blood. In: Gupta, R.C. (Ed.), Toxicology of Organophosphate and Carbamate Compounds. Academic Press/Elsevier, Amsterdam, pp. 199–208.

Renou, J., Mercey, G., Verdelet, T., et al., 2013. Syntheses and *in vitro* evaluations of uncharged reactivators for human acetylcholinesterase inhibited by organophosphorus nerve agents. Chem. Biol. Interact. 203, 81–84.

Reymond, C., Jaffre, N., Taudon, N., et al., 2018. Superior efficacy of HI-6 dimethanesulfonate over pralidoxime methylsulfate against Russian VX poisoning in cynomolgus monkeys (*Macca fascicularis*). Toxicology 410, 96–105.

Richardson, R.J., Worden, R.M., Wijeyesakere, S.J., et al., 2015. Neuropathy target esterase as a biomarker and biosensor of delayed neuropathic agents. In: Gupta, R.C. (Ed.), Handbook of Toxicology of Chemical Warfare Agents, second ed. Academic Press/Elsevier, Amsterdam, pp. 935–952.

Rocha, E.S., Santos, M.D., et al., 1999. Low concentrations of the organophosphate VX affect spontaneous and evoked transmitter release from hippocampal neurons: toxicological relevance of cholinesterase-dependent actions. Toxicol. Appl. Pharmacol. 159, 31–40.

Rochu, D., Chabriere, E., Masson, P., 2007. Human paraoxonase: a promising approach for pre-treatment and therapy of organophosphorus poisoning. Toxicology 233, 47–59.

Rosman, Y., Eisenkraft, A., Milk, N., et al., 2014. Lessons learned from the Syrian sarin attack: evaluation of a clinical syndrome through social media. Ann. Intern. Med. 160, 644–648.

Schier, J.G., Ravikumar, P.R., Nelson, L.S., et al., 2004. Preparing for chemical terrorism: stability of injectable atropine sulfate. Acad. Emerg. Med. 11, 329–334.

Seto, Y., 2015. On-site detection of chemical warfare agents. In: Gupta, R.C. (Ed.), Handbook of Toxicology of Chemical Warfare Agents, second ed. Academic Press/ Elsevier, Amsterdam, pp. 897–914.

Sharma, R., Gupta, B., Singh, N., et al., 2015. Development and structural modifications of cholinesterase reactivators against chemical warfare agents in last decade: a mini review. Med. Chem. 15, 58–72.

Shih, T.-M., 1982. Time course effects of soman on acetylcholine and choline levels in six discrete areas of the rat brain. Neuropharmacology 78, 170–175.

Shih, T.M., McDonough Jr., J.H., 1997. Neurochemical mechanisms in soman-induced seizures. J. Appl. Toxicol. 17, 255–264.

Shih, T.M., Duniho, S.M., McDonough, J.H., 2003. Control of nerve agent-induced seizures is critical for neuroprotection and survival. Toxicol. Appl. Pharmacol. 188, 69–80.

Sirin, G.S., Zhou, Y., Lior-Hoffmann, L., et al., 2012. Aging mechanism of soman inhibited acetylcholinesterase. J. Phys. Chem. B 116, 12199–12207.

Sit, R.K., Kovarik, Z., Hrvat, N.M., et al., 2018. Pharmacology, pharmacokinetics, and tissue disposition of zwitterionic hydroxyiminoacetamido alkylamines as reactivating antidotes for organophosphate exposure. J. Pharmacol. Exp. Ther. 367, 363–372.

Sivam, S.P., Hoskins, B., Ho, I.K., 1984. An assessment of comparative acute toxicity of diisopropyl fluorophosphate, tabun, sarin, and soman in relation to cholinergic and GABAergic enzyme activities in rats. Fundam. Appl. Toxicol. 4, 531–538.

Škrbić, R., Stojiljković, M.P., Ćetković, S.S., et al., 2017. Naloxone antagonizes soman-induced central respiratory depression in rats. Basic Clin. Pharmacol. Toxicol. 120, 615–620.

Solberg, Y., Belkin, M., 1997. The role of excitotoxicity in organophosphorus nerve agents central poisoning. Trends Pharmacol. Sci. 18, 183–185.

Sparenborg, S., Brennecke, L.H., Jaax, N.K., et al., 1992. Dizocilpine (MK-801) arrests status epilepticus and prevents brain damage induced by soman. Neuropharmacology 31, 357–368.

Sterri, S.H., Fonnum, F., 2015. The role of carboxylesterases in therapeutic intervention of nerve gases poisoning. In: Gupta, R.C. (Ed.), Handbook of Toxicology of Chemical Warfare Agents, second ed. Academic Press/Elsevier, Amsterdam, pp. 1099–1106.

Stojiljković, M.P., Škrbić, R., Jokanović, M., et al., 2019. Prophylactic potential of memantine against soman poisoning in rats. Toxicology 416, 62–74.

Stolbach, A., Bebarta, V., Beuhler, M., et al., 2018. ACMT position statement: alternative or contingency countermeasures for acetylcholinesterase inhibiting agents. J. Med. Toxicol. 14, 261–263.

Tan, C., Nasir, M.Z.M., Ambrosi, A., et al., 2017. 3D printed electrodes for detection of nitroaromatic explosives and nerve agents. Anal. Chem. 89, 8995–9001.

Tattersall, J.E., 1990. Effects of organophosphorus anticholinesterases on nicotinic ion channels at adult mouse muscle endplates. Br. J. Pharmacol. 101, 349–357.

Tsuchihashi, H., Katagi, M., Nishikawa, M., et al., 1998. Identification of metabolites of nerve agent VX in serum collected from a victim. J. Anal. Toxicol. 22, 383–388.

Tu, A.T., 2002. Chemical Terrorism: Horrors in Tokyo Subway and Matsumoto City. Alaken Inc., Fort Collins, CO, USA, pp. 1–240.

Tuovinen, K., 2004. Organophosphate-induced convulsions and prevention of neuropathological damages. Toxicology 196, 31–39.

United States Department of Health and Human Services (Food and Drug Administration) (US FDA), 2016. Memorandum Expiry During Extension Update for AtroPen (atropine), CANA (diazepam), Morphine Sulfate, and Pralidoxime Chloride. Auto-Injectors. http://www.fda.gov/downloads/Drugs/DrugsSafety/UCM496442.pdf.

Van Der Schans, M.J., 2015. Laboratory analysis of chemical warfare agents, adducts, and metabolites in biomedical samples. In: Gupta, R.C. (Ed.), Handbook of Toxicology of Chemical Warfare Agents, second ed. Academic Press/Elsevier, Amsterdam, pp. 915–923.

Vanova, N., Pejchal, J., Herman, D., et al., 2018. Oxidative stress in organophosphate poisonings: role of standard antidotal therapy. J. Appl. Toxicol. 38, 1058–1070.

Wade, J.V., Samson, F.E., Nelson, S.R., et al., 1987. Changes in extracellular amino acids during soman- and kainic acid-induced seizures. J. Neurochem. 49, 645–650.

Walker, B., Nidiri, J., 2002. Current concepts: organophosphate toxicity. Inhal. Toxicol. 14, 975–990.

Watson, A., Bakshi, K., Opresko, D., et al., 2006. Cholinesterase inhibitors as chemical warfare agents: community preparedness guidelines. In: Gupta, R.C. (Ed.), Toxicology of Organophosphate and Carbamate Compounds. Academic Press/Elsevier, Amsterdam, pp. 47–68.

Watson, A., Opresko, D., Young, R.A., et al., 2015. Organophosphate nerve agents. In: Gupta, R.C. (Ed.), Handbook of Toxicology of Chemical Warfare Agents, second ed. Academic Press/Elsevier, Amsterdam, pp. 87–109.

Weger, N., Szinicz, L., 1981. Therapeutic effects on new oximes, benactyzine and atropine in soman poisoning: part 1. Effects of various oximes in soman, and VX poisoning in dogs. Fundam. Appl. Toxicol. 1, 161–163.

Wei, Z., Liu, Y., Wang, Y., et al., 2016. Novel nonquternary reactivators showing reactivation efficiency for soman-inhibited human acetylcholinesterase. Toxicol. Lett. 246, 1–6.

Wiener, S.W., Hoffman, R.S., 2004. Nerve agents: a comprehensive review. J. Intensive Care Med. 19, 22–37.

Wilhelm, C.M., Snider, T.H., Babin, M.C., et al., 2018. Evaluating the broad-spectrum efficacy of the acetylcholinesterase oximes reactivators MMB4 DMS, HLö7 DMS, and 2-PAM Cl against phorate oxon, sarin, and VX in the Hartley guinea pig. Neurotoxicology 68, 142–148.

Worek, F., Koller, M., Thiermann, H., et al., 2016a. Reactivation of nerve agent-inhibited human acetylcholinesterase by obidoxime, HI-6 and obidoxime + HI-6: kinetic in vitro study with simulated nerve agent toxicokinetics and oxime pharmacokinetics. Toxicology 350–352, 25–30.

Worek, F., Wille, T., Koller, M., et al., 2016b. Toxicology of organophosphorus compounds in view of an increasing terrorist threat. Arch. Toxicol. 90, 2131–2145.

Zaja-Milatovic, S., Gupta, R.C., 2015. Excitotoxicity, oxidative stress, and neuronal injury. In: Gupta, R.C. (Ed.), Handbook of Toxicology of Chemical Warfare Agents, second ed. Academic Press/Elsevier, Amsterdam, pp. 709–724.

Zaja-Milatovic, S., Gupta, R.C., Aschner, M., Milatovic, D., 2009. Protection of DFP-induced oxidative damage and neurodegeneration by antioxidants and NMDA receptor antagonist. Toxicol. Appl. Pharmacol. 240, 124–131.

Zhao, Y., Yang, M., Fu, Q., et al., 2018. A nanozyme- and ambient light-based smartphone platform for simultaneous detection of dual biomarkers from exposure to organophosphorus pesticides. Anal. Chem. 90, 7391–7398.
Zimmer, L.A., Ennis, M., Shipley, M.T., 1997. Soman-induced seizures rapidly activates astrocytes and microglia in discrete brain regions. J. Comp. Neurol. 378, 482–492.

Further reading

Albuquerque, E.X., Akaike, A., Shaw, K.-P., et al., 1984. The interaction of anticholinesterase agents with the acetylcholine receptor-ionic channel complex. Fundam. Appl. Toxicol. 4, S27–S33.
Meshul, C.K., Boyne, A.F., Deshpande, S.S., et al., 1985. Comparison of the ultrastructural myopathy induced by anticholinesterase agents at the end plates of rat soleus and extensor muscles. Exp. Neurol. 89, 96–114.

CHAPTER THREE

Neurotoxicology of pyrethroid insecticides

David M. Soderlund*
Department of Entomology, Cornell University, Geneva, NY, United States
*Corresponding author: e-mail address: dms6@cornell.edu

Contents

Advances in Neurotoxicology, Volume 4
ISSN 2468-7480
https://doi.org/10.1016/bs.ant.2019.11.002

1. Introduction

Pyrethroids are a class of synthetic insecticides derived structurally from the pyrethrins (Fig. 1), the insecticidal constituents of the natural insecticide pyrethrum. The development of pyrethroids was guided by efforts to retain or improve on the insecticidal activity and low mammalian toxicity of the pyrethrins while increasing environmental stability, and in particular photostability, to permit their use in agricultural and other applications where residual activity is required. Discoveries in the 1960s and 1970s of novel synthetic compounds with improved properties led to a proliferation

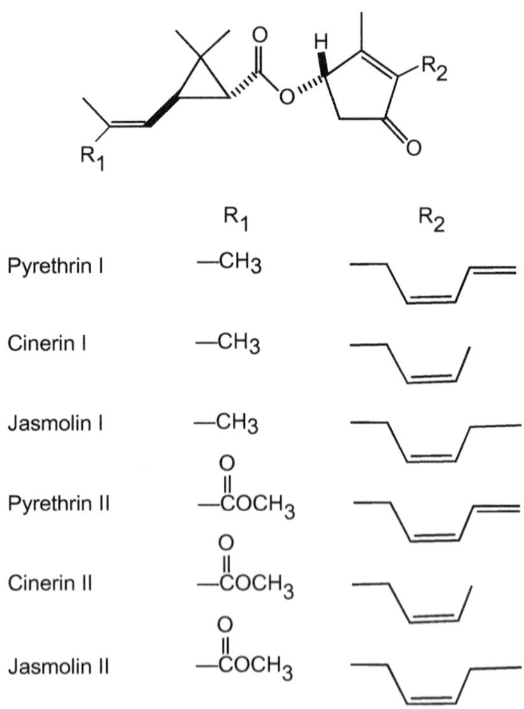

Fig. 1 Structures of the six natural pyrethrins.

of commercial pyrethroid-based insecticide products that continue to play important roles in arthropod pest and disease vector control today (Elliott, 1989). This review focuses on the development and current status of pyrethroids as an insecticide class, the neurotoxicology of pyrethroids in mammals and humans, and the molecular neuropharmacology of pyrethroid action on mammalian neuronal voltage-gated sodium channels in relation to their neurotoxic effects.

2. Pyrethroid insecticides

2.1 Development and use

Allethrin, tetramethrin and resmethrin (see Fig. 2 for pyrethroid structures), the first synthetic pyrethroids to enter commercial use, did not offer improved photostability when compared to the pyrethrins and were therefore limited to applications where residual activity was not required. However, the discoveries of permethrin, deltamethrin and fenvalerate Fig. 2 in the 1970s identified the first synthetic pyrethroids that were potent insecticides and stable enough in sunlight for agricultural use and other residual applications (Elliott, 1989). At the time of its discovery, deltamethrin was substantially more potent than all other known insecticides. Permethrin, deltamethrin, and fenvalerate (as the resolved isomer esfenvalerate) remain in widespread use.

The success of permethrin, deltamethrin, and fenvalerate catalyzed intensive research and development, which led to the discovery and worldwide registration of more than 20 pyrethroid insecticides by 1995 (Casida and Quistad, 1998). Prior to introduction of photostabilized compounds for agricultural applications in the late 1970s the use of pyrethroid-based insecticides was insignificant compared to other insecticide classes, but by the late 1980s pyrethroid products had captured nearly 20% of the world insecticide market (Soderlund, 2017; Sparks and Lorsbach, 2017). Initially introduced as replacements for organochlorine, organophosphate, and carbamate insecticides against pests of cotton, pyrethroids were subsequently registered for the management of a wide variety of agricultural and household pests as well as human disease vectors. Bifenthrin, tefluthrin, and etofenprox are examples of pyrethroids with unique properties (acaricidal activity, vapor phase activity in soil, and reduced toxicity to fish, respectively) that allowed the expansion of pyrethroid uses.

Pyrethroids are uniquely durable among modern insecticide classes (Soderlund, 2017; Sparks and Lorsbach, 2017). Over the past two decades

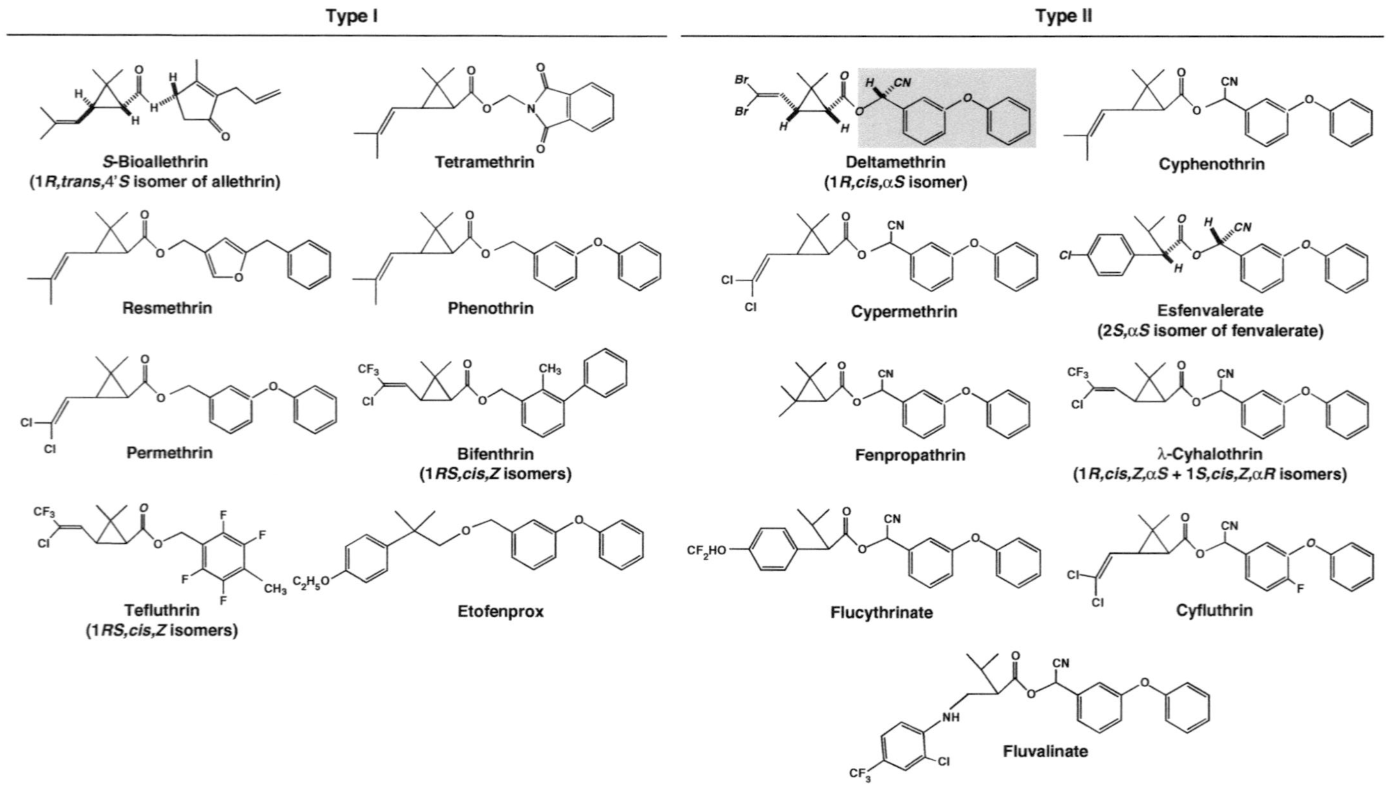

Fig. 2 Structures of Type I and Type II pyrethroids. Stereochemistry is indicated for compounds that are used as resolved isomers or defined isomer mixtures. The α-cyano-3-phenoxybenzyl moiety that defines the Type II structural class is shaded in the structure of deltamethrin.

pyrethroid use has remained relatively constant, representing approximately 15–20% of the world insecticide market. By contrast, during this period the use of organochlorine, organophosphate and carbamate insecticides declined by more than 50%. At the same time, the introduction in the mid-1990s of transgenic cotton varieties that express insecticidal proteins significantly reduced pyrethroid use on this crop, and the neonicotinoid and diamide classes of insecticides were developed and gained significant market share. Despite these shifts in the insecticide market and the development of pyrethroid resistance in many pest populations, pyrethroids remain second only to neonicotinoids in their share of the worldwide insecticide market four decades after their introduction for agricultural use.

2.2 Structure-activity relationships

The pyrethroid insecticides in use today were discovered by replacing the structural elements of the pyrethrins with novel structural moieties that conserved the molecular shape and physical properties of the pyrethrins (Elliott, 1995). This process yielded new compounds with pyrethrin-like insecticidal activity which, in turn, served as templates for further structural elaboration. Most initial discovery efforts held the chrysanthemic acid moiety of pyrethrin I constant and evaluated esters of this acid with a variety of alcohols, which led to resmethrin and phenothrin. Replacement of methyls by chlorines in the acid moiety of phenothrin yielded permethrin, the first photostabilized pyrethroid. Inclusion of an α-cyano substituent in the 3-phenoxybenzyl moiety (as in cypermethrin and deltamethrin) significantly enhanced insecticidal activity. Esters of α-cyano-3-phenoxybenzyl alcohols are considered to be a separate structural class of pyrethroids (designated Type II), whereas the diverse group of pyrethroids lacking this moiety are designated as Type I pyrethroids.

The pyrethroid structure-activity space was further expanded by the discovery that the substituted cyclopropanecarboxylate acid moiety of the pyrethrins and many synthetic pyrethroids could be replaced by an α-isopropylphenylacetic acid moiety, as in fenvalerate. Further exploration of non-cyclopropanecarboxylate structures led to compounds such as etofenprox, which lacks the central ester linkage common to other pyrethroids. Pyrethrin I and etofenprox, despite their structural divergence, conform to a single operationally-defined structure-activity relationship for insecticidal activity that is based on the physical properties and overall shape of the entire molecule (Fig. 3). Moreover, it is clear from the structural

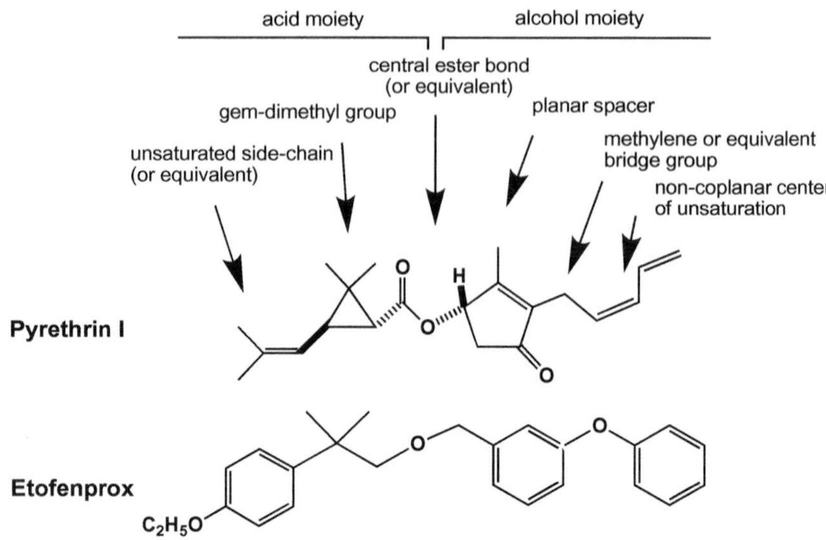

Fig. 3 Essential elements of the pyrethroid structure-activity relationship illustrated by the structures of pyrethrin I and etofenprox.

diversity of pyrethroid insecticides that there is no chemical toxophore or reactive entity that is required for pyrethroid-like activity.

The insecticidal activity of pyrethroids is also stereospecific. The chrysanthemic acid moiety of pyrethrin I contains two chiral centers at carbon-1 and carbon-3 of the cyclopropane ring, producing two pairs of diastereomers (Fig. 4). Only the chrysanthemic acid and related 3-substituted cyclopropanecarboxylic acid isomers with the *R* configuration at C-1, or the equivalent configuration in other chiral acid moieties, are insecticidal. By contrast, the insecticidal activity of the 1*S* enantiomers is negligible (Elliott et al., 1974). The enantiomers at C-3 yield a pair of diastereomers, which are commonly designated *trans* or *cis* based on the orientation of the C-1 and C-3 substituents in relation to the plane of the cyclopropane ring (Fig. 4). An unsaturated substituent at C-3 (either *trans* or *cis*) is required both for high insecticidal activity and to confer the profound stereospecificity for 1*R* isomers (Elliott et al., 1974). The dependence of insecticidal activity on the absolute configuration of the carbon bearing the acid moiety carbonyl extends to non-cyclopropane acids (e.g., esfenvalerate; Fig. 2).

Stereoisomerism is less common in the alcohol moieties of pyrethroids. Nevertheless, when a chiral center is present at the carbon bearing the alcohol hydroxyl group (e.g., allethrin, deltamethrin; see Fig. 2) only one epimer of the resulting ester exhibits high insecticidal activity even when esterified

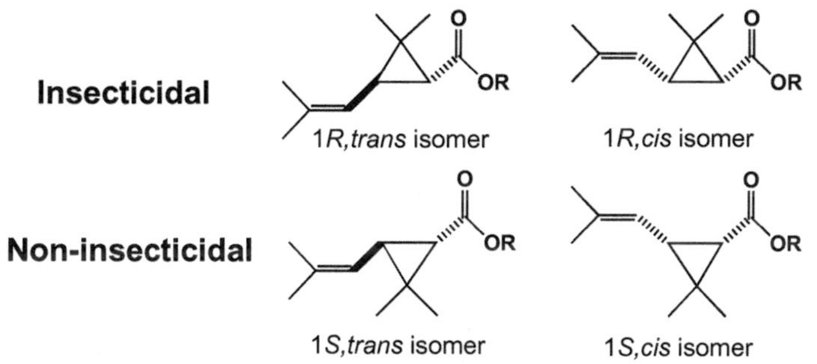

Fig. 4 Structures of the insecticidal 1*R* and non-insecticidal 1*S* chrysanthemic acid moieties of pyrethroids.

to a chiral acid moiety with the insecticidally active 1*R* or equivalent configuration. The consistent stereospecificity for insecticidal activity among pyrethroids implies that these compounds act by the binding of the intact molecule to a chiral receptor.

2.3 Mechanism of insecticidal action

The rapid paralytic (knockdown) activity of the pyrethrins and early synthetic pyrethroids implied an action on the insect nervous system. Electrophysiological studies of the pyrethrins and allethrin (reviewed in Soderlund, 1995) showed that these compounds caused an initial hyperexcitatory effect, evident as the production of bursts of action potentials following a single stimulus, followed by the block of action potential conduction. The effects of pyrethrins and allethrin in these studies were similar to those produced by DDT.

A study of the action of pyrethroids and DDT analogs on resting membrane potentials and neuronal excitability in crayfish giant axon preparations (Lund and Narahashi, 1983) revealed effects of pyrethroid structure on the relative importance of hyperexcitatory and blocking effects. DDT analogs and many Type I pyrethroids induced a negative afterpotential and repetitive bursts of action potentials following a single stimulus but had little or no effect on the resting membrane potential. By contrast, many Type II pyrethroids did not cause repetitive activity but instead produced a use-dependent block of nerve conduction coupled with a depolarization of the nerve membrane. Some compounds from both structural classes of pyrethroids exhibited intermediate actions, producing bursts of action potentials of declining amplitude coupled with use-dependent depolarization and subsequent nerve block.

Studies on voltage-clamped crayfish giant axons correlated the effects of DDT analogs and pyrethroids on neuronal excitability with the modification of the transient sodium currents carried by voltage-gated sodium channels (VGSCs) in these preparations (Lund and Narahashi, 1983). All compounds caused a slowing of sodium current inactivation during a depolarizing pulse and caused a slowly-decaying tail current following membrane repolarization. However, they differed significantly in the kinetics of tail current decay. Tail currents caused by DDT analogs and Type I pyrethroids decayed rapidly, whereas those caused by Type II pyrethroids were much more persistent. These results suggested that the principal difference between Type I and Type II compounds lies in the relative lifetimes of pyrethroid-modified open channels. Type I compounds cause a transient prolongation of sodium permeability, which results in a negative afterpotential and the induction of repetitive nerve firing but does not produce sufficient sodium influx to compromise the resting potential. By contrast, Type II compounds recruit VGSCs into persistent pyrethroid-modified open states, which results in use-dependent depolarization, inactivation of unmodified channels, and ultimate block of nerve conduction.

Definitive confirmation that VGSCs are the critical targets for the insecticidal action of pyrethroids comes from studies of the molecular genetics of pyrethroid resistance traits in insects that involve reduced nerve sensitivity (reviewed in Dong et al., 2014; Soderlund, 2005; Soderlund and Knipple, 2003). In the house fly, the *kdr* (knockdown resistance) and *super-kdr* (enhanced knockdown resistance) genetic traits confer resistance to DDT and all pyrethroids by reducing the sensitivity of the fly nervous system to these compounds. Genetic linkage analyses mapped these traits close to the principal VGSC gene of the house fly, designated *Vssc1*. Moreover, *kdr*-like resistance traits in other insect species have been mapped to *Vssc1*-orthologous VGSC sequences.

Sequence analyses of *Vssc1* alleles in susceptible, *kdr*, and *super-kdr* fly strains identified a single point mutation common to both resistant strains and a second mutation found only in the highly-resistant *super-kdr* strain. Insertion of these mutations into the susceptible *Vssc1* coding sequence, followed by functional analysis of susceptible and specifically-mutated channels in vitro, showed that these mutations reduced the pyrethroid sensitivity of sodium channels sufficiently to account for the resistance afforded by the *kdr* and *super-kdr* traits. Subsequent studies with other arthropod populations exhibiting *kdr*-like resistance identified additional sodium channel point mutations associated with resistance, and the functional significance of some of these mutations was confirmed by site-directed mutagenesis and functional analysis of the expressed channels. Taken together, these results provide convincing evidence that VGSCs are the targets for the insecticidal action of pyrethroids.

2.4 Regulatory considerations

The Food Quality Protection Act (FQPA) of 1996 significantly altered the regulatory framework for pesticides in the United States. Three new regulatory principles were embodied in the mandates of the FQPA. First, risk assessments for pesticides must include all routes of exposure rather than only dietary exposure. Second, cumulative risk assessments are required for groups of chemicals having a "common mechanism of toxicity," a significant departure from the practice of assessing risks of individual chemicals in isolation. Third, there is a default presumption of the greater sensitivity of fetuses and neonates to the deleterious effects of pesticides in the absence of evidence to the contrary.

The new regulatory mandates of the FQPA identified gaps in the available data for exposure, toxicity, and mechanism of action for many insecticide classes including the pyrethroids. Much of the research on pyrethroid neurotoxicology published in the past two decades was undertaken specifically to address the issues of common mechanism and differential sensitivity. In particular, significant research addressed the issue of whether pyrethroids represented a single common mechanism group as defined by the FQPA.

In 2011, the U.S. Environmental Protection Agency (USEPA) published their cumulative risk assessment for pyrethroids (USEPA, 2011). This assessment included the pyrethrins and 15 synthetic pyrethroids with registered uses, representing both Type I and Type II structures, in a single common mechanism group for the purposes of cumulative risk assessment. Four additional pyrethroids registered for use in the United States were recognized as members of the common mechanism group but were not included in the risk assessment either because they were not toxic to mammals at high doses, showed no evidence for human exposure either through the diet or resulting from residential exposure, or had limited registrations for non-agricultural use. Surprisingly, etofenprox was omitted from the pyrethroid common mechanism group despite its structural and functional relationship to the pyrethroids that were included in this group by the USEPA.

3. Acute neurotoxic effects of pyrethroids in mammals

3.1 Acute toxicity

Pyrethroids are regarded as relatively safe insecticides when compared to organochlorine, organophosphate and carbamate insecticides. Table 1 summarizes the oral toxicity to rats of several pyrethroids that are registered for use in the United States. The LD_{50} values for most pyrethroids following oral administration in vegetable oils lie between 50 and 500 mg/kg. Sublethal pyrethroid intoxication following a single dose is transient. Signs of toxicity are observed from minutes to a few hours after dosing, depending on the route of administration, and animals that survive acute intoxication at near-lethal doses recover and appear normal 1–14 days after treatment (Soderlund et al., 2002).

Table 1 Acute toxicity to rats, structural classification, and signs of intoxication for selected pyrethroid insecticides.

	Oral LD_{50} (mg/kg)[a]			
Compound	**Males**	**Females**	**Structural class**[b]	**Intoxication syndrome**[c]
S-Bioallethrin	370	320	Type I	T
Bifenthrin	70	54	Type I	T
Cyfluthrin	155	160	Type II	CS
λ-Cyhalothrin	79	56	Type II	CS
Cypermethrin	297	372	Type II	CS
Deltamethrin	95	87	Type II	CS
Esfenvalerate	87	87	Type II	CS
Fenpropathrin	71	67	Type II	T/CS
Permethrin	1200	1200	Type I	T
Pyrethrins	710	320	Type I	T
Resmethrin	1695	1640	Type I	T
Tefluthrin	22	35	Type I	T

[a]Data from Soderlund et al. (2002).
[b]See Fig. 2 for structures.
[c]Intoxication syndromes as described by Verschoyle and Aldridge (1980), supported by behavioral results from the functional observational battery study of Weiner et al. (2009); T/CS indicates elements of both syndromes were observed.

The toxicity of any pyrethroid varies widely depending on the route of administration and the vehicle used for dosing. Pyrethroids are poorly absorbed through the skin and therefore exhibit very low systemic toxicities following dermal exposure (Clark, 1995). Similarly, oral administration of pyrethroids in aqueous suspensions gives much lower acute toxicities than administration in vegetable oils or organic vehicles (Soderlund et al., 2002), presumably because the low aqueous solubility of pyrethroids limits their bioavailability following dosing in aqueous media. By contrast, routes of administration that favor efficient delivery to the central nervous system (e.g., intravenous or intracerebral injection) render pyrethroids much more toxic with LD_{50} values for the most toxic compounds in the 1–10 mg/kg range (Gray and Soderlund, 1985).

Pyrethroids are rapidly metabolized in mammals in vivo by hydrolysis of the central ester bond and oxidative attack at several sites to give complex arrays of primary and secondary metabolites, which are less toxic and more readily excreted than the parent pyrethroid (Kaneko, 2010). Compounds that interfere with pyrethroid biotransformation can act as pyrethroid synergists in toxicity studies, thereby confirming the significance of biotransformation in limiting pyrethroid toxicity. Pretreatment with the esterase inhibitor *S*,*S*,*S*-tributyl phosphorotrithioate (DEF) or the cytochrome P450 monooxygenase inhibitor piperonyl butoxide increases the intraperitoneal toxicity of many pyrethroids to mice (Gray and Soderlund, 1985). Differences between pyrethroids in the degree of synergism observed in these studies suggest that the relative importance of hydrolytic and oxidative attack in limiting toxicity varies from pyrethroid to pyrethroid and from isomer to isomer of any given pyrethroid.

Acute pyrethroid intoxication by any route of administration is correlated with levels of the parent compound that are achieved in the central nervous system (CNS). The CNS concentrations of pyrethroids obtained following intracerebral injection of severely toxic or lethal doses correspond closely to those found after the peripheral administration of much larger but equally toxic doses (Gray and Soderlund, 1985). This relationship is exemplified in the toxicity of deltamethrin to mice by different routes of administration with various synergist pretreatment regimes (Ruzo et al., 1979). Whereas the observed LD_{50} values for deltamethrin varied over a more than 30-fold range, the levels of deltamethrin in the CNS at death following equitoxic doses varied over only a 2.6-fold range. These results provide further confirmation that targets in the CNS mediate the acute toxicity of pyrethroids.

3.2 Structure-toxicity relationships

The structural features required for pyrethroid-like insecticidal activity in most cases also confer measurable neurotoxicity to mammals. Determination of the acute toxicity of the resolved optical isomers of several pyrethroids following intravenous administration to rats (Verschoyle and Aldridge, 1980) and intracerebral administration to mice (Lawrence and Casida, 1982) showed that in mammals, as in insects, neurotoxicity depends on the 1*R* stereochemical configuration at cyclopropane C-1 or its steric equivalent at the corresponding position in compounds lacking the cyclopropanecarboxylate moiety (see Fig. 3). The corresponding 1*S* cyclopropanecarboxylates and their steric equivalents are without measurable toxicity even when administered at high doses directly to the CNS (Gray and Soderlund, 1985). In mammals, as in insects, the presence of an α-cyano substituent in *S* configuration in the 3-phenoxybenzyl alcohol moiety also greatly enhances acute neurotoxicity. In the case of mammalian toxicity, this effect is highly stereospecific: the α-*R* epimers of compounds with otherwise appropriate stereochemistry in the acid moiety have no demonstrable toxicity when injected directly into the brain (Ghiasuddin and Soderlund, 1985).

Despite the broad conservation of pyrethroid structure-activity relationships for both insecticidal activity and mammalian toxicity, some insecticidal compounds appear to lack measurable effects on the mammalian CNS. Cismethrin (the 1*R*,*cis* isomer of resmethrin), bioresmethrin (the 1*R*,*trans* isomer of resmethrin), and the 1*R*,*trans* isomer of ethanomethrin (Fig. 5) are closely related compounds and effective insecticides. Cismethrin is also toxic to mice by both intraperitoneal and intracerebral injection, and the intraperitoneal toxicity can be synergized by pretreatment with either DEF or piperonyl butoxide (Table 2). By contrast, bioresmethrin has no measurable toxicity to mice by either intraperitoneal or intracerebral injection, and neither DEF nor piperonyl butoxide increases its intraperitoneal toxicity (Table 2). These results suggest that the targets for pyrethroid intoxication in the mammalian CNS are intrinsically insensitive to bioresmethrin. 1*R*,*trans*-Ethanomethrin, a close structural analog of bioresmethrin, also lacks measurable toxicity by the intraperitoneal route. However, the intraperitoneal toxicity of this compound is strongly synergized by both DEF and piperonyl butoxide, and its intracerebral toxicity is similar to that of cismethrin (Table 2).

Small structural changes that confer intrinsic selectivity are not limited to resmethrin and its close analogs. Similar to bioresmethrin, the 1*R*,*trans* and

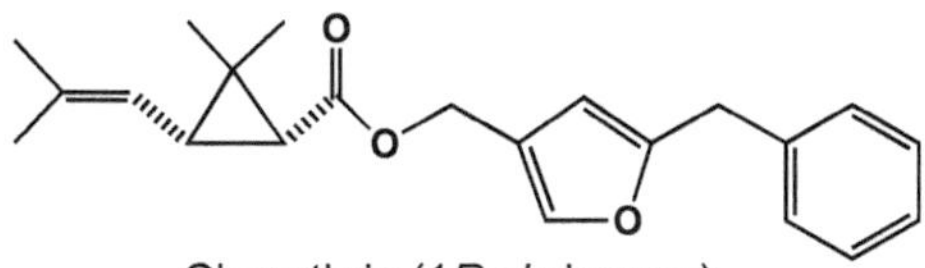

Cismethrin (1*R*,*cis* isomer)

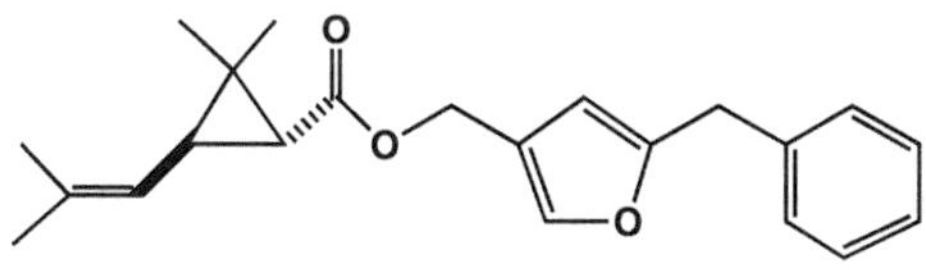

Bioresmethrin (1*R*,*trans* isomer)

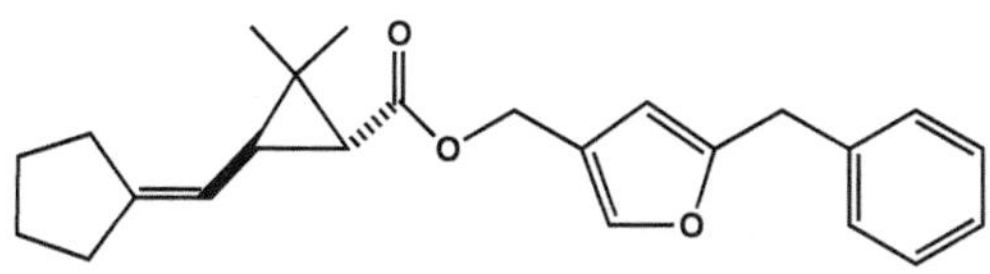

[1*R*,*trans*]-Ethanomethrin

Fig. 5 Structures of cismethrin, bioresmethrin, and [1*R*,*trans*]-ethanomethrin.

Table 2 Intraperitoneal (IP) toxicity to mice in the absence or presence of DEF (*S,S,S*-tributyl phosphorotrithioate) or PB (piperonyl butoxide), and intracerebral (IC) toxicity to mice of cismethrin, bioresmethrin, and [1*R*,*trans*]-ethanomethrin.

			Synergism factor[a]		
Compound[c]	**Isomer**	**IP LD_{50} (mg/kg)[a]**	**DEF**	**PB**	**IC LD_{50} (mg/kg)[b]**
Cismethrin	1R,*cis*	320	11X	25X	0.6
Bioresmethrin	1R,*trans*	>1500	None	None	>120
Ethanomethrin	1R,*trans*	>1000	>188X	>60X	0.2

[a]Data from Soderlund and Casida (1977).
[b]Data from Lawrence and Casida (1982).
[c]See Fig. 5 for structures.

1*R*,*cis* isomers of phenothrin and the 1*RS*, *trans* isomers of permethrin are also without measurable toxicity to mice following either intraperitoneal or intracerebral treatment. However, inclusion of an α-cyano substituent in the 3-phenoxybenzyl alcohol moiety of these compounds (to give the corresponding isomers of cyphenothrin and cypermethrin) confers significant mammalian toxicity by both intraperitoneal and intracerebral administration (Soderlund and Casida, 1977). It is surprising that the intrinsic insensitivity of mammalian targets to [1*R*,*cis*]-phenothrin does not extend to [1*R*,*cis*]-permethrin, which is toxic to mice by intracerebral injection

(Lawrence and Casida, 1982) despite differing from 1*R*,*cis*-phenothrin only by the replacement of methyl groups by chlorines in the cyclopropane C-3 substituent. Taken together, these results show that small differences in structure and stereochemistry in both the acid and alcohol moieties of pyrethroids can differentiate between the receptors for pyrethroid intoxication in insects and mammals.

3.3 Syndromes of intoxication

Two distinct intoxication syndromes are associated with the acute toxicity of pyrethroids to mammals. Verschoyle and Barnes (1972) provided the first systematic description of the signs of pyrethroid intoxication in rats following oral and intravenous dosing and noted that pyrethrins, bioallethrin and resmethrin produced the same signs of intoxication by either route of administration. This syndrome included hypersensitivity and aggression followed by stimulus-induced bouts of general tremor, convulsive twitching, coma, and death. The principal difference observed between oral and intravenous dosing was the speed of onset of intoxication. A subsequent study (Barnes and Verschoyle, 1974) identified a markedly different syndrome of intoxication following either oral or intravenous administration of deltamethrin, which involved salivation without lacrimation followed by jerking leg movements and progressive writhing convulsions (choreoathetosis).

A comparative study of the signs of intoxication of 36 pyrethroids following intravenous administration (Verschoyle and Aldridge, 1980) clarified the relationship between pyrethroid structure and intoxication syndromes and established the taxonomy of pyrethroid intoxication in mammals. Of 18 pyrethroids in the Type I structural class, 15 produced a syndrome of intoxication like that first described for pyrethroids (Verschoyle and Barnes, 1972), which was designated the T (tremor) syndrome. The remaining three Type I structures were not toxic at high doses. By contrast, 12 of 17 compounds in the Type II class caused a syndrome of intoxication like that observed for deltamethrin (Barnes and Verschoyle, 1974), which was designated the CS (choreoathetosis with salivation) syndrome. Four of the remaining Type II compounds produced the T syndrome, whereas one Type II compound and one Type II-like compound in which the α-cyano group was replaced by an α-ethynyl group produced elements of both intoxication syndromes. A second study of the intraperitoneal toxicity of 29 pyrethroids to mice (Lawrence and Casida, 1982) confirmed the effects of pyrethroid structure on signs of intoxication defined by intravenous administration to rats.

Although the first descriptions of the T and CS syndromes suggested that the signs of intoxication were independent of the route of administration, it is not clear that this generalization extends to all pyrethroids or to all routes of administration. For example, comparison of the results of regulatory neurotoxicity studies for nine pyrethroids (reviewed in Soderlund et al., 2002) showed that the behavioral signs of intoxication following oral exposure were not well correlated with the signs characteristic of the T and CS syndromes following intravenous dosing. The relevance of the T/CS classification system to the signs of intoxication following oral exposure was re-examined in a controlled behavioral neurotoxicity study of 12 pyrethroids (Weiner et al., 2009). This study employed a modified functional observational battery that was optimized to facilitate distinction of the T and CS syndromes. The results of this study confirmed that the T and CS syndromes of intoxication were also evident following oral administration, but the magnitude and pattern of response traits varied between compounds in each symptomatic group. This study also confirmed that some pyrethroids exhibit intermediate signs of intoxication that contain elements of both the T and CS syndromes.

3.4 Age-related differences in pyrethroid toxicity

Identifying the relationship between developmental stage and pyrethroid toxicity has important regulatory implications, but information on differences in pyrethroid sensitivity between immature and adult animals is limited. Neonatal rats were 6- to 17-fold more sensitive than adults to permethrin, cypermethrin and deltamethrin (Cantalamessa, 1993; Sheets et al., 1994). The use of synergists to block metabolic detoxication (Cantalamessa, 1993) and the measurement of concentrations in whole brain at equitoxic doses (Sheets et al., 1994) suggested that the greater sensitivity of young animals was due to age-dependent differences in detoxication rather than the intrinsically greater sensitivity of the neonatal CNS. A subsequent study failed to identify age-related differences in the sensitivity rats to cismethrin or permethrin at either high (lethal) or low (behaviorally active) doses (Sheets, 2000). The same study also found differential sensitivity to deltamethrin and cypermethrin only at high doses. The age-related differences in pyrethroid sensitivity observed at high doses in these studies may be due to the reduced ability of the incompletely-developed detoxication enzyme systems of the neonatal liver to metabolize a large bolus of insecticide. A recently-developed physiologically-based toxicokinetic model (Song et al., 2019) confirmed the importance of age-dependent differences in pyrethroid detoxication as the basis for the differential sensitivity of juvenile and adult rats to pyrethroids.

3.5 Pyrethroid intoxication in humans

Despite decades of extensive worldwide pyrethroid use, there are relatively few documented instances of pyrethroid intoxication in humans (Bradberry et al., 2005). The most comprehensive source of information on human intoxication by pyrethroids is a review of 573 cases of acute pyrethroid poisoning that were reported in the Chinese medical literature during the period 1983–88 (He et al., 1989). These cases encompassed both occupational exposure, the result of mishandling during agricultural uses, and accidental exposure, usually the result of ingestion of formulated insecticide products. All but seven of these cases involved deltamethrin, fenvalerate, and cypermethrin. The most common signs of systemic pyrethroid poisoning included dizziness, headache, nausea, anorexia, and fatigue. More serious cases exhibited coarse muscle fasciculations, disturbance of consciousness, coma, and convulsive attacks. Seven deaths were reported, of which one was due to atropine intoxication following misdiagnosis of organophosphate insecticide poisoning. All patients received symptomatic and supportive therapies and most recovered within 6 days of exposure, although recovery of patients experiencing convulsions required up to 55 days. Other reports of pyrethroid exposure and intoxication incidents (Box and Lee, 1996; Gotoh et al., 1998; Lessenger, 1992) are in substantial agreement with the findings reported in the Chinese literature (He et al., 1989).

Although lethal or serious systemic poisonings by pyrethroids in humans are rare, information collected by poison control centers shows a significant incidence of pyrethroid exposure. During a 4-year period (2001–04) the Toxic Exposure Surveillance System of the American Association of Poison Control Centers received 55,610 reports of exposure to pyrethroid insecticides, of which nearly one-third were exposures of children younger than 5 years of age (Litovitz et al., 2002; Watson et al., 2003, 2004, 2005). Fewer than 5% of these exposures were characterized as producing toxic effects classified by the authors as moderate or major, and only three exposures were fatal. In the two fatalities for which details were reported, pyrethroid exposure was accompanied by exposure to an organic solvent (xylene or ethylene glycol), presumably as a component of the formulated insecticide, making it unclear whether death was due to insecticide or solvent intoxication (Bradberry et al., 2005). A second survey (Hudson et al., 2014) of pyrethroid-related illnesses and injuries over a 9 year period (2000–08), based on data from 11 states, identified nearly 5000 cases of acute pyrethrin/pyrethroid-related illness or injury. Respiratory effects were the most common symptoms, and the most common exposure scenarios involved spills or splashes of concentrated pesticide, improper storage, and failure to evacuate during pesticide application.

The experience of workers involved in the handling of technical or formulated pyrethroids, during either manufacture or use, provides insight into the local effects of pyrethroids in humans following cutaneous exposure. Two reviews (Clark, 1995; Vijverberg and van den Bercken, 1990) summarize occupational exposure data from published clinical reports and unpublished reports from industrial sources. The most frequently-reported symptom in worker exposure studies was paresthesia, which was characterized by numbness, itching, burning, or tingling of the skin following dermal exposure to a pyrethroid. These sensations generally occurred in the absence of erythema, edema, vesiculation, or other signs of overt skin irritation and were usually limited to the directly exposed areas of the skin. Pyrethroid-induced paresthesia was transient and reversible within hours after exposure, but in some instances it lasted for up to 48 h. No clinical signs of acute pyrethroid intoxication were observed in any of these cases.

Effects observed during occupational exposure are supported by the results of a study with human volunteers. The effects of four pyrethroids were evaluated by application to the earlobe of volunteers (Flannigan and Tucker, 1985). In this study, permethrin caused the least pronounced and flucythrinate the most pronounced paresthesia, whereas cypermethrin and fenvalerate were intermediate and approximately equal in effect. In each case, paresthesia developed within 30 min of exposure, peaked by 8 h, and dissipated by 24–32 h after exposure. The results of worker exposure data coupled with the results of controlled experiments with human volunteers show that paresthesia is a local effect. It occurs only at the site of dermal exposure, is not correlated with the appearance of a rash or other signs of classical skin irritation, and is not associated with any signs of systemic intoxication. Based on these observations and the proposed origins of paresthesia in the sensory nervous system (Rizzo et al., 1996), pyrethroid-induced paresthesia has been postulated to be a direct excitatory effect of pyrethroids on small sensory nerve fibers in the skin rather than a response due to classical skin irritation.

4. Neurotoxic effects following sublethal acute or chronic exposure

4.1 Effects on locomotor activity

Changes in locomotor activity have been widely employed as an index of pyrethroid intoxication following sublethal acute exposure (reviewed in Wolansky and Harrill, 2008). Crofton and Reiter (1984, 1988a,b) compared the effects of nine commercial or experimental pyrethroids on locomotor activity in rats following oral administration using a figure-eight maze, measured at the time of peak effect (1.5–2 h after dosing). All nine compounds caused a dose-dependent decrease in motor activity, and no differences in response were found between compounds identified as producing the T syndrome of intoxication and those producing the CS syndrome. A subsequent comparative study of 11 pyrethroids (Wolansky et al., 2006) confirmed the universal reduction in motor activity by all pyrethroid regardless of symptomatic class and established relative potencies based on doses required to produce a 30% reduction in motor activity. More recent studies established robust correlations between dose, blood and brain levels, and reduction in locomotor activity for bifenthrin (Scollon et al., 2011) and two different isomer mixtures of cyhalothrin (Moser et al., 2016). Characterization of pyrethroid effects on locomotor activity has been extended to the correlation of blood and brain concentrations of environmentally relevant mixtures of pyrethroids with reduction in motor activity as a means of assessing the additive neurotoxic effects of pyrethroids in cumulative risk assessments (Hughes et al., 2016; Starr et al., 2012).

4.2 Effects on acoustic startle response

Unlike locomotor activity, which is reduced by all pyrethroids, the acoustic startle response partially differentiates between the T and CS symptomatic classes of pyrethroids. Crofton and Reiter (1984, 1988a,b) assessed the effects of nine pyrethroids on the acoustic startle response of rats, measured as the latency to onset and peak amplitude of startle responses to standard acoustical stimuli 1.5–2 h after dosing. Compounds identified as producing the T syndrome of intoxication increased the amplitude of the startle response without affecting its latency, whereas compounds identified as producing the CS syndrome of intoxication produced varied effects. In the latter group, cypermethrin, deltamethrin, cyfluthrin and flucythrinate decreased the amplitude and increased the latency of the startle response whereas fenvalerate increased the amplitude and did not affect latency and also, unlike any other compound tested, increased sensitization to background acoustical stimuli. By contrast fluvalinate had no effect on any measure of the acoustic startle response. In another study (Hijzen and Slangen, 1988), cypermethrin, fenfluthrin, and permethrin increased the amplitude of the acoustic startle response in rats whereas deltamethrin decreased the amplitude of this response. The effects of permethrin and deltamethrin in this study are generally consistent with previous results, but the increase in startle amplitude caused by cypermethrin is an unexpected and contradictory finding. A recent study suggests the effects of deltamethrin on the acoustic startle response is more pronounced in juvenile rats compared to adults at doses that produce equivalent brain concentrations (Williams et al., 2019).

4.3 Effects on conditioned behaviors

There are only a few reports of the impact of pyrethroids on conditioned behaviors, learning, and memory. Intraperitoneal administration of allethrin, deltamethrin, fenvalerate or permethrin to rats caused a dose-dependent reduction in the frequency of a previously-learned behavior (i.e., bar-pressing reinforced by food) (Bloom et al., 1983; Stein et al., 1987). Oral administration of cypermethrin, deltamethrin or permethrin to rats caused similar reductions in food-reinforced learned behavior (Glowa, 1986; Peele and Crofton, 1987). Daily oral administration of deltamethrin to rats for 15 days reduced learning and memory measured in a Y-maze using a negatively-reinforced visual discrimination response (Husain et al., 1996). Similarly, repeated intraperitoneal administration of deltamethrin to mice impaired memory retention in passive avoidance and Y-maze tests (Nieradko-Iwanicka and Borzecki, 2015).

4.4 Developmental neurotoxicity

The reported exposure of children and pregnant women to low levels of pyrethroids has raised concerns about the effect of chronic, low-level exposure to pyrethroids on the developing nervous system. Mice exposed to bioallethrin and deltamethrin during the second and third postnatal weeks, a period of rapid brain development, exhibited increases in motor activity without habituation and a downregulation of muscarinic receptor expression in the brain at 4 months (Ahlbom et al., 1994; Eriksson and Fredriksson, 1991; Eriksson and Nordberg, 1990; Eriksson et al., 1993). However, attempts to replicate these results in another laboratory failed to reproduce the effects of either bioallethrin or deltamethrin on behavior or muscarinic receptor expression (Muhammad et al., 2003). A review of 22 developmental neurotoxicity studies in rodents (Shafer et al., 2005) did not identify a consistent pattern of pyrethroid-dependent effects on the developing rodent CNS.

Recent studies provide additional information on the effects of pyrethroids on neuronal development in mice exposed in utero. Exposure of female mice to deltamethrin during gestation at doses below the developmental no-effect level caused male offspring to exhibit hyperactivity and impulsive-like behaviors that were interpreted as features of attention deficit hyperactivity disorder (ADHD) (Richardson et al., 2015). A subsequent study from the same laboratory involving exposure of pregnant females to low doses of deltamethrin found significant downregulation of the expression of genes for multiple sodium channel subunits and brain-derived neurotrophic factor in offspring at 10–11 months of age, a time when behavioral abnormalities were still observed (Magby and Richardson, 2017). Also, low doses of either cypermethrin or deltamethrin administered to pregnant female mice caused a reduction in neuronal proliferation, cell maturation and differentiation, and increased apoptosis in the cerebral cortex of offspring when measured at the end of embryonic development (Guo et al., 2018). A causal relationship between the behavioral effects of deltamethrin following in utero exposure and either the long-term sodium channel downregulation or alteration in neuronal development caused by deltamethrin remains to be established.

It is not clear whether these recently-described effects of deltamethrin on mouse neuronal development and behavior are relevant to pyrethroid exposure and developmental effects in human populations. A large survey of postnatal pyrethroid exposure in children found no association between levels of pyrethroid metabolites in the urine of children aged 6–15 and parental reports of ADHD or learning disability (Quirós-Alcalá et al., 2014). By contrast, a study of associations between residential proximity to pesticide use and the occurrence of neurodevelopmental disorders found that children of mothers residing near pyrethroid insecticide applications just prior to conception or during the last trimester of pregnancy were at greater risk for autism spectrum disorders and developmental delay (Shelton et al., 2014).

4.5 Pyrethroids and neurodegenerative disease

Epidemiological studies consistently identify chronic exposure to herbicides and insecticides as a risk factor for idiopathic Parkinson's disease (PD), but they are unable to pinpoint the chemicals that constitute the greatest risk (Brown et al., 2006; van der Mark et al., 2012). Efforts to understand the potential involvement of pyrethroids in PD have focused on the effects of these compounds on the dopaminergic neurons of the substantia nigra, the locus of the neurodegeneration underlying PD (reviewed in Baltazar et al., 2014; Bloomquist et al., 2002; Moretto and Colosio, 2011). Deltamethrin preferentially stimulated the release of dopamine from preloaded mouse striatal synaptosomes. Both deltamethrin and permethrin increased dopamine uptake into striatal synaptosomes and caused a slowly-developing upregulation of expression of the dopamine transporter protein. Although long-term low dose exposure to permethrin produced persistent changes in striatal dopamine transport, it did not cause degeneration of dopaminergic neurons, nor did it enhance the neurodegenerative effects of 1-methyl-4-phenyl-1,2,3,6-tetrahydropyridine (MPTP), a chemical that causes PD-like neurodegeneration in humans and animal models. Unlike permethrin, cypermethrin produced signs of neurodegeneration in the striatum following long-term exposure in adult rats; moreover, postnatal pre-exposure to cypermethrin increases subsequent adult sensitivity to the neurodegenerative effects of this compound. The effects of cypermethrin in on the striatum have been attributed in part to oxidative stress rather than a direct effect on dopaminergic neurotransmission.

5. Action of pyrethroids on mammalian voltage-gated sodium channels

5.1 Structure and heterogeneity of mammalian voltage-gated sodium channels (VGSCs)

The compelling evidence for effects on VGSCs as the mechanism of insecticidal activity of pyrethroids (summarized above) and the strong conservation of VGSC structure, function and pharmacology across animal taxa (Goldin, 2002) implicate VGSCs in the CNS as important target sites for the acute neurotoxic effects of pyrethroids in mammals. Potential VGSC targets for pyrethroids in mammals comprise nine α subunit isoforms ($Na_v1.1$–$Na_v1.9$) that exhibit unique patterns of developmental and anatomical expression and varied functional and pharmacological properties (Goldin, 2001). The α subunits are large pseudotetrameric proteins that form the ion pore and contain receptor domains for a variety of natural and synthetic neurotoxins, including pyrethroids. Four α subunits ($Na_v1.1$, $Na_v1.2$, $Na_v1.3$, and $Na_v1.6$) are abundantly expressed in either the embryonic or adult brain and represent possible targets for pyrethroid neurotoxicity in the CNS. The $Na_v1.3$ and $Na_v1.6$ isoforms are of particular interest because they are the most abundantly-expressed isoforms in the embryonic ($Na_v1.3$) and adult ($Na_v1.6$) brain (Felts et al., 1997; Shah et al., 2001; Whitaker et al., 2000, 2001).

Additional diversity among mammalian VGSCs results from the coassembly in the nerve membrane of an α subunit with one or two small auxiliary β subunits that modulate channel gating and regulate expression (Isom, 2001). VGSCs in the adult brain are heterotrimeric complexes of one α subunit and two β subunits that differ in structure and their mode of association (covalent or noncovalent) with the α subunit (Hartshorne and Catterall, 1984). Although there are four β subunits in mammals, the ubiquitous expression of the β1 and β2 subunits in the adult brain implies that the majority of brain VGSCs are $\alpha + \beta1 + \beta2$ complexes (Schaller and Caldwell, 2003; Shah et al., 2001; Whitaker et al., 2000). However, the subunit composition of individual native sodium channel complexes is not known.

5.2 Pyrethroid action on VGSCs in mammalian neurons

Given the significance of VGSCs as targets for pyrethroid action, there are surprisingly few studies available in the literature of the action of pyrethroids on VGSCs in mammalian neurons (Ginsburg and Narahashi, 1993; Song et al., 1996; Tabarean and Narahashi, 1998; Tatebayashi and Narahashi, 1994; Wu et al., 2009; Yang and Li, 2015). In these studies, which involved the measurement of macroscopic sodium currents under voltage clamp, all neuroactive pyrethroids retarded the kinetics of sodium channel activation, inactivation, and deactivation. This prolongation is evident during a depolarization step as the induction of a slowly-developing and persistent sodium current and after repolarization as a slowly-decaying sodium tail current. Type I compounds produce tail currents that decay relatively rapidly, whereas Type II compounds produce extremely persistent tail currents. In some cases, these effects on channel kinetics are accompanied by pyrethroid-dependent shifts in the voltage dependence of activation and inactivation gating. The effects of pyrethroids on VGSCs in mammalian neurons are therefore fully consistent with the effects on VGSCs in invertebrate nerve preparations.

Studies of pyrethroid action on sodium currents in dorsal root ganglion (DRG) neurons provide evidence for differences in pyrethroid sensitivity between functionally distinct VGSC populations expressed in a single cell. DRG neurons contain two populations of sodium channels that are distinguished by their gating properties and their differential sensitivity to block by tetrodotoxin (TTX). The TTX-resistant current in these cells is much more sensitive to than the TTX-sensitive current to allethrin (Ginsburg and Narahashi, 1993), tetramethrin (Song et al., 1996; Tatebayashi and Narahashi, 1994) and deltamethrin (Tabarean and Narahashi, 1998). The apparent use-dependent modification of TTX-R and TTX-S channels in DRG neurons by deltamethrin was attributed to the accumulation of persistently open channels rather than preferential modification of channels in the open state (Tabarean and Narahashi, 2001).

The effects of pyrethroids on macroscopic sodium currents are in good agreement with more limited studies at the single-channel level. In patch clamp experiments tetramethrin, a Type I compound, increased the mean open time of individual sodium channels approximately 10-fold (Yamamoto et al., 1983). By contrast, the Type II compounds deltamethrin and fenvalerate increase the mean open times of individual sodium channels in patch clamp studies up to 200-fold and produced channels that remained open on repolarization of the membrane patch (Chinn and Narahashi, 1986; Holloway et al., 1989). Deltamethrin also delayed channel opening in response to a depolarizing pulse.

Voltage clamp and patch clamp experiments have also been employed to assess the combined effects of two different pyrethroids on sodium channels. In voltage clamp experiments, initial exposure of DRG neurons to either fenvalerate (Song et al., 1996) or deltamethrin (Tabarean and Narahashi, 1998) produced characteristic modifications of sodium currents, including the induction of slowly-decaying tail currents. Subsequent application of tetramethrin to these preparations antagonized the effects of fenvalerate or deltamethrin, so that the persistent tail current associated with these compounds disappeared and was replaced by the rapidly-decaying tail current typical of the action of tetramethrin in this experimental system. Patch clamp studies of the interaction of tetramethrin and deltamethrin at single sodium channels (Motomura and Narahashi, 2001) confirmed the results of voltage clamp studies by documenting the displacement of deltamethrin by subsequently-applied tetramethrin. These results imply that pyrethroids that produce divergent kinetic modifications of sodium channels exert their effects either at a shared binding site, partially overlapping binding sites, or at sites that are allosterically coupled so that occupancy of one site precludes occupancy of the second.

An alternate approach to characterizing the effects of pyrethroids on CNS neurons in vitro involves the use of microelectrode array (MEA) chips to measure spontaneous electrical activity of neural networks formed by dissociated neurons in culture (Johnstone et al., 2010). Spontaneous activity (measured as the mean firing rate) was inhibited by permethrin and deltamethrin in networks formed from cultured rat hippocampal neurons (Meyer et al., 2008) and mouse frontal cortex and spinal cord neurons (Shafer et al., 2008), an effect attributed to an action on the VGSCs responsible for action potentials in these networks. In MEA recordings from neural networks formed by rat cortical neurons in culture, the inhibitory potency of an environmentally relevant mixture of five pyrethroids was consistent with that predicted by dose additivity based on the potencies of the five pyrethroids measured individually and the proportional representation of the five compounds in the mixture (Johnstone et al., 2017).

MEA recordings from rat cortical neural networks defined the relative potency of six pyrethroids as inhibitors of mean firing rates (Baskar and Murthy, 2018) and found a strong correlation between potencies in MEA assays and the potencies of the same compounds as inhibitors of locomotor activity in behavioral assays with rats (Wolansky et al., 2006).

5.3 Pyrethroid action on defined VGSC complexes expressed in *Xenopus* oocytes

Individual CNS neurons express multiple VGSC isoforms and contain multiple functionally and pharmacologically distinct VGSC heteromultimeric complexes (Felts et al., 1997; Whitaker et al., 2000, 2001). Thus, the relative sensitivity of different isoforms and subunit complexes to pyrethroids, and therefore the relative importance of these isoforms and complexes as targets in intoxication, cannot be determined in studies using native neurons. This difficulty can be overcome by using in vitro systems for the heterologous expression and functional characterization of VGSC complexes of defined subunit structure.

The *Xenopus* oocyte expression system is widely employed for the reconstitution and study of both ligand-gated and voltage-gated ion channels (Goldin, 2006). When injected with synthetic mRNA the oocyte efficiently translates the message, performs post-translational modifications on the nascent protein, and inserts the mature protein into the cell membrane. The functional and pharmacological properties of sodium channels expressed in oocytes can be assessed in voltage clamp assays of macroscopic sodium currents. The oocyte expression system is particularly well-suited to studies in which channel structure is the principal experimental variable. Expression in oocytes, when coupled with site-directed mutagenesis, also facilitates the testing of specific hypotheses regarding the effects of channel structure on both functional and pharmacological properties and the role of specific domains and amino acid residues in drug and insecticide binding to the channel.

Most of the studies of pyrethroid action on sodium channels in mammalian neurons summarized above were performed under voltage clamp conditions by perfusing cells with pyrethroids at hyperpolarized membrane potentials and assessing the effects of pyrethroids on sodium current kinetics during and after a depolarizing pulse. This approach is biased toward the detection of closed-state modification, and therefore most of these studies did not consider the contribution of other channel states to pyrethroid action. However, in some cases pyrethroid action on sodium channels in voltage-clamped squid and crayfish axons and rat DRG neurons is enhanced by repetitive depolarization (Brown and Narahashi, 1992; de Weille et al., 1988; Salgado and Narahashi, 1993; Tabarean and Narahashi, 2001). Moreover, modification by cypermethrin and deltamethrin of cloned insect sodium channels expressed in *Xenopus* oocytes is absolutely dependent on repeated depolarization (Smith et al., 1998; Vais et al., 2000b). In light of these results, most subsequent studies of the action of pyrethroids on mammalian VGSC complexes expressed in oocytes have considered both resting and use-dependent channel modification.

Initial studies of mammalian VGSCs expressed in *Xenopus* oocytes examined the rat $Na_v1.2$ (adult splice variant; previously designated rat brain IIa) isoform, which is abundantly expressed in the adult rat brain, and the impact of coexpression with the rat β1 subunit. In the absence of the β1 subunit [*1R,cis,αS*]-cypermethrin produced a slowly-decaying tail current typical of pyrethroid effects of sodium channels in other systems, but these effects were small and observable only at very high pyrethroid concentrations (Smith and Soderlund, 1998). Expression with the β1 subunit increased the sensitivity of $Na_v1.2$ channels to cypermethrin (Smith and Soderlund, 1998). By contrast, $Na_v1.2$ + β1 channels were not detectably modified by the high concentrations of cismethrin or [1*R,cis*]-permethrin (Smith and Soderlund, 1998; Vais et al., 1997).

A subsequent study (Meacham et al., 2008) showed that the β1 subunit increased the sensitivity of $Na_v1.2$ channels to deltamethrin and four other Type II pyrethroids (cypermethrin, β-cyfluthrin, esfenvalerate and fenpropathrin). However, $Na_v1.2+\beta1$ channels were not detectably modified by the Type I compounds permethrin and tetramethrin. All of these studies considered only closed-state modification of $Na_v1.2+\beta1$ channels pyrethroids. However, Vais et al. (2000a) showed that repeated depolarizing pulses were required to detect modification of $Na_v1.2+\beta1$ channels by deltamethrin and concluded that deltamethrin acted by binding exclusively to open channels. The reason for the lack of resting-state modification by deltamethrin in this study, which contradicts the results obtained elsewhere with $Na_v1.2+\beta1$ channels and several Type II pyrethroids (Meacham et al., 2008; Smith and Soderlund, 1998), is not known.

The rat $Na_v1.8$ isoform is more sensitive to pyrethroids than the $Na_v1.2$ isoform. $Na_v1.8$ channels, which are resistant to block by TTX, are expressed in the peripheral nervous system, primarily in sensory neurons (Catterall et al., 2005), and are likely to carry the TTX-resistant, pyrethroid-sensitive component of sodium current found in rat DRG neurons (Ginsburg and Narahashi, 1993). Whereas actions on peripheral nerve channels such as $Na_v1.8$ are not relevant to the systemic neurotoxic effects of pyrethroids, they may underlie the paresthesia that can occur following dermal exposure to pyrethroids (Soderlund et al., 2002).

Two studies (Choi and Soderlund, 2006; Smith and Soderlund, 2001) examined the effects of 11 pyrethroids, comprising both Type I and Type II structures, on $Na_v1.8$ channels expressed in oocytes in the absence of auxiliary β subunits. All 11 compounds produced clearly detectable closed-state modification. Manual subtraction of scaled control sodium currents from currents obtained following pyrethroid exposure permitted reconstruction and kinetic characterization of currents carried only by the pyrethroid-modified channel population. Fig. 6A shows example control and cismethrin-modified currents (top) and the reconstructed cismethrin-modified current derived from these traces (bottom). All of the pyrethroid-modified currents activated and inactivated much more slowly than currents carried by unmodified channels. There was a strong positive correlation between the kinetics of channel activation during depolarization and the kinetics of deactivation following repolarization (Fig. 6B).

This study also evaluated the effects of repetitive depolarization on the extent of pyrethroid modification. Use-dependent enhancement was restricted to the four compounds with the slowest kinetics of activation

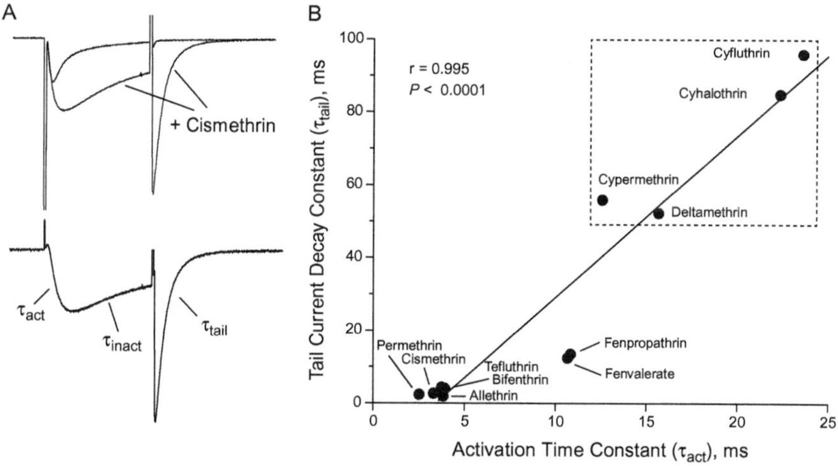

Fig. 6 Action of pyrethroids on rat $Na_v1.8$ sodium channels expressed in *Xenopus* oocytes. (A) Sodium currents recorded from an oocyte expressing $Na_v1.8$ sodium channels before and after exposure to 100 μM cismethrin (top) and the net cismethrin-modified current (bottom) calculated by subtraction of the control current from the composite current measured in the presence of cismethrin. (B) Correlation between the time constants of activation and tail current decay, obtained from traces such as those shown in panel A, for rat $Na_v1.8$ sodium channels modified by pyrethroids. The compounds enclosed in the rectangle exhibited statistically-significant use-dependent enhancement of channel modification. *The figure is adapted from the data of Choi, J.-S., Soderlund, D.M., 2006. Structure-activity relationships for the action of 11 pyrethroid insecticides on rat $Na_v1.8$ sodium channels expressed in Xenopus oocytes. Toxicol. Appl. Pharmacol. 211, 233–244.*

and deactivation (see Fig. 6B). It is possible, therefore, that the use-dependent effects of these four compounds are due to the accumulation of persistently open channels upon repeated depolarization rather than to enhanced binding of these compounds to the open state of the $Na_v1.8$ channel.

Further evidence for differences in pyrethroid sensitivity between rat VGSC isoforms comes from studies of the $Na_v1.2$, $Na_v1.3$, $Na_v1.6$, and $Na_v1.7$ isoforms expressed in oocytes as heterotrimeric $\alpha + \beta1 + \beta2$ complexes using tefluthrin as a pharmacological probe (Tan and Soderlund, 2009, 2010, 2011a). All four isoforms exhibited resting modification by tefluthrin that was enhanced by repeated depolarization (Fig. 7A). The $Na_v1.2$ and $Na_v1.7$ isoforms were relatively resistant to modification by tefluthrin, whereas the $Na_v1.3$ and $Na_v1.6$ isoforms were more sensitive. The difference in sensitivity of $Na_v1.2$ and $Na_v1.6$ channels to

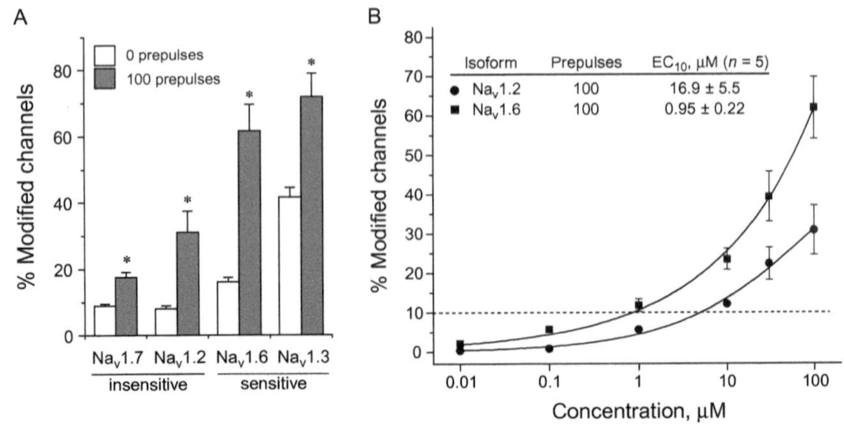

Fig. 7 Differential sensitivity of rat voltage-gated sodium channel isoforms to tefluthrin. (A) Resting (0 prepulses) and use-dependent (100 depolarizing prepulses) modification oocytes by 100 μM tefluthrin of four sodium channel isoforms expressed as α + β1 + β2 complexes in *Xenopus*. Asterisks indicate significant (paired *t*-tests, $P < 0.05$) use-dependent enhancement of channel modification. (B) Concentration dependence of use-dependent modification oocytes by tefluthrin of rat Na$_v$1.2 and Na$_v$1.6 sodium channels expressed as α + β1 + β2 complexes in *Xenopus* oocytes by tefluthrin. *The figure is adapted from the data of Tan, J., Soderlund, D.M., 2009. Human and rat Na$_v$1.3 voltage-gated sodium channels differ in inactivation properties and sensitivity to the pyrethroid insecticide tefluthrin. Neurotoxicology 30, 81–89; Tan, J., Soderlund, D.M., 2010. Divergent actions of the pyrethroid insecticides S-bioallethrin, tefluthrin, and deltamethrin on rat Na$_v$1.6 sodium channels. Toxicol. Appl. Pharmacol. 247, 229–237; and Tan, J., Soderlund, D.M., 2011a. Action of tefluthrin on rat Na$_v$1.7 voltage-gated sodium channels expressed in Xenopus oocytes. Pestic. Biochem. Physiol. 101, 21–26.*

use-dependent modification by 100 μM tefluthrin shown in Fig. 7A corresponds to an approximately 17-fold difference in potency based on EC$_{10}$ values (Fig. 7B) (Tan and Soderlund, 2010).

Studies with Na$_v$1.6 + β1 + β2 channels (Tan and Soderlund, 2010) also compared the effects of *S*-bioallethrin, tefluthrin and deltamethrin on the kinetics of channel deactivation and the relative importance of resting and use-dependent modification. Consistent with results in other assay systems, tail currents induced by *S*-bioallethrin decayed rapidly and those induced by deltamethrin were very persistent, whereas those induced by tefluthrin decayed with intermediate kinetics. Modification of Na$_v$1.6 + β1 + β2 channels by *S*-bioallethrin was not affected by repeated depolarization, whereas modification by both tefluthrin and deltamethrin was enhanced by repeated depolarization. In the case of deltamethrin, resting modification was barely detectable. Tefluthrin was approximately 10-fold more potent than deltamethrin on $Na_v1.6 + \beta1 + \beta2$ channels based on EC_{10} values obtained from partial concentration-effect curves for use-dependent modification. The high potency of tefluthrin against this putative CNS target is consistent with the high acute toxicity of this compound to rats (see Table 1).

Additional studies with the rat $Na_v1.6$ isoform in the oocyte expression system explored the effects of the β1 and β2 subunits on resting and use-dependent channel modification by tefluthrin (Tan and Soderlund, 2011b). Coexpression of the $Na_v1.6$ α subunit with the β subunits, individually or in combination, did not affect the sensitivity of $Na_v1.6$ channels to resting modification by tefluthrin. Trains of depolarizing prepulses did not affect channel modification during a test depolarization in the absence of β subunits, but coexpression with the β1 subunit enhanced the extent of channel modification twofold following repeated depolarization. The β2 subunit also slightly enhanced modification following repeated depolarization, but coexpression with both β subunits caused enhanced modification following repeated depolarization that was indistinguishable from that found with $Na_v1.6 + \beta1$ channels. A more limited set of experiments with the rat $Na_v1.3$ isoform showed that coexpression with the β1 and β2 subunits was required to observe enhanced modification following repeated depolarization. These results establish an important role for sodium channel auxiliary subunits in modulating channel responses to pyrethroids in the oocyte expression system and, in particular, conferring sensitivity to use-dependent modification.

The conventional interpretation of use-dependent channel modification is that it signifies preferential binding to the open state of the channel, but in the case of pyrethroids use-dependent effects could also result from the accumulation of persistently open channels following binding to the resting state (Smith and Soderlund, 2001; Tabarean and Narahashi, 2001). The binding of pyrethroids to open sodium channels was assessed directly in a series of experiments employing the rat $Na_v1.6$ sodium channel isoform containing a tripeptide mutation (IFM/QQQ) that is known to prevent fast inactivation (Fig. 8A) (McCavera and Soderlund, 2012). Expression of mutated rat $Na_v1.6$ channels (designated $Na_v1.6Q3$) with the rat β1 and β2 subunits produced noninactivating sodium currents (Fig. 8B), permitting the assessment of the duration of channel opening on the extent of channel modification by *S*-bioallethrin, tefluthrin and deltamethrin. Modification by tefluthrin was clearly detectable following short depolarizations and was increased by long depolarizations (Fig. 8C and D). Deltamethrin caused little modification of $Na_v1.6Q3$ channels following short depolarizations, but prolonged

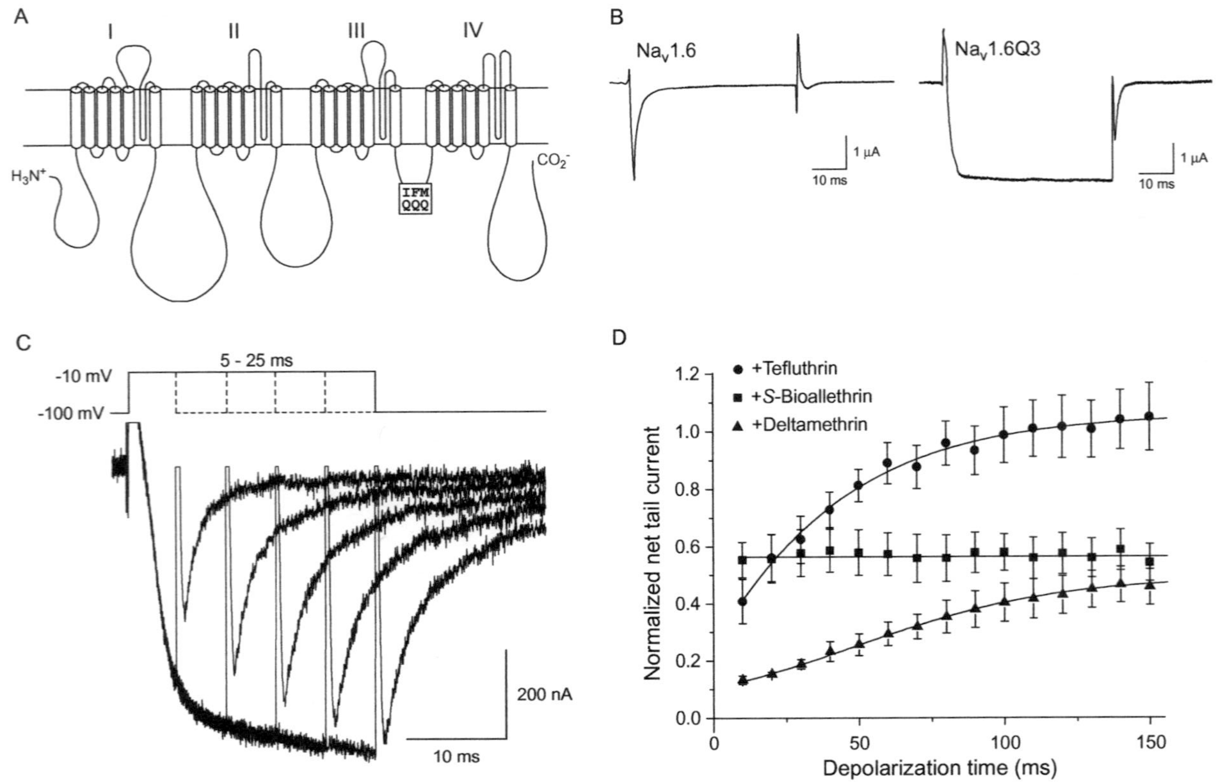

Fig. 8 See legend on opposite page.

depolarizations caused a progressive increase in channel modification (Fig. 8D). By contrast the extent of modification by *S*-bioallethrin following short depolarizations was not increased by prolonged depolarization (Fig. 8D). These studies provide direct evidence for the preferential binding of deltamethrin and tefluthrin (but not *S*-bioallethrin) to $Na_v1.6Q3$ channels in the open state.

5.4 Pyrethroid action on defined VGSC complexes expressed in mammalian cell lines

Although the *Xenopus* oocyte system readily permits the manipulation of channel structure as an experimental variable, the biophysical and pharmacological properties of channels expressed in the oocyte membrane environment may differ from those in native neurons or mammalian cell expression systems due to species differences in membrane composition and post-translational modification (Goldin, 2006). The HEK293 cell line, derived from human embryonic kidney cells by transformation with sheared adenovirus type 5 DNA (Graham et al., 1977), is widely employed as an alternative to *Xenopus* oocytes for the heterologous expression of a variety of proteins, including mammalian voltage-gated sodium channels (Thomas and Smart, 2005). Despite their origin HEK293 cells exhibit some characteristics of neurons, expressing more than 60 neuronal genes including neurofilament proteins and neuroreceptor and ion channel subunits (Shaw et al., 2002; Thomas and Smart, 2005).

Fig. 8 Action of *S*-bioallethrin, deltamethrin and tefluthrin on inactivation-deficient rat $Na_v1.6$ sodium channels expressed as $\alpha + \beta1 + \beta2$ complexes in *Xenopus* oocytes. (A) Diagram of the extended transmembrane structure of a voltage-gated sodium channel α subunit protein showing the location of the IFM/QQQ mutation introduced in to the $Na_v1.6Q3$ channel. (B) Representative sodium current trace recorded from an oocytes expressing $Na_v1.6$ channels (left) or $Na_v1.6Q3$ channels during a 40-ms step depolarization from -100 to -10 mV. (C) Effect of increasing depolarization time on the amplitude tefluthrin-induced sodium tail currents. Traces were recorded from a single oocyte expressing $Na_v1.6Q3$ channels either before or after equilibration with tefluthrin (100 μM) during and after depolarizations of 5–25 ms in 5-ms increments from a holding potential of -100 mV to a test potential of -10 mV. (D) Normalized net tail currents measured following equilibration of oocytes expressing $Na_v1.6Q3$ sodium channels with *S*-bioallethrin (100 μM), tefluthrin (100 μM) or deltamethrin (10 μM) and step depolarizations of 10–150 ms in 10-ms increments. Values are means ± SE of six determinations with different cells and are corrected for control tail currents measured in the same cell prior to pyrethroid exposure. *The figure is adapted from the data of McCavera, S.J., Soderlund, D.M., 2012. Differential state-dependent modification of inactivation-deficient $Na_v1.6$ sodium channels by the pyrethroid insecticides S-bioallethrin, tefluthrin and deltamethrin. Neurotoxicology. 33, 384–390.*

Moreover, some clonal populations of HEK293 cells exhibit small endogenous voltage-gated sodium currents that are associated primarily with the expression of the human $Na_v1.7$ sodium channel isoform (He and Soderlund, 2010).

The high-level expression of rat $Na_v1.6 + \beta1 + \beta2$ channel complexes in stably-transformed clonal HEK293 cell lines permitted the assessment of the action of pyrethroids on channels in this expression system and facilitated comparisons between identical $Na_v1.6 + \beta1 + \beta2$ channel complexes expressed in oocytes and in mammalian cells (He and Soderlund, 2011). Assays of macroscopic sodium currents using the whole-cell patch clamp technique evaluated the effects of tefluthrin and deltamethrin in this system. Tefluthrin (Fig. 9) and deltamethrin produced concentration-dependent hyperpolarizing shifts in the voltage dependence of channel activation and steady-state inactivation. In addition to their effects on the voltage dependence of inactivation, both compounds caused

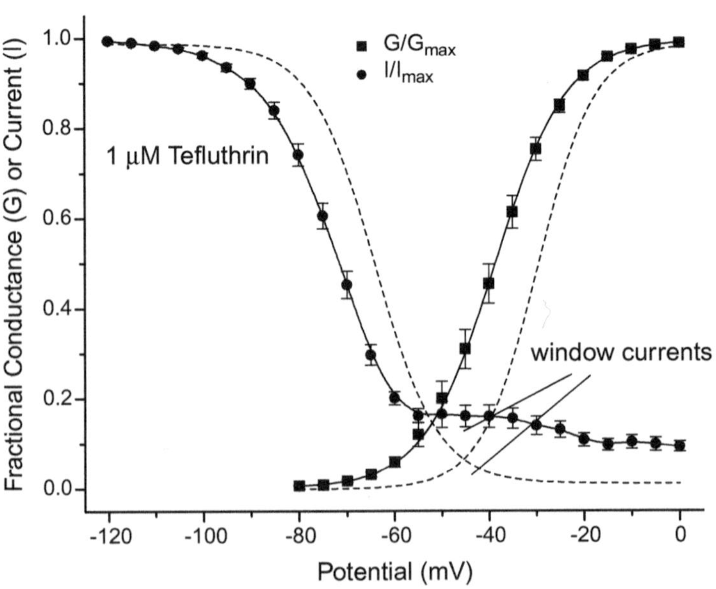

Fig. 9 Effect of tefluthrin on the voltage dependence of activation (G/G_{max}, squares) and steady-state activation (I/I_{max}, circles) of $Na_v1.6 + \beta1 + \beta2$ sodium channels expressed in HEK cells. Dashed lines show curves obtained for activation and steady-state inactivation in the absence of tefluthrin. *The figure is adapted from He, B., Soderlund, D.M., 2011. Differential state-dependent modification of rat $Na_v1.6$ sodium channels expressed in human embryonic kidney (HEK293) cells by the pyrethroid insecticides tefluthrin and deltamethrin. Toxicol. Appl. Pharmacol. 257, 377–387.*

concentration-dependent increases in the fraction of sodium current that was resistant to inactivation following strong depolarizing prepulses. The combined effects on voltage dependence and resistance to inactivation produced large window currents (Fig. 9), creating a population of persistently activated channels at membrane potentials at which unmodified channels are either unresponsive or inactivated. The pyrethroid-dependent shifts in the voltage dependence of activation and inactivation and the production of a population of inactivation-resistant channels seen in assays of $Na_v1.6 + \beta1 + \beta2$ channels in HEK293 cells were not found in prior studies of the same channel complex in oocytes.

The relative importance of resting and use-dependent modification of rat $Na_v1.6$ sodium channels by pyrethroids also differed between expression systems. In oocytes (Tan and Soderlund, 2010) modification of $Na_v1.6$ channels by deltamethrin required repetitive depolarization, indicating that this compound binds almost exclusively to open $Na_v1.6$ channels in this system. Tefluthrin produced readily detectable modification of resting $Na_v1.6$ channels, but repetitive depolarization significantly enhanced the extent of modification, increasing the apparent affinity of channels for this compound by ~15-fold. This result indicates that tefluthrin binds to both resting and open $Na_v1.6$ channels but exhibits preference for binding to the open state. In contrast to the results obtained in oocytes, both tefluthrin and deltamethrin produced detectable resting modification of $Na_v1.6$ channels expressed in HEK293 cells, but only modification by deltamethrin was further enhanced by repeated depolarization (He and Soderlund, 2011). Based on results with these two pyrethroids, $Na_v1.6$ channels expressed in oocytes appear to exhibit a preference for open-state modification, whereas the same channels expressed in HEK293 cells may undergo more extensive modification in the resting state.

A subsequent study (He and Soderlund, 2016) examined the effects of tefluthrin and deltamethrin on the rat $Na_v1.6$ channel expressed in HEK293 cells either alone or with the rat $\beta1$ subunit. Tefluthin and deltamethrin caused large shifts in the voltage dependence of activation and inactivation and produced a population of inactivation-resistant channels with both $Na_v1.6$ monomeric channels and $Na_v1.6 + \beta1$ channels, a result consistent with the effects of these compounds on $Na_v1.6 + \beta1 + \beta2$ channels. However, repeated depolarization had no effect on the extent of modification of $Na_v1.6$ or $Na_v1.6 + \beta1$ channels. Thus it appears that open-state modification of $Na_v1.6$ channels expressed in HEK293 cells requires the $\beta2$ subunit, unlike in oocytes where open-state modification requires the β1 subunit. The results obtained with rat $Na_v1.6$ channels expressed in HEK293 cells alone or in combination with the β1 or β2 auxiliary subunits highlight the importance of cellular context as a determinant of the actions of pyrethroids on sodium channels in heterologous expression systems.

The only other study of the action of pyrethroids on sodium channels expressed in HEK293 cells compared the sensitivities of the human $Na_v1.1$ and $Na_v1.6$ isoforms, expressed in the absence of β subunits, to deltamethrin (James et al., 2017). Deltamethrin modified the human $Na_v1.1$ channel principally in the resting state, whereas some use-dependent effects were observed with the human $Na_v1.6$ channel. The $Na_v1.1$ isoform also appeared to be more sensitive to deltamethrin modification than the $Na_v1.6$ isoform. These studies represent the only information on a mammalian $Na_v1.1$ sodium channel isoform in any expression system. However, the interpretation of the use-dependent actions of deltamethrin on the $Na_v1.1$ and $Na_v1.6$ isoforms in this study is complicated by the use of only α subunits expressed in the absence of β subunits and repetitive depolarization protocols that differed markedly from those used by others to examine use-dependent actions on channels expressed in oocytes and HEK293 cells.

6. Pyrethroid receptors on VGSCs

6.1 Pharmacological characterization of pyrethroid receptors

Sodium channels are the site of action of a large structural variety of naturally-occurring neurotoxins (Catterall et al., 2007; Mattei and Legros, 2014). This coevolutionary convergence of strategies for chemical predation and defense on a single molecular target underscores both the fundamental significance of VGSCs in the normal function of animal nervous systems and the disruptive potential of agents that modify normal VGSC function. The sites of action of these neurotoxins, identified in functional and radioligand binding assays and by site-directed mutagenesis, identify six distinct binding domains for naturally-occurring neurotoxins on the VGSC α subunit protein (Table 3). These binding sites are operationally defined by competitive interactions among toxins considered to act at the same domain and allosteric interactions between toxins acting at different domains. Some of these sites have been mapped by site-directed mutagenesis to specific physical domains of the sodium channel α subunit protein.

Table 3 Discrete neurotoxin receptor domains on voltage-gated sodium channels.

Site[a]	Neurotoxins	Effects
Site 1	Tetrodotoxin	Channel block
	Saxitoxin	
	μ-Conotoxin	
Site 2	Batrachotoxin	Enhanced activation
	Veratridine	Persistent activation
	Aconitine	Repetitive nerve firing
	Grayanotoxins	
Site 3	α-Scorpion toxins	Slowed inactivation
	Sea anemone toxins	Prolonged channel opening
	δ-Astracotoxins	
Site 4	β-Scorpion toxins	Enhanced activation
Site 5	Brevetoxins	Enhanced activation
	Ciguatoxins	Block of inactivation
Site 6	δ-Conotoxins	Prolonged channel opening
Site 7	Pyrethrins	Persistent activation
	DDT and analogs	Repetitive nerve firing
	Pyrethroids	Nerve depolarization

[a]Sites 1–6 after Catterall et al. (2007).

Studies of pyrethroid action on radiosodium uptake and radioligand binding in brain preparations elucidated the relationship between the pyrethroid receptor site and other identified binding domains of the sodium channel. Pyrethroids alone do not affect radiosodium uptake into brain synaptosomes, but they allosterically enhance sodium uptake that is stimulated by veratridine or batrachotoxin (Bloomquist and Soderlund, 1988; Ghiasuddin and Soderlund, 1985). Type II pyrethroids also allosterically enhanced the binding of [^{3}H]batrachotoxinin A-20-α-benzoate (BTX-B) to brain sodium channels and synergistically enhanced the allosteric effects of α-scorpion toxins and brevetoxins on BTX-B binding, but most Type I compounds and DDT analogs were without effect (Brown et al., 1988; Lombet et al., 1988; Rubin et al., 1993; Trainer et al., 1993). Comparisons of deltamethrin, its neurotoxic 1*R*,*trans*,α*S* epimer (designated NRDC 158A), and structurally related nontoxic isomers showed that only deltamethrin and NRDC 158A enhanced BTX-B binding to mouse brain sodium channels (Rubin and Soderlund, 1993). The results of these studies imply the existence of a pyrethroid binding site on sodium channels that is distinct from Sites 1–6 (Table 3) and is allosterically coupled to Sites 2, 3 and 5. Binding studies using [^{3}H]RU58487, a pyrethroid photoaffinity radioligand, demonstrated high-affinity saturable binding to brain sodium channels that exhibits the allosteric coupling to other binding domains predicted by previous studies (Trainer et al., 1997). Moreover, [^{3}H]RU58487 specifically labeled the sodium channel α subunit protein as demonstrated by immunoprecipitation. These studies provide direct evidence for a unique pyrethroid binding site (Table 3, Site 7) that is intrinsic to the α subunit protein and is allosterically coupled to sites for other sodium channel ligands.

Modification of VGSC function by pyrethroids is stereospecific for neurotoxic isomers, but there is indirect evidence that nontoxic isomers also bind to VGSCs without altering channel function. A study of the action of the four isomers of tetramethrin on sodium currents in squid giant axon preparations under voltage clamp showed that the nontoxic 1*S* isomers of tetramethrin antagonized the effects of the toxic 1*R* isomers by binding to an allosteric site (Lund and Narahashi, 1982). Similarly, nontoxic isomers of deltamethrin antagonized deltamethrin's enhancement of BTX-B binding to sodium channels in mouse brain synaptosomes (Rubin and Soderlund, 1993). Recent studies show that the nontoxic 1*S*,*cis* isomer of permethrin has no effect on sodium currents carried by insect sodium channels expressed in *Xenopus* oocytes, but it competitively inhibits the action of [1*R*,*cis*] permethrin in these preparations (Du et al., 2010, 2013; Tan et al., 2005). Taken together these findings suggest that nontoxic isomers of pyrethroids bind to VGSCs, thereby interfering with the action of toxic isomers without modifying channel function.

6.2 Molecular model of pyrethroid receptors on insect VGSCs

Point mutations in insect sodium channel genes that are associated with pyrethroid resistance provide unique insight into the sodium channel domains that are important for pyrethroid action and may therefore be involved in pyrethroid binding. Comparisons of partial or complete coding sequences of sodium channel genes from pyrethroid-susceptible and pyrethroid-resistant strains of arthropod species identified more than 50 mutations or combinations of mutations at more than 30 sequence positions that are associated with pyrethroid resistance (Dong et al., 2014). Studies employing site-directed mutagenesis and expression in *Xenopus* oocytes confirmed that 18 single amino acid substitutions at 13 sequence positions (Fig. 10) reduce the sensitivity of insect sodium channels to pyrethroids (Dong et al., 2014). The majority of confirmed resistance-associated mutations are found in or adjacent to the intracellular linker between transmembrane helices S4 and S5 (designated L4–5) or within transmembrane helices S5 and S6 of homology domain II. Models of the structure of sodium channels and related voltage-gated potassium channels place the L4–5, S5 and S6 segments of each homology domain in close proximity to each other at or near the inner mouth of the channel, with the S6 helices involved in activation gating of the channel (Lipkind and Fozzard, 2000; Shrivastava et al., 2004; Zhao et al., 2004). The cluster of confirmed resistance mutations in the L4–5, S5 and S6 segments of homology domain II led to conceptual models postulating a pyrethroid receptor formed primarily by residues in these segments of homology domain II with some contribution from the S6 helices and associated regions of homology domains I and III (Lee and Soderlund, 2001; Tan et al., 2005; Vais et al., 2003).

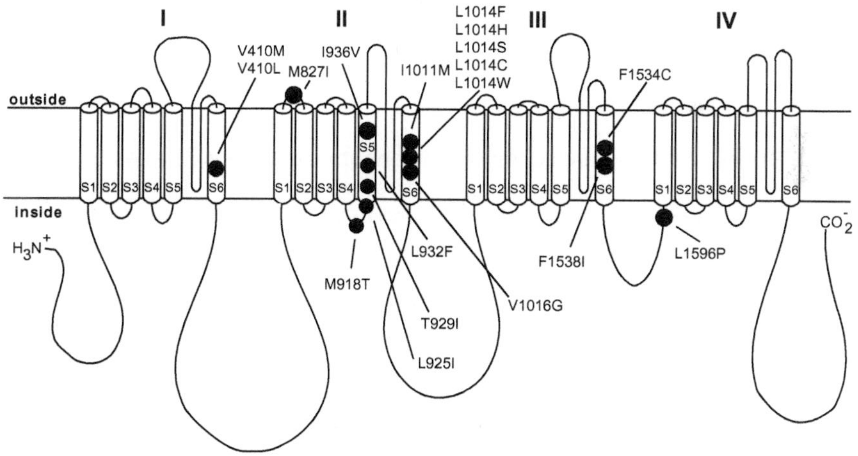

Fig. 10 Diagram of the extended transmembrane structure of a voltage-gated sodium channel α subunit protein showing the location (numbered according the house fly Vssc1 protein sequence) and identity of point mutations identified in pyrethroid-resistant arthropod populations that reduce the sensitivity to pyrethroids of insect sodium channels expressed in *Xenopus* oocytes. *The figure is based on the summary of resistance-associated mutations in Dong, K., Du, Y., Rinkevich, F., Nomura, Y., Peng, X., Wang, L., Silver, K., Zhorov, B.S., 2014. Molecular biology of insect sodium channels and pyrethroid resistance. Insect Biochem. Mol. Biol. 50, 1–17.*

In the absence of a crystal structure for any animal sodium channel, O'Reilly et al. (2006) constructed a homology model of the pore region of the house fly Vssc1 sodium channel based on the crystal structure of the homologous rat $K_v1.2$ potassium channel, which was crystallized in the open configuration (Long et al., 2005). Computer-aided docking of pyrethroid ligands and the predicted impact of resistance mutations on ligand docking parameters led to the identification of a putative receptor for pyrethroids and DDT that includes four resistance-associated residues in the L4–5 and S5 segments of homology domain II and one resistance-associated residue in the S6 segment of homology domain III (Fig. 11) (O'Reilly et al., 2006).

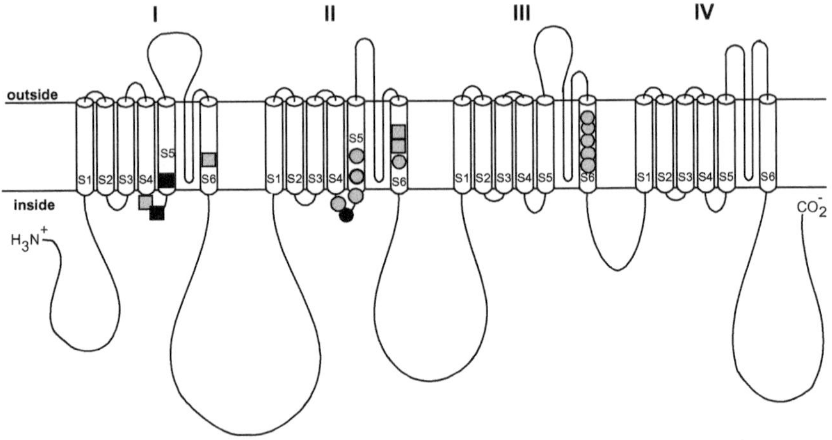

Fig. 11 Diagram of the extended transmembrane structure of a voltage-gated sodium channel α subunit protein showing locations of pyrethroid-sensing residues in two putative pyrethroid receptor sites on insect voltage-gated sodium channels. Pyrethroid-sensing residues in putative receptor Site 1 are indicated by circles, and pyrethroid-sensing residues in putative receptor Site 2 are indicated by squares. The three black symbols identify sites of amino acid sequence polymorphisms between insect and mammalian voltage-gated sodium channels that confer reduced pyrethroid sensitivity on mammalian channels. *The figure is based on the data of O'Reilly, A.O., Khambay, B.P.S., Williamson, M.S., Field, L.M., Wallace, B.A., Davies, T.G.E., 2006. Modelling insecticide binding sites at the voltage-gated sodium channel. Biochem. J. 396, 255–263; Usherwood, P.N.R., Davies, T.G.E., Mellor, I.R., O'Reilly, A.O., Peng, F., Vais, H., Khambay, B.P.S., Field, L.M., Williamson, M.S., 2007. Mutations in DIIS5 and the DIIS4-5 linker of Drosophila melanogaster sodium channel define binding domains for pyrethroids and DDT. FEBS Lett. 581, 5485–5492; Du, Y., Lee, J.-E., Nomura, Y., Zhang, T., Zhorov, B., Dong, K., 2009. Identification of a cluster of residues in transmembrane 6 of domain III of the cockroach sodium channel essential for the action of pyrethroid insecticides. Biochem. J. 419, 377–385; Du, Y., Nomura, Y., Satar, G., Hu, Z., Nauen, R., He, S.Y., Zhorov, B., Dong, K., 2013. Molecular evidence for dual pyrethroid-receptor sites on a mosquito sodium channel. Proc. Natl. Acad. Sci. U. S. A. 110, 11785–11790.*

Site-directed mutagenesis studies in these regions identified additional pyrethroid-sensing residues that contribute to the proposed pyrethroid receptor (Du et al., 2009; Usherwood et al., 2007).

A subsequent study (Du et al., 2013) created a similar homology model of the pore region of the mosquito (*Aedes aegypti*) sodium channel and investigated the potential role of pyrethroid resistance mutations lying outside the pyrethroid receptor (designated receptor Site 1) described by O'Reilly et al. (2006). Additional site-directed mutagenesis experiments and computer-aided docking of pyrethroid ligands identified a second putative pyrethroid receptor (designated receptor Site 2) associated with amino acid residues in the L4–5, S5, and S6 regions of homology domain I and the S6 domain of homology domain II (Fig. 11). A three-dimensional model of the pore region of the mosquito sodium channel (Du et al., 2013) places receptor Site 1 at the interface of homology domains II and III and receptor Site 2 at the interface of homology domains I and II. The existence of a second pyrethroid receptor on insect VGSCs is consistent with Hill analyses of the action of deltamethrin on insect sodium channels expressed in *Xenopus* oocytes, which implied the existence of more than one pyrethroid binding site per channel (Vais et al., 2000b).

7. Molecular basis of the differential sensitivity of VGSCs to pyrethroids

7.1 Differential sensitivity of rat VGSC isoforms

The differential sensitivity of rat VGSC isoforms to pyrethroids has been mapped to a single amino acid polymorphism. Alignment of the complete amino acid sequences of four pyrethroid-sensitive isoforms (house fly Vssc1 and rat $Na_v1.3$, $Na_v1.6$ and $Na_v1.8$) and the two pyrethroid-insensitive isoforms (rat $Na_v1.2$ and $Na_v1.7$) identified a single sequence position at which the amino acid sequence was conserved among all four sensitive isoform sequences but differed in the two insensitive isoform sequences (Fig. 12) (Tan and Soderlund, 2011a). This residue, a conserved valine in sensitive isoforms, was replaced with an isoleucine in the $Na_v1.2$ and $Na_v1.7$ sequences. The site of this polymorphism lies within a block of highly conserved sequence in a region of the channel that forms part of putative pyrethroid receptor Site 2. This residue (Val410 in Vssc1) has been identified as the site of multiple point mutations resulting in replacement by leucine, alanine, glycine or methionine in sodium channel sequences from pyrethroid-resistant insects (Dong et al., 2014). The significance of this

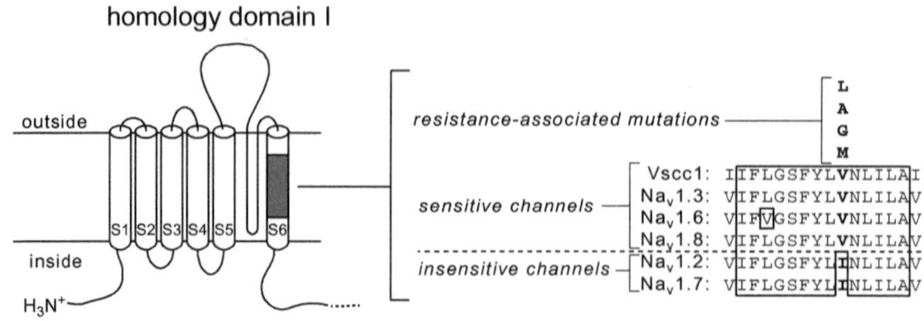

Fig. 12 Identification of a sodium channel amino acid sequence polymorphism associated with differential pyrethroid sensitivity of sodium channel isoforms. Left: diagram of sodium channel homology domain I showing the location of a conserved block of amino acid sequence in transmembrane segment 6. Right: Aligned amino acid sequences in transmembrane segment 6 from four pyrethroid-sensitive channels (house fly Vssc1, rat Na$_v$1.3, rat Na$_v$1.6 and rat Na$_v$1.8) and two pyrethroid-resistant channels (rat Na$_v$1.2 and rat Na$_v$1.7). Residues enclosed in the solid line are identical in all six sequences. Polymorphic residues aligning with V410 of Vssc1 are shown in boldface type; substitutions at the valine residue aligning with V410 of Vssc1 that are associated with pyrethroid resistance in insects are shown above the aligned sequences. *The figure is from Tan, J., Soderlund, D.M., 2011a. Action of tefluthrin on rat Na$_v$1.7 voltage-gated sodium channels expressed in Xenopus oocytes. Pestic. Biochem. Physiol. 101, 21–26.*

polymorphism in determining relative pyrethroid sensitivity was confirmed in studies with the rat Na$_v$1.4 isoform, which also contains an isoleucine at this position and is therefore expected to be pyrethroid-resistant. Replacement of I433 in the Na$_v$1.4 sequence with valine increased the sensitivity of this isoform to deltamethrin (Oliveira et al., 2013).

7.2 Differential sensitivity of insect and mammalian VGSCs

The favorable selective toxicity of pyrethroids is due in part to the generally lower pyrethroid sensitivity of mammalian sodium channels. Direct comparisons of the pyrethroid sensitivity of insect and mammalian sodium channel isoforms in the common expression environment of the *Xenopus* oocyte showed that insect channels are much more sensitive to pyrethroids (Vais et al., 2001). The difference in sensitivity found in this study is exaggerated somewhat by the use of a pyrethroid-insensitive rat sodium channel isoform (Na$_v$1.2) as the basis for comparison, but more sensitive rat isoforms such as Na$_v$1.3, Na$_v$1.6 or Na$_v$1.8 are also at least an order of magnitude less sensitive to pyrethroid modification than insect channels (Smith et al., 1997; Soderlund and Lee, 2001; Tan and Soderlund, 2009, 2010).

Amino acid sequence alignments of insect and mammalian VGSCs have provided clues to the molecular basis for the reduced pyrethroid sensitivity of mammalian VGSCs. A single amino acid polymorphism, which is located in the L4–5 region of homology domain II and is a component of pyrethroid receptor Site 1 on insect sodium channels (O'Reilly et al., 2006), confers reduced pyrethroid sensitivity on mammalian VGSC isoforms (see Fig. 11). All pyrethroid-sensitive insect channels contain a methionine at this position (M918 of the house fly Vssc1 sequence). By contrast all mammalian channels, regardless of their relative pyrethroid sensitivity, contain an isoleucine at the homologous position. Replacement of this isoleucine residue with methionine in rat $Na_v1.2$ (Vais et al., 2000a), $Na_v1.4$ (Wang et al., 2001) or $Na_v1.8$ (Soderlund and Lee, 2001) channels significantly increased the sensitivity of these channels to pyrethroids.

Similarly, two amino acid polymorphisms located at pyrethroid-sensing residues in models of pyrethroid receptor Site 2 in insect sodium channels may also contribute to reduced sensitivity of mammalian VGSCs to pyrethroids (see Fig. 11) (Du et al., 2013). Pyrethroid-susceptible insects have conserved valine and isoleucine residues at positions 252 and 263 (house fly Vssc1 sequence numbering), respectively, whereas mammalian VGSCs have conserved leucine at positions homologous to V252 of Vssc1 and either methionine or threonine at positions homologous to I263 of Vssc1. The single L252V or M263I mutations slightly increased the sensitivity of rat $Na_v1.2$ or $Na_v1.4$ channels to deltamethrin, but the effects of the two mutations were additive. Moreover, combination of the L252V/M263I double mutation with the previously-described I918M mutation produced mammalian channels with greatly increased sensitivity to deltamethrin.

7.3 Differential sensitivity of rat and human VGSCs

The uncertain fidelity of rodent models in predicting intoxication risks in humans remains a central problem for both experimental and regulatory toxicology. Sodium channel α subunit sequences are highly conserved, so that orthologous subunits in rats and humans are >95% identical at the level of amino acid sequence (Goldin, 2001). However, this degree of sequence conservation still results in as many as 100 amino acid sequence differences between orthologous channel proteins and could conceivably produce species differences in functional and pharmacological properties. Heterologous expression systems offer the opportunity to compare directly the properties and pyrethroid sensitivities of orthologous rat and human sodium channels.

The only study of this type to date compared the sensitivities to tefluthrin of orthologous rat and human $Na_V1.3 + \beta1 + \beta2$ sodium channels in *Xenopus* oocytes (Tan and Soderlund, 2009). Modification of rat $Na_V1.3$ channels by 100 μM tefluthrin was fourfold greater than that measured in parallel assays with human $Na_V1.3$ channels. Human $Na_V1.3$ channels were also less sensitive to tefluthrin than rat $Na_V1.2$ channels, which are considered to be relatively insensitive to pyrethroids (see Fig. 7). The rat and human channels also differed in their kinetics of fast inactivation and voltage dependence of steady-state inactivation. These two sequences differ at only 54 of 1951 amino acid residues (Tan and Soderlund, 2009), and none of these differences align with any residues identified as putative determinants of pyrethroid sensitivity shown in Figs. 10 and 11. Therefore, additional structural determinants of pyrethroid sensitivity remain to be identified.

8. Correlation of effects on VGSCs with mammalian toxicity

The insecticidal properties of pyrethroids derive from their ability to alter the function of VGSCs. Although the discovery and development of pyrethroids involved the optimization of insecticidal activity using whole insects, optimization for high intrinsic activity on insect sodium channels was implicit throughout this process. The strong conservation of the structure, function and pharmacology of voltage-gated sodium between insects and mammals implicate VGSCs the likely targets for pyrethroid neurotoxicity in mammals as well.

Three lines of evidence provide more direct evidence that VGSCs are sites of action for the acute neurotoxic effects of pyrethroids in mammals. First, the structural and stereochemical requirements identified in the optimization of insecticidal activity are conserved both in assays with mammalian VGSCs in a variety of in vitro systems and in the acute toxicity of pyrethroids to rodents. In particular, the absolute stereospecificity for 1*R* cyclopropanecarboxylates and esters of α-*S*-cyano-3-phenoxybenzyl alcohol that is found in insecticidal activity is conserved in the acute toxicity of pyrethroids to mammals.

Second, qualitative differences in sodium channel modification between the two structural subclasses of pyrethroids are generally correlated with the production of different intoxication syndromes (Bloomquist, 1993; Soderlund et al., 2002; Weiner et al., 2009). Most Type I compounds, which induce burst discharges in invertebrate neurons following a single stimulus and rapidly-decaying sodium tail currents under voltage clamp conditions, produce the T syndrome of intoxication in rodents. Most Type II compounds, which cause use-dependent nerve block resulting from a degradation of the resting membrane potential in invertebrate neurons and long-lived sodium tail currents under voltage clamp conditions, produce the CS syndrome of intoxication. A small number of compounds from both the Type I and Type II structural classes produce intermediate effects on neuronal excitability and tail current decay and also produce signs of intoxication that include elements of both the T and CS syndromes.

Third, recent studies have identified amino acid sequence polymorphisms between mammalian and insect VGSCs, located within the two putative pyrethroid receptor domains of insect VGSCs, that reduce the sensitivity of mammalian VGSCs to pyrethroids (Du et al., 2013). These intrinsic differences in pyrethroid sensitivity between insect and mammalian VGSCs are thought to contribute in part to the favorable selective toxicity of pyrethroids in practice.

Despite the evidence connecting effects on sodium channels with toxicity and the high acute toxicity of pyrethroids when introduced directly to the mammalian CNS, the apparent potency of pyrethroids as modifiers of mammalian VGSCs in vitro is relatively modest. Biochemical and electrophysiological studies of pyrethroid action require relatively high concentrations of insecticide to produce populations of pyrethroid-modified channels large enough to be detected by the methods of assay, and threshold concentrations for detectable effects are typically in the range of 10–100 nM. However, the concentrations required to disrupt normal action potential generation in neurons are equal to or lower than these threshold concentrations because only a small fraction of the available population of channels must be modified in order to disrupt electrical signaling. This relationship was demonstrated directly by Song and Narahashi (1996) in a study of the action of tetramethrin on sodium channels in rat Purkinje neurons. In this study, micromolar concentrations of tetramethrin were required to produce sufficient channel modification (5–25%) to be detected readily under voltage clamp conditions, whereas disruption of action potential generation in the same cells was achieved at 100 nM tetramethrin, a concentration calculated to modify fewer than 1% of sodium channels based on voltage clamp data. Similarly, concentrations of deltamethrin in the micromolar range are required to produce detectable modification various rat sodium channel isoforms expressed in oocytes or mammalian cell lines under voltage clamp conditions (Choi and Soderlund, 2006; He and Soderlund, 2011; Meacham et al., 2008; Tan and Soderlund, 2009, 2010; Vais et al., 2000a), but deltamethrin concentrations in the nanomolar range are sufficient to disrupt normal electrical activity in primary cultures of mouse frontal cortex and spinal cord neurons (Shafer et al., 2008). Thus, studies performed in some in vitro systems tend to underestimate significantly the potency of pyrethroids as disruptors of VGSC function and neuronal signaling in intact nervous systems.

9. Conclusions

Pyrethroids have been important pest management tools in agriculture, public health, and a variety of household applications for more than three decades. Pyrethroids are neurotoxic to both insects and mammals, and the empirical structure-activity rules that govern insecticidal activity also hold for toxicity to mammals and other nontarget organisms. Most pyrethroids exhibit low to moderate acute oral toxicity to mammals. Systemic toxicity is limited by inefficient absorption by some routes and by rapid metabolic detoxication, but the intrinsic toxicity of pyrethroids, which is observed upon direct administration to the CNS, is significant. The Type I and Type II structural subclasses of pyrethroids produce divergent syndromes of acute intoxication, which has complicated efforts to forge causal connections between effects on neuronal target sites in vitro and the physiological consequences of poisoning in vivo.

The insecticidal properties of pyrethroids result from their ability to alter the function of voltage-gated sodium channels in insect neuronal membranes, thereby disrupting electrical signaling in the nervous system. The structure, function and pharmacology of voltage-gated sodium channels are highly conserved between insects and mammals. Therefore, it is not surprising that pyrethroids also alter the function of mammalian sodium channels. For the Type I and Type II structural subclasses of pyrethroids, qualitative differences in sodium channel modification are generally correlated with the production of different intoxication syndromes, implying that actions on sodium channels are sufficient to account for the acute toxicity of this insecticide class.

Pyrethroids modify sodium channel function by binding to unique receptors on the pore-forming α subunits of voltage-gated sodium channels. Homology models of the pore-forming domains of insect VGSC subunits, coupled with computer-aided ligand docking experiments, have identified two pyrethroid receptor domains on the VGSC α subunit. These two receptor sites include locations of several amino acid polymorphisms identified in pyrethroid-resistant arthropod populations that reduce the pyrethroid sensitivity of insect VGSCs in heterologous expression systems. Moreover, these receptor domains include sequence locations at which polymorphisms between mammalian VGSC isoforms determine their relative pyrethroid sensitivity and locations at which polymorphisms between insect and mammalian VGSCs result in the lower sensitivity of all mammalian isoforms compared to insect VGSCs.

Studies employing in vitro assay systems have described actions of pyrethroids on a variety of putative biochemical and physiological target sites in mammals other than neuronal VGSCs. Pyrethroid actions on these putative alternative targets are beyond the scope of this chapter, but other reviews (Bloomquist, 1993; Costa, 2015; Ray and Fry, 2006; Shafer and Meyer, 2004; Soderlund, 2010, 2012; Soderlund et al., 2002) summarize this research and provide critical analyses of the relevance of targets other than neuronal VGSCs to pyrethroid intoxication in mammals. The consensus of these analyses is that actions at a small number of putative alternative targets—VGSCs expressed in skeletal and cardiac muscle, voltage-gated calcium channels, and γ-aminobutyric acid$_A$ receptors—may contribute to intoxication by some pyrethroids, particularly at very high doses. However, there is no evidence that these targets supersede the significance of neuronal VGSCs as the principal sites of action of pyrethroids as a class in both target arthropod species and nontarget organisms.

References

Ahlbom, J., Fredriksson, A., Erikkson, P., 1994. Neonatal exposure to a type-I pyrethroid (bioallethrin) induces dose-response changes in brain muscarinic receptors and behaviour in neonatal and adult mice. Brain Res. 645, 318–324.

Baltazar, M.T., Dinis-Oliveira, R.J., de Lourdes Bastros, M., Tsatsakis, A.M., Duarte, J.A., Carvalho, F., 2014. Pesticides exposure as etiological factor of Parkinson's disease and other neurodegenerative diseases—a mechanistic approach. Toxicol. Lett. 230, 85–103.

Barnes, J.M., Verschoyle, R.D., 1974. Toxicity of new pyrethroid insecticide. Nature 248, 711.

Baskar, M.K., Murthy, P.B., 2018. Acute *in vitro* neurotoxicity of some pyrethroids using microelectrode arrays. Toxicol. In Vitro 47, 165–177.

Bloom, A.S., Staatz, C.G., Dieringer, T., 1983. Pyrethroid effects on operant responding and feeding. Neurobehav. Toxicol. Teratol. 5, 321–324.

Bloomquist, J.R., 1993. Neuroreceptor mechanisms in pyrethroid mode of action and resistance. In: Roe, M., Kuhr, R.J. (Eds.), Reviews in Pesticide Toxicology. Toxicology Communications, Raleigh, NC, pp. 181–226.

Bloomquist, J.R., Soderlund, D.M., 1988. Pyrethroid insecticides and DDT modify alkaloid-dependent sodium channel activation and its enhancement by sea anemone toxin. Mol. Pharmacol. 33, 543–550.

Bloomquist, J.R., Barlow, R.L., Gillette, J.S., Li, W., Kirby, M.L., 2002. Selective effects of insecticides on nigrostriatal dopaminergic nerve pathways. Neurotoxicology 23, 537–544.

Box, S.A., Lee, M.R., 1996. A systemic reaction following exposure to a pyrethroid insecticide. Hum. Exp. Toxicol. 15, 389–390.

Bradberry, S.M., Cage, S.A., Proudfoot, A.T., Vale, J.A., 2005. Poisoning due to pyrethroids. Toxicol. Rev. 24, 93–106.

Brown, L.D., Narahashi, T., 1992. Modulation of nerve membrane sodium channel activation by deltamethrin. Brain Res. 584, 71–76.

Brown, G.B., Gaupp, J.E., Olsen, R.W., 1988. Pyrethroid insecticides: stereospecific allosteric interaction with the batrachotoxinin-A benzoate binding site of mammalian voltage-sensitive sodium channels. Mol. Pharmacol. 34, 54–59.

Brown, T.P., Rumsby, P.C., Capleton, A.C., Rushton, L., Levy, L.S., 2006. Pesticides and Parkinson's disease—is there a link? Environ. Health Perspect. 114, 156–164.

Cantalamessa, F., 1993. Acute toxicity of two pyrethroids, permethrin and cypermethrin, in neonatal and adult rats. Arch. Toxicol. 67, 510–513.

Casida, J.E., Quistad, G.B., 1998. Golden age of insecticide research: past, present, or future? Annu. Rev. Entomol. 43, 1–16.

Catterall, W.A., Goldin, A.L., Waxman, S.G., 2005. International Union of Pharmacology. XLVII. Nomenclature and structure-function relationships of voltage-gated sodium channels. Pharmacol. Rev. 57, 397–409.

Catterall, W.A., Cestele, S., Yarov-Yarovoy, V., Yu, F.H., Konoki, K., Scheuer, T., 2007. Voltage-gated ion channels and gating modifier toxins. Toxicon 49, 124–141.

Chinn, K., Narahashi, T., 1986. Stabilization of sodium channel states by deltamethrin in mouse neuroblastoma cells. J. Physiol. 380, 191–207.

Choi, J.-S., Soderlund, D.M., 2006. Structure-activity relationships for the action of 11 pyrethroid insecticides on rat $Na_V1.8$ sodium channels expressed in *Xenopus* oocytes. Toxicol. Appl. Pharmacol. 211, 233–244.

Clark, J.M., 1995. Effects and mechanisms of action of pyrethrin and pyrethroid insecticides. In: Chang, L.W., Dyer, R.S. (Eds.), Handbook of Neurotoxicology. Marcel Dekker, Inc., New York, pp. 511–546.

Costa, L.G., 2015. The neurotoxicity of organochlorine and pyethroid pesticides. In: Lotti, M., Bleecker, M.L. (Eds.), Handbook of Clinical Neurology. Elsevier, New York, pp. 135–148.

Crofton, K.M., Reiter, L.W., 1984. Effects of two pyrethroid insecticides on motor activity and acoustic startle response in the rat. Toxicol. Appl. Pharmacol. 75, 318–328.

Crofton, K.M., Reiter, L.W., 1988a. The effects of type I and II pyrethroids on motor activity and acoustic startle response in the rat. Fundam. Appl. Toxicol. 10, 624–634.

Crofton, K.M., Reiter, L.W., 1988b. Pyrethroid insecticides and the gamma-aminobutyric acid receptor complex. Motor activity and acoustic startle response in the rat. J. Pharmacol. Exp. Ther. 243, 946–954.

de Weille, J.R., Vijverberg, H.P.M., Narahashi, T., 1988. Interactions of pyrethroids and octylguanidine with sodium channels of squid giant axons. Brain Res. 445, 1–11.

Dong, K., Du, Y., Rinkevich, F., Nomura, Y., Peng, X., Wang, L., Silver, K., Zhorov, B.S., 2014. Molecular biology of insect sodium channels and pyrethroid resistance. Insect Biochem. Mol. Biol. 50, 1–17.

Du, Y., Lee, J.-E., Nomura, Y., Zhang, T., Zhorov, B., Dong, K., 2009. Identification of a cluster of residues in transmembrane 6 of domain III of the cockroach sodium channel essential for the action of pyrethroid insecticides. Biochem. J. 419, 377–385.

Du, Y., Song, W., Groome, J., Nomura, Y., Luo, N., Dong, K., 2010. A negative charge in transmembrane segment 1 of Domian II of the cockroach sodium channel is critical for channel gating and action of pyrethroid insecticides. Toxicol. Appl. Pharmacol. 247, 53–59.

Du, Y., Nomura, Y., Satar, G., Hu, Z., Nauen, R., He, S.Y., Zhorov, B., Dong, K., 2013. Molecular evidence for dual pyrethroid-receptor sites on a mosquito sodium channel. Proc. Natl. Acad. Sci. U. S. A. 110, 11785–11790.

Elliott, M., 1989. The pyrethroids: early discovery, recent advances and the future. Pestic. Sci. 27, 337–351.

Elliott, M., 1995. Chemicals in insect control. In: Casida, J.E., Quistad, G.B. (Eds.), Pyrethrum Flowers: Production, Chemistry, Toxicology, and Uses. Oxford University Press, New York, pp. 3–31.

Elliott, M., Farnham, A.W., Janes, N.F., Needham, P.H., Pulman, D.A., 1974. Insecticidally active conformations of pyrethroids. In: Kohn, G.K. (Ed.), Mechanism of Pesticide Action. American Chemical Society, Washington, DC, pp. 80–91.

Eriksson, P., Fredriksson, A., 1991. Neurotoxic effects of two different pyrethroids, bioallethrin and deltamethrin on immature and adult mice: changes and behavioral and muscarinic receptor variables. Toxicol. Appl. Pharmacol. 108, 78–85.

Eriksson, P., Nordberg, A., 1990. Effects of two pyrethroids, bioallethrin and deltamethrin on subpopulations of muscarinic and nicotinic receptors in the neonatal brain. Toxicol. Appl. Pharmacol. 102, 456–463.

Eriksson, P., Johansson, U., Ahlbom, J., Fredriksson, A., 1993. Neonatal exposure to DDT induces increased susceptibility to pyrethroid (bioallethrin) exposure at adult age. Changes in muscarinic receptor and behavioural variables. Toxicology 77, 21–30.

Felts, P.A., Yokoyama, S., Dib-Hajj, S., Black, J.A., Waxman, S.G., 1997. Sodium channel α-subunit mRNAs I, II, III, NaG, Na6 and hNE (PN1): different expression patters in developing rat nervous system. Mol. Brain Res. 45, 71–82.

Flannigan, S.A., Tucker, S.B., 1985. Variation in cutaneous sensation between synthetic pyrethroid insecticides. Contact Dermatitis 13, 140–147.

Ghiasuddin, S.M., Soderlund, D.M., 1985. Pyrethroid insecticides: potent, stereospecific enhancers of mouse brain sodium channel activation. Pestic. Biochem. Physiol. 24, 200–206.

Ginsburg, K.S., Narahashi, T., 1993. Differential sensitivity of tetrodotoxin-sensitive and tetrodotoxin-resistant sodium channels to the insecticide allethrin in rat dorsal root ganglion neurons. Brain Res. 627, 239–248.

Glowa, J.R., 1986. Acute and sub-acute effects of deltamethrin and chlordimeform on schedule-controlled responding in the mouse. Neurobehav. Toxicol. Teratol. 8, 97–102.

Goldin, A.L., 2001. Resurgence of sodium channel research. Annu. Rev. Physiol. 63, 871–894.

Goldin, A.L., 2002. Evolution of voltage-gated Na^+ channels. J. Exp. Biol. 205, 575–584.

Goldin, A.L., 2006. Expression of ion channels in *Xenopus* oocytes. In: Clare, J.J., Trezise, D.J. (Eds.), Expression and Analysis of Recombinant Ion Channels. Wiley VCH Verlag GmbH & Co., Weinheim, pp. 1–25.

Gotoh, Y., Kawakami, M., Matsumoto, N., Okada, Y., 1998. Permethrin emulsion ingestion: clinical manifestations and clearance of isomers. Clin. Toxicol. 36, 57–61.

Graham, F.L., Smiley, J., Russell, W.C., Nairn, R., 1977. Characteristics of a human cell line transformed by DNA from human adenovirus type 5. J. Gen. Virol. 36, 59–77.

Gray, A.J., Soderlund, D.M., 1985. Mammalian toxicology of pyrethroids. In: Hutson, D.H., Roberts, T.R. (Eds.), Insecticides. Wiley, New York, pp. 193–248.

Guo, J., Xu, J., Zhang, J., An, L., 2018. Alteration of mice cerebral cortex development after prenatal exposure to cypermethrin and deltamethrin. Toxicol. Lett. 287, 1–9.

Hartshorne, R.P., Catterall, W.A., 1984. The sodium channel from rat brain: purification and subunit composition. J. Biol. Chem. 259, 1667–1685.

He, B., Soderlund, D.M., 2010. Human embryonic kidney (HEK293) cells express endogenous voltage-gated sodium currents and $Na_v1.7$ sodium channels. Neurosci. Lett. 469, 268–272.

He, B., Soderlund, D.M., 2011. Differential state-dependent modification of rat Nav1.6 sodium channels expressed in human embryonic kidney (HEK293) cells by the pyrethroid insecticides tefluthrin and deltamethrin. Toxicol. Appl. Pharmacol. 257, 377–387.

He, B., Soderlund, D.M., 2016. Effects of the β1 auxiliary subunit on modification of rat $Na_v1.6$ sodium channels expressed in HEK293 cells by the pyrethroid insecticides tefluthrin and deltamethrin. Toxicol. Appl. Pharmacol. 291, 58–69.

He, F., Wang, S., Liu, L., Chen, S., Zhang, Z., Sun, J., 1989. Clinical manifestations and diagnosis of acute pyrethroid poisoning. Arch. Toxicol. 63, 54–58.

Hijzen, T.H., Slangen, J.L., 1988. Effects of type I and type II pyrethroids on the acoustic startle response in rats. Toxicol. Lett. 40, 141–152.

Holloway, S.F., Salgado, V.L., Wu, C.H., Narahashi, T., 1989. Kinetic properties of single sodium channels modified by fenvalerate in mouse neuroblastoma cells. Pflugers Arch. 414, 613–621.

Hudson, N.L., Kasner, E.J., Beckman, J., Mehler, L., Schwartz, A., Higgins, S., Bonnar-Prado, J., Lackovic, M., Mulay, P., MItchell, Y., Larios, L., Walker, R., Waltz, J., Moraga-McHaley, S., Roisman, R., Calvert, G.M., 2014. Characteristics of acute pesticide-related illnesses and injuries associated with pyrethrin and pyrethroid exposures—11 states, 2000–2008. Am. J. Ind. Med. 57, 15–30.

Hughes, M.F., Ross, D.G., Starr, J.M., Scollon, E.J., Wolansky, M.J., Crofton, K.M., DeVito, M.J., 2016. Environmentally relevant pyrethroid mixtures: a study on the correlation of blood and brain concentrations of a mixture of pyrethroid insecticides to motor activity in the rat. Toxicology 359–360, 19–28.

Husain, R., Adhami, V.M., Seth, P.K., 1996. Behavioral, neurochemical, and neuromorphological effects of deltamethrin in adult rats. J. Toxicol. Environ. Health 48, 515–526.

Isom, L.L., 2001. Sodium channel β subunits: anything but auxiliary. Neuroscientist 7, 42–54.

James, T.F., Nenov, M., Tapia, C.M., Lecchi, M., Koshy, S., Green, T.A., Laezza, F., 2017. Consequences of acute $Na_v1.1$ exposure to deltamethrin. Neurotoxicology 60, 150–160.

Johnstone, A.F.M., Gross, G.W., Weiss, D.G., Schroeder, O.H.-U., Gramowski, A., Shafer, T.J., 2010. Microelectrode arrays: a physiologically based neurotoxicity testing platform for the 21st century. Neurotoxicology 31, 331–350.

Johnstone, A.F.M., Strickland, J.D., Crofton, K.M., Gennings, C., Shafer, T.J., 2017. Effects of an environmentally-relevant mixture of pyrethroid insecticides on spontaneous activity in primary cortical networks on microelectrode arrays. Neurotoxicology 60, 234–239.

Kaneko, H., 2010. Pyrethroid chemistry and metabolism. In: Krieger, R.I. (Ed.), Hayes' Handbook of Pesticide Toxicology. Elsevier, New York, pp. 1635–1663.

Lawrence, L.J., Casida, J.E., 1982. Pyrethroid toxicology: mouse intracerebral structure-toxicity relationships. Pestic. Biochem. Physiol. 18, 9–14.

Lee, S.H., Soderlund, D.M., 2001. The V410M mutation associated with pyrethroid resistance in *Heliothis virescens* reduces the pyrethroid sensitivity of house fly sodium channels expressed in *Xenopus* oocytes. Insect Biochem. Mol. Biol. 31, 19–29.

Lessenger, J.E., 1992. Five office workers inadvertently exposed to cypermethrin. J. Toxicol. Environ. Health 35, 261–267.

Lipkind, G.M., Fozzard, H.A., 2000. KcsA crystal structure as framework for a molecular model of the Na^+ channel pore. Biochemistry 39, 8161–8170.

Litovitz, T.L., Klein-Schwartz, W., Rodgers, G.C.J., Cobaugh, D.J., Youniss, J., Omslaer, J.C., May, M.E., Woolf, A.D., Benson, B.E., 2002. 2001 annual report of the American Association of Poison Control Centers Toxic Exposure Surveillance System. Am. J. Emerg. Med. 20, 391–452.

Lombet, A., Mourre, C., Lazdunski, M., 1988. Interactions of insecticides of the pyrethroid family with specific binding sites on the voltage-dependent sodium channel from mammalian brain. Brain Res. 459, 44–53.
Long, S.B., Campbell, E.B., MacKinnon, R., 2005. Crystal structure of a mammalian voltage-dependent Shaker family K^+ channel. Science 309, 897–903.
Lund, A.E., Narahashi, T., 1982. Dose-dependent interaction of the pyrethroid isomers with sodium channels of squid axonal membranes. Neurotoxicology 3, 11–24.
Lund, A.E., Narahashi, T., 1983. Kinetics of sodium channel modification as the basis for the variation in the nerve membrane effects of pyrethroids and DDT analogs. Pestic. Biochem. Physiol. 20, 203–216.
Magby, J.P., Richardson, J.R., 2017. Developmental pyrethroid exposure causes long-term decreases of neuronal sodium channel expression. Neurotoxicology 60, 274–279.
Mattei, C., Legros, C., 2014. The voltage-gated sodium channel: a major target of marine neurotoxins. Toxicon 91, 84–95.
McCavera, S.J., Soderlund, D.M., 2012. Differential state-dependent modification of inactivation-deficient $Na_V1.6$ sodium channels by the pyrethroid insecticides *S*-bioallethrin, tefluthrin and deltamethrin. Neurotoxicology 33, 384–390.
Meacham, C.A., Brodfuehrer, P.D., Watkins, J.A., Shafer, T.J., 2008. Developmentally-regulated sodium channel subunits are differentially sensitive to α-cyano containing pyrethroids. Toxicol. Appl. Pharmacol. 231, 273–281.
Meyer, D.A., Carter, J.M., Johnstone, A.F.M., Shafer, T.J., 2008. Pyrethroid modulation of spontaneous neuronal excitability and neurotransmission in hippocampal neurons in culture. Neurotoxicology 29, 213–225.
Moretto, A., Colosio, C., 2011. Biochemical and toxicological evidence of neurological effects of pesticides: the example of Parkinson's disease. Neurotoxicology 32, 383–391.
Moser, V.C., Liu, Z., Schlosser, C., Spanogle, T.L., Chandrasekaran, A., McDaniel, K.L., 2016. Locomotor activity and tissue levels following acute administration of lambda- and gamma-cyhalothrin in rats. Toxicol. Appl. Pharmacol. 313, 97–103.
Motomura, H., Narahashi, T., 2001. Interaction of tetramethrin and deltamethrin at the single sodium channel in rat hippocampal neurons. Neurotoxicology 22, 329–339.
Muhammad, B.Y., Verschoyle, R.D., Ray, D.E., 2003. Developmental toxicity of pyrethroids. Arch. Toxicol. 77, 48–49.
Nieradko-Iwanicka, B., Borzecki, A., 2015. Subacute poisoning of mice with deltamethrin produces memory impairment, reduced locomotor activity, liver damage, and changes in blood morphology in the mechanism of oxidative stress. Pharmacol. Rep. 67, 535–541.
Oliveira, E.E., Du, Y., Nomura, Y., Dong, K., 2013. A residue in the transmembrane segment 6 of domain I in insect and mammalian sodium channels regulate differential sensitivities to pyrethroid insecticides. Neurotoxicology 38, 42–50.
O'Reilly, A.O., Khambay, B.P.S., Williamson, M.S., Field, L.M., Wallace, B.A., Davies, T.G.E., 2006. Modelling insecticide binding sites at the voltage-gated sodium channel. Biochem. J. 396, 255–263.
Peele, D.B., Crofton, K.M., 1987. Pyrethroid effects on schedule-controlled behavior: time and dosage relationships. Neurotoxicol. Teratol. 9, 387–394.
Quirós-Alcalá, L., Mehta, S., Eskenazi, B., 2014. Pyrethroid exposure and parental report of learning disability and attention deficit/hyperactivity disorder in U.S. children: NHANES 1999–2002. Environ. Health Perspect. 122, 1336–1342.
Ray, D.E., Fry, J.R., 2006. A reassessment of the neurotoxicity of pyrethroid insecticides. Pharmacol. Ther. 111, 174–193.
Richardson, J.R., Taylor, M.M., Shalat, S.L., Guillot III, T.S., Caudle, W.M., Hossain, M.M., Mathews, T.A., Jones, S.R., Cory-Schlechta, D.A., Miller, G.W., 2015. Developmental pesticide exposure reproduces features of attention deficit hyperactivity disorder. FASEB J. 29, 1960–1972.

Rizzo, M.A., Kocsis, J.D., Waxman, S.G., 1996. Mechanisms of paresthesiae, dysthesiae, and hyperesthesiae: role of Na^+ channel heterogeneity. Eur. Neurol. 36, 3–12.
Rubin, J.G., Soderlund, D.M., 1993. Binding of [^{3}H]batrachotoxinin A-20-α-benzoate and [^{3}H]saxitoxin to receptor sites associates with sodium channels in trout brain synaptoneurosomes. Comp. Biochem. Physiol. 105C, 231–238.
Rubin, J.G., Payne, G.T., Soderlund, D.M., 1993. Structure-activity relationships for pyrethroids and DDT analogs as modifiers of [^{3}H]batrachotoxinin A 20-α-benzoate binding to mouse brain sodium channels. Pestic. Biochem. Physiol. 45, 130–140.
Ruzo, L.O., Engel, J.L., Casida, J.E., 1979. Decamethrin metabolites from oxidative, hydrolytic and conjugative reactions in mice. J. Agric. Food Chem. 27, 725–731.
Salgado, V.L., Narahashi, T., 1993. Immobilization of sodium channel gating charge in crayfish giant axons by the insecticide fenvalerate. Mol. Pharmacol. 43, 626–634.
Schaller, K.L., Caldwell, J.H., 2003. Expression and distribution of voltage-gated sodium channels in the cerebellum. Cerebellum 2, 2–9.
Scollon, E.J., Starr, J.M., Crofton, K.M., Wolansky, M.J., DeVito, M.J., Hughes, M.F., 2011. Correlation of tissue concentrations of the pyrethroid bifenthrin with neurotoxicity in the rat. Toxicology 290, 1–6.
Shafer, T.J., Meyer, D.A., 2004. Effects of pyrethroids on voltage-sensitive calcium channels: a critical evaluation of strengths, weaknesses, data needs, and relationship to assessment of cumulative neurotoxicity. Toxicol. Appl. Pharmacol. 196, 303–318.
Shafer, T.J., Meyer, D.A., Crofton, K.M., 2005. Developmental neurotoxicity of pyrethroid insecticides: critical review and future research needs. Environ. Health Perspect. 113, 123–136.
Shafer, T.J., Rijal, S.O., Gross, G.W., 2008. Complete inhibition of spontaneous activity in neuronal networks in vitro by deltamethrin and permethrin. Neurotoxicology 29, 203–212.
Shah, B.S., Stevens, E.B., Pinnock, R.D., Dixon, A.K., Lee, K., 2001. Developmental expression of the novel voltage-gated sodium channel auxiliary subunit β3, in rat CNS. J. Physiol. 534, 763–776.
Shaw, G., Morse, S., Ararat, M., Graham, F.L., 2002. Preferential transformation of human neuronal cells by human adenoviruses and the origin of HEK 293 cells. FASEB J. 16, 869–871.
Sheets, L.P., 2000. A consideration of age dependent differences in susceptibility to organophosphorus and pyrethroid insecticides. Neurotoxicology 21, 57–64.
Sheets, L.P., Doherty, J.D., Law, M.W., Reiter, L.W., Crofton, K.M., 1994. Age-dependent differences in the susceptibility of rats to deltamethrin. Toxicol. Appl. Pharmacol. 126, 186–190.
Shelton, J.F., Geraghty, E.M., Tancredi, D.J., Delwiche, L.D., Schmidt, R.J., Ritz, B., Hansen, R.L., Hertz-Picciotto, I., 2014. Neurodevelopmental disorders and parental residential proximity to agricultural pesticides: the CHARGE study. Environ. Health Perspect. 122, 1103–1109.
Shrivastava, I.H., Durell, S.R., Guy, H.R., 2004. A model of voltage gating developed using the KvAP channel crystal structure. Biophys. J. 87, 2255–2270.
Smith, T.J., Soderlund, D.M., 1998. Action of the pyrethroid insecticide cypermethrin on rat brain IIa sodium channels expressed in *Xenopus* oocytes. Neurotoxicology 19, 823–832.
Smith, T.J., Soderlund, D.M., 2001. Potent actions of the pyrethroid insecticides cismethrin and cypermethrin on rat tetrodotoxin-resistant peripheral nerve (SNS/PN3) sodium channels expressed in *Xenopus* oocytes. Pestic. Biochem. Physiol. 70, 52–61.
Smith, T.J., Lee, S.H., Ingles, P.J., Knipple, D.C., Soderlund, D.M., 1997. The L1014F point mutation in the house fly *Vssc1* sodium channel confers knockdown resistance to pyrethroids. Insect Biochem. Mol. Biol. 27, 807–812.

Smith, T.J., Ingles, P.J., Soderlund, D.M., 1998. Actions of the pyrethroid insecticides cismethrin and cypermethrin on house fly *Vssc1* sodium channels expressed in *Xenopus* oocytes. Arch. Insect Biochem. Physiol. 38, 126–136.

Soderlund, D.M., 1995. Mode of action of pyrethrins and pyrethroids. In: Casida, J.E., Quistad, G.B. (Eds.), Pyrethrum Flowers: Production, Chemistry, Toxicology, and Uses. Oxford University Press, New York, pp. 217–233.

Soderlund, D.M., 2005. Sodium channels. In: Gilbert, L., Iatrou, K., Gill, S. (Eds.), Comprehensive Molecular Insect Science. Elsevier, New York, pp. 1–24.

Soderlund, D.M., 2010. Toxicology and mode of action of pyrethroid insecticides. In: Krieger, R.I. (Ed.), Hayes' Handbook of Pesticide Toxicology. Elsevier, New York, pp. 1665–1686.

Soderlund, D.M., 2012. Molecular mechanisms of pyrethroid insecticide neurotoxicity: recent advances. Arch. Toxicol. 86, 165–181.

Soderlund, D.M., 2017. Targeting voltage-sensitive sodium channels for insect control: past, present and future. In: Gross, A.D., Ozoe, Y., Coats, J.R. (Eds.), Advances in Agrochemicals: Ion Channels and G Protein-Coupled Receptors (GPCRs) as Targets for Pest Control. Volume 1: Ion Channels and Gap Junctions. American Chemical Society, Washington, DC, pp. 37–51.

Soderlund, D.M., Casida, J.E., 1977. Substrate specificity of mouse-liver microsomal enzymes in pyrethroid metabolism. In: Elliott, M. (Ed.), Synthetic Pyrethroids. American Chemical Society, Washington, DC, pp. 162–172.

Soderlund, D.M., Knipple, D.C., 2003. The molecular biology of knockdown resistance to pyrethroid insecticides. Insect Biochem. Mol. Biol. 33, 563–577.

Soderlund, D.M., Lee, S.H., 2001. Point mutations in homology domain II modify the sensitivity of rat $Na_v1.8$ sodium channels to the pyrethroid cismethrin. Neurotoxicology 22, 755–765.

Soderlund, D.M., Clark, J.M., Sheets, L.P., Mullin, L.S., Piccirillo, V.J., Sargent, D., Stevens, J.T., Weiner, M.L., 2002. Mechanisms of pyrethroid toxicity: implications for cumulative risk assessment. Toxicology 171, 3–59.

Song, J.-H., Narahashi, T., 1996. Modulation of sodium channels of rat cerebellar Purkinje neurons by the pyrethroid tetramethrin. J. Pharmacol. Exp. Ther. 277, 445–453.

Song, J.-H., Nagata, K., Tatebayashi, H., Narahashi, T., 1996. Interactions of tetramethrin, fenvalerate and DDT at the sodium channel in rat dorsal root ganglion neurons. Brain Res. 708, 29–37.

Song, G., Moreau, M., Efremenko, A., Lake, B.G., Wu, H., Bruckner, J.V., White, C.A., Osimitz, T.G., Creek, M.R., Hinderliter, P.M., Clewell, H.J., Yoon, M., 2019. Evaluation of age-related pyrethroid pharmacokinetic differences in rats: physiologically-based pharmacokinetic model development using *in vitro* data and *in vitro* to *in vivo* extrapolation. Toxicol. Sci. 169, 365–379.

Sparks, T.C., Lorsbach, B.A., 2017. Agrochemical discovery—building the next generation of insect control agents. In: Gross, A.D., Ozoe, Y., Coats, J.R. (Eds.), Advances in Agrochemicals: Ion Channels and G Protein-Coupled Receptors (GPCRs) as Targets for Pest Control. Volume 1: Ion Channels and Gap Junctions. American Chemical Society, Washington, DC, pp. 1–17.

Starr, J.M., Scollon, E.J., Hughes, M.F., Ross, D.G., Graham, S.E., Crofton, K.M., Wolansky, M.J., DeVito, M.J., Tornero-Velez, R., 2012. Environmentally relevant mixtures in cumulative assessments: an acute study of toxicokinetics and effects on motor activity in rats exposed to a mixture of pyrethroids. Toxicol. Sci. 130, 309–318.

Stein, E.A., Washburn, M., Walczak, C., Bloom, A.S., 1987. Effects of pyrethroid insecticides on operant responding maintained by food. Neurotoxicol. Teratol. 9, 27–31.

Tabarean, I.V., Narahashi, T., 1998. Potent modulation of tetrodotoxin-sensitive and tetrodotoxin-resistant sodium channels by the type II pyrethroid deltamethrin. J. Pharmacol. Exp. Ther. 284, 958–965.
Tabarean, I.V., Narahashi, T., 2001. Kinetics of modulation of tetrodotoxin-sensitive and tetrodotoxin-resistant sodium channels by tetramethrin and deltamethrin. J. Pharmacol. Exp. Ther. 299, 988–997.
Tan, J., Soderlund, D.M., 2009. Human and rat $Na_v1.3$ voltage-gated sodium channels differ in inactivation properties and sensitivity to the pyrethroid insecticide tefluthrin. Neurotoxicology 30, 81–89.
Tan, J., Soderlund, D.M., 2010. Divergent actions of the pyrethroid insecticides *S*-bioallethrin, tefluthrin, and deltamethrin on rat $Na_v1.6$ sodium channels. Toxicol. Appl. Pharmacol. 247, 229–237.
Tan, J., Soderlund, D.M., 2011a. Action of tefluthrin on rat $Na_v1.7$ voltage-gated sodium channels expressed in *Xenopus* oocytes. Pestic. Biochem. Physiol. 101, 21–26.
Tan, J., Soderlund, D.M., 2011b. Coexpression with auxiliary β subunits modulates the action of tefluthrin on rat $Na_v1.6$ and $Na_v1.3$ sodium channels. Pestic. Biochem. Physiol. 101, 256–264.
Tan, J., Liu, Z., Wang, R., Huang, Z.Y., Chen, A.C., Gurevitz, M., Dong, K., 2005. Identification of amino acid residues in the insect sodium channel critical for pyrethroid binding. Mol. Pharmacol. 67, 513–522.
Tatebayashi, H., Narahashi, T., 1994. Differential mechanism of action of the pyrethroid tetramethrin on tetrodotoxin-sensitive and tetrodotoxin-resistant sodium channels. J. Pharmacol. Exp. Ther. 270, 595–603.
Thomas, P., Smart, T.G., 2005. HEK293 cell line: a vehicle for the expression of recombinant proteins. J. Pharmacol. Toxicol. Methods 51, 187–200.
Trainer, V.L., Moreau, E., Baden, D.G., Catterall, W.A., 1993. Neurotoxin binding and allosteric modulation at receptor sites 2 and 5 on purified and reconstituted rat brain sodium channels. J. Biol. Chem. 268, 17114–17119.
Trainer, V.L., McPhee, J.C., Boutelet-Bochan, H., Baker, C., Scheuer, T., Babin, D., Demoute, J.-P., Guedin, D., Catterall, W.A., 1997. High affinity binding of pyrethroids to the α subunit of brain sodium channels. Mol. Pharmacol. 51, 651–657.
USEPA, 2011. Pyrethrins/Pyrethroid Cumulative Risk Assessment. U. S. Environmental Protection Agency, Office of Pesticide Programs, p. 88. October 4, 2011.
Usherwood, P.N.R., Davies, T.G.E., Mellor, I.R., O'Reilly, A.O., Peng, F., Vais, H., Khambay, B.P.S., Field, L.M., Williamson, M.S., 2007. Mutations in DIIS5 and the DIIS4-5 linker of *Drosophila melanogaster* sodium channel define binding domains for pyrethroids and DDT. FEBS Lett. 581, 5485–5492.
Vais, H., Williamson, M.S., Hick, C.A., Eldursi, N., Devonshire, A.L., Usherwood, P.N.R., 1997. Functional analysis of a rat sodium channel carrying a mutation for insect knock-down resistance (*kdr*) to pyrethroids. FEBS Lett. 413, 327–332.
Vais, H., Atkinson, S., Eldursi, N., Devonshire, A.L., Williamson, M.S., Usherwood, P.N.R., 2000a. A single amino acid change makes a rat neuronal sodium channel highly sensitive to pyrethroid insecticides. FEBS Lett. 470, 135–138.
Vais, H., Williamson, M.S., Goodson, S.J., Devonshire, A.L., Warmke, J.W., Usherwood, P.N.R., Cohen, C.J., 2000b. Activation of *Drosophila* sodium channels promotes modification by deltamethrin: reductions in affinity caused by knock-down resistance mutations. J. Gen. Physiol. 115, 305–318.
Vais, H., Williamson, M.S., Devonshire, A.L., Usherwood, P.N.R., 2001. The molecular interactions of pyrethroid insecticides with insect and mammalian sodium channels. Pest Manag. Sci. 57, 877–888.
Vais, H., Atkinson, S., Pluteanu, F., Goodson, S.J., Devonshire, A.L., Williamson, M.S., Usherwood, P.N.R., 2003. Mutations of the *para* sodium channel of *Drosophila melanogaster* identify putative binding sites for pyrethroids. Mol. Pharmacol. 64, 914–922.

van der Mark, M., Brouwer, M., Kromhout, H., Nijssen, P., Huss, A., Vermeulen, R., 2012. Is pesticide use related to Parkinson's disease? Some clues to heterogeneity in study results. Environ. Health Perspect. 120, 340–347.

Verschoyle, R.D., Aldridge, W.N., 1980. Structure-activity relationships of some pyrethroids in rats. Arch. Toxicol. 45, 325–329.

Verschoyle, R.D., Barnes, J.M., 1972. Toxicity of natural and synthetic pyrethrins to rats. Pestic. Biochem. Physiol. 2, 308–311.

Vijverberg, H.P.M., van den Bercken, J., 1990. Neurotoxicological effects and the mode of action of pyrethroid insecticides. CRC Crit. Rev. Toxicol. 21, 105–126.

Wang, S.-Y., Barile, M., Wang, G.K., 2001. A phenylalanine residue at segment D3-S6 in Nav1.4 voltage-gated Na^+ channels is critical for pyrethroid action. Mol. Pharmacol. 60, 620–628.

Watson, W.A., Litovitz, T.L., Rodgers, G.C.J., Klein-Schwartz, W., Youniss, J., Rose, S.R., Borys, D., May, M.E., 2003. 2002 annual report of the American Association of Poison Control Centers Toxic Exposure Surveillance System. Am. J. Emerg. Med. 21, 353–421.

Watson, T.F., Litovitz, T.L., Klein-Schwartz, W., Rodgers, G.C.J., Youniss, J., Reid, N., Rouse, W.G., Rembert, R.S., Borys, D., 2004. 2003 annual report of the American Association of Poison Control Centers Toxic Exposure Surveillance System. Am. J. Emerg. Med. 22, 335–404.

Watson, W.A., Litovitz, T.L., Rodgers, G.C.J., Klein-Schwartz, W., Reid, N., Youniss, J., Flanagan, A., Wruk, K.M., 2005. 2004 annual report of the American Association of Poison Control Centers Toxic Exposure Surveillance System. Am. J. Emerg. Med. 23, 589–666.

Weiner, M.L., Nemec, M., Sheets, L., Sargent, D., Breckenridge, C., 2009. Comparative functional observational battery of twelve commercial pyrethroid insecticides in male rats following acute oral exposure. Neurotoxicology 30, S1–S16.

Whitaker, W.R.J., Clare, J.J., Powell, A.J., Chen, Y.H., Faull, R.L.M., Emson, P.C., 2000. Distribution of voltage-gated sodium channel α-subunit and β-subunit mRNAs in human hippocampal formation, cortex, and cerebellum. J. Comp. Neurol. 422, 123–139.

Whitaker, W.R.J., Faull, R.L.M., Waldvogel, H.J., Plumpton, C.J., Emson, P.C., Clare, J.J., 2001. Comparative distribution of voltage-gated sodium channel proteins in human brain. Mol. Brain Res. 88, 37–53.

Williams, M.T., Gutierrez, A., Vorhees, C.V., 2019. Effects of acute deltamethrin exposure in adult and developing Sprague Dawley rats on acoustic startle response in relation to deltamethrin brain and plasma concentrations. Toxicol. Sci. 168, 61–69.

Wolansky, M.J., Harrill, J.A., 2008. Neurobehavioral toxicology of pyrethroid insecticides in adult animals: a critical review. Neurotoxicol. Teratol. 30, 55–78.

Wolansky, M.J., Gennings, C., Crofton, K.M., 2006. Relative potencies for acute effects of pyrethroids on motor function in rats. Toxicol. Sci. 89, 271–277.

Wu, S.-N., Wu, Y.-H., Chen, B.-S., Liu, Y.-C., 2009. Underlying mechanism of action of tefluthrin, a pyrethroid insecticide, on voltage-gated ion currents and on action currents in pituitary tumor (GH_3) cells and GnRH-secreting (GT1-7) neurons. Toxicology 2009, 70–77.

Yamamoto, D., Quandt, F.N., Narahashi, T., 1983. Modification of single sodium channels by tetramethrin. Brain Res. 274, 344–349.

Yang, L., Li, L., 2015. Actions of the pyrethroid insectiide bifenthrin on sodium channels expressed in rat cerebral cortical neurons. Toxicol. Mech. Methods 25, 63–69.

Zhao, Y., Yarov-Yarovoy, V., Scheuer, T., Catterall, W.A., 2004. A gating hinge in Na^+ channels: a molecular switch for electrical signaling. Neuron 41, 859–865.

CHAPTER FOUR

Neurotoxicity of Neonicotinoids

Arturo Anadón*, Irma Ares, Marta Martínez, María-Rosa Martínez-Larrañaga, María-Aránzazu Martínez
Department of Pharmacology and Toxicology, Faculty of Veterinary Medicine, Universidad Complutense de Madrid, Madrid, Spain
*Corresponding author: e-mail address: aanadon@ucm.es

Contents

1. Introduction

Neonicotinoids are the most important new class of synthetic insecticides, with similar chemical structures to nicotine, which have been used in plant and animal health over the last decade to control sucking insects, and have become one of the most widespread pesticides. Neonicotinoids are registered in 120 countries with a global market share of more than 25% but their use is firmly increasing (Jeschke et al., 2011), being among the most

Advances in Neurotoxicology, Volume 4
ISSN 2468-7480
https://doi.org/10.1016/bs.ant.2019.11.005

effective insecticides with a high intrinsic acute and residual activity for control of sucking insect pests such as aphids, whiteflies, lead- and planthoppers, thrips, some microlepidoptera, and a number of coleopteran pests, particularly as insecticides of crops (i.e., vegetables, pome and some fruits, citrus, rice, cotton, corn, canola, potato, sugar beet, oilseed rape, and soybean, among many others). Moreover, neonicotinoids are used as ectoparasiticides in domestic animals for controlling fleas (e.g., *Ctenocephalides felis*, *Ctenocephalides canis*) (imidacloprid, nitenpyram), lice (e.g., *Linognathus setosus*, *Trichodectus canis*) and flies (e.g., *Musca domestica*) in cats and dogs. In addition to plant and health protection use, applications of neonicotinoid insecticides in non-agricultural fields has also expanded in recent years, such as household sectors, lawn, and garden for controlling termites (e.g., *Reticulitermes* spp.) (e.g., imidacloprid) and turf pests such as white grubs and cockroaches (e.g., *Blattella germanica*). New bait gel formulations containing imidacloprid are also used for the control of cockroaches, and thiamethoxam, for broad-spectrum control of ants (Jeschke et al., 2011). These insecticides have activity against some chewing insect species including termites, soil insects, and fleas on pets. In Table 1 is listed the spectrum of activity (insect order, family and insect species) of neonicotinoids used in plant and animal health.

Neonicotinoids are applied as foliar uses (spray), soil uses, or seed treatment. It is important to note that use of imidacloprid and clothianidin as seed treatments on some crops poses risks to small birds, and ingestion of even a few treated seeds could cause mortality or reproductive impairment to sensitive bird species (Gibbons et al., 2015). Today approximately 60% of all neonicotinoid applications are soil/seed treatments, and most spray applications are especially targeted against pests attacking crops such as cereals, corn, rice, vegetables, sugar beet, potatoes, cotton, and others (Elbert et al. , 2008). In plants, the neonicotinoids are metabolized, although is known that, for example, the metabolism of imidacloprid is strongly influenced by the method of application. Whereas in foliar application most of the residues remain on the leaf surface as parent compound, nearly all of the imidacloprid administered to plants by soil application or seed treatment is metabolized more or less completely, depending on plant species and time (Nauen et al., 1998).

The main neonicotinoids presently on the market are imidacloprid, thiamethoxam, clothianidin, thiacloprid, dinotefuran, acetamiprid, nitenpyram and sulfoxaflor. The majority of neonicotinoids and their metabolites are highly persistent in soil, aquatic sediments and water and they accumulate in

Table 1 Spectrum of activity of Neonicotinoids used in plant and animal health.

Use	Order	Family	Scientific name (insect species)	Common name
Plant health	Hemiptera	Aphididae	*Aphis gossypii*	Aphids
			Myzus persicae	
			Phorodon humili	
			Rhopalosiphum padi	
			Eriosoma lanigerum	
			Rhopalosiphum maidis	
		Cicadellidae	*Nephotettix cincticeps*	Leafhoppers
			Idioscopus clypealis	
		Delphacidae	*Nilaparvata lugens*	Planthoppers
		Aleyrodidae	*Bemisia tabaci*	Whiteflies
			Trialeurodes vaporariorum	
		Blissidae	*Blissus leucopterus*	
		Pentatomidae	*Nezara viridula*	Stink bug
	Coleoptera	Chrysomelidae	*Leptinotarsa decemlineata*	Beetles
			Chaetocnema pulicaria	
			Colaspis brunnea	
		Scarabaeidae	*Popillia japonica*	
			Heteronychus arator	
		Brachyceridae	*Lissorhoptrus oryzophilus*	Weevils
	Lepidoptera	Tortricidae	*Cydia pomonella*	Moths
		Gracillariidae	*Phyllocnistis citrella*	
		Lyonetiidae	*Leucoptera coffeella*	
	Thysanoptera	Thripidae	*Thrips tabaci*	Thrips
	Diptera	Anthomyiidae	*Delia platura*	Flies
Animal health	Siphonaptera	Pulicidae	*Ctenocephalides felis*	Fleas
			Ctenocephalides canis	
	Phthiraptera	Linognathidae	*Linognathus setosus*	Lice
		Trichodectidae	*Trichodectes canis*	
	Diptera	Muscidae	*Musca domestica*	Flies

soil after repeated uses. They are used on maize, fruit trees, potatoes and many other crops. Imidacloprid [1-(6-chloro-3-pyridylmethyl)-N-nitroimidazolidin-2-ylideneamine, CAS number 138261-41-3] was the first product of this class of insecticides, was discovered in 1984 and commercialized in 1991 and it was used in foliar application and seed treatments (Tomizawa and Casida, 2003) and later on topical used on pets. Imidacloprid is a neonicotinoid insecticide belonging to the chloronicotinyl nitroguanidine chemical family. Since the discovery of imidacloprid, several other chemical compounds analogues with the 6-chloro-3-pyridylmethyl moiety that confers added insecticidal potency (chlorine atom) and selective action have been developed for commercial use, including in this first-generation chloropyridyls group acetamiprid [(*E*)-N^1-[(6-chloro-3-pyridyl)methyl]-N^2-cyano-N^1-methylacetamide, CAS number 135410-20-7], nitenpyram [(E)-1-N′-[(6-chloropyridin-3-yl)methyl]-1-N′-ethyl-1-*N*-methyl-2-nitroethene-1,1-diamine, CAS number 150824-47-8] and thiacloprid [(2Z)-3-(6-chloropyridin-3-yl)methyl-1,3-thiazolidin-2-ylidene cyanamide, CAS number 111988-49-9] (Sheets, 2001). For related analogues, the binding affinity to insect receptor is decreased when chlorine atom is not present. The replacement of the chloropyridinyl moiety with a chlorothiazolyl group led to development of a second generation of neonicotinoid insecticide (e.g., imidaclothiz [N-[1-[(2-chloro-1,3-thiazol-5-yl)methyl]-4,5-dihydroimidazol-2-yl]nitramide, CAS number 105843-36-5], clothianidin [(E)-1-(2-Chlor-1,3-thiazol-5-ylmethyl)-3-methyl-2-nitroguanidine, CAS number 210880-92-5], thiamethoxam [3-[(2-Chloro-1,3-thiazol-5-yl)methyl]-5-methyl-N-nitro-1,3,5-oxadiazinan-4-imine, CAS number 153719-23-4]. Dinotefuran [2-methyl-1-nitro-3-[(tetrahydro-3-furanyl) methyl] guanidine, CAS number 165252-70-0] was the first representative of the furanicotinyl subclass of neonicotinoid insecticides (third-generation furanyls), and sulfoxaflor (fourth-generation sulfoximines) (Wang et al., 2018) (see Table 2).

Neonicotinoids termed "systemic insecticides" have systemic properties, meaning that they are absorbed and transported throughout the plant, resulting in pesticide residues in all parts of the growing plant (leaves, flowers, roots, stems), including pollen and nectar; that imply translocation to pollen, nectar and guttation droplets. Today, the "systemic insecticides" neonicotinoids imidacloprid, thiamethoxam, and clothianidin are widely used for seed treatment in different crops such as cotton, corn, cereals, sugar beet, oilseed rape, and others (Jeschke et al., 2011).

Table 2 Chemical structures of some of the most important neonicotinoid insecticides.

Neonicotinoids/Chemical Structure	Nitromethylene (CH-NO2)	Nitroimine (N-NO2)	Cyanoimine (N-CN)
First-generation chloropyridyls	**Nitenpyram** 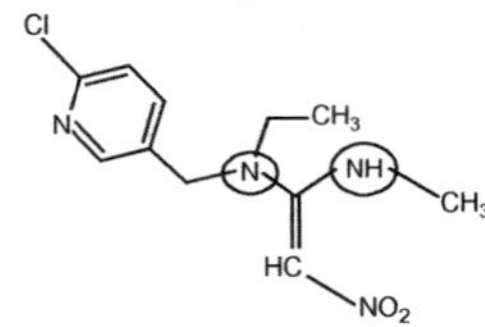 **Paichongding** 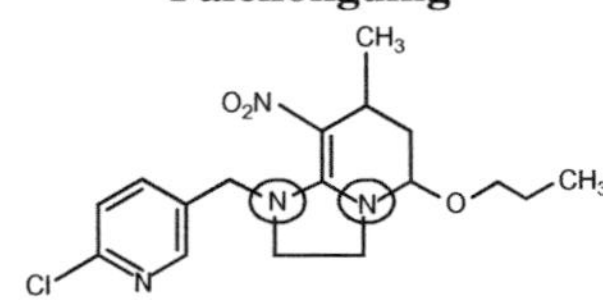 **Cycloxaprid** 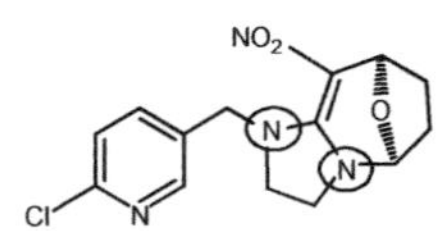	**Imidacloprid** 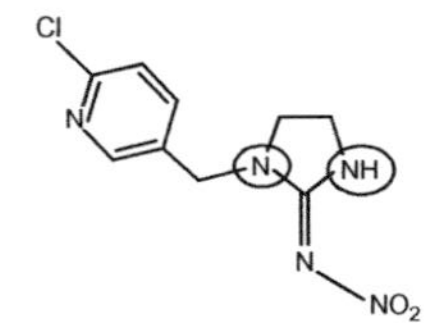	**Thiacloprid** 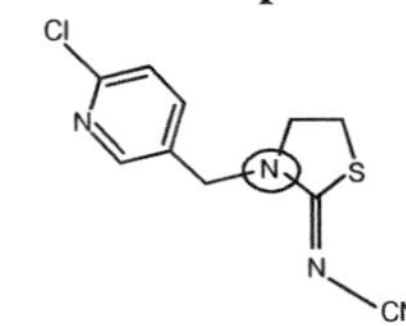 **Acetamiprid** 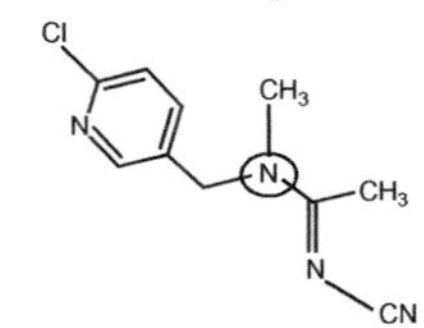

Continued

Table 2 Chemical structures of some of the most important neonicotinoid insecticides.—cont'd

Neonicotinoids/Chemical Structure	Nitromethylene (CH-NO2)	Nitroimine (N-NO2)	Cyanoimine (N-CN)
Second-generation chlorothiazoles		**Imidaclothiz** 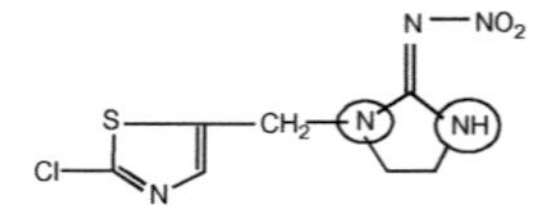 **Thiamethoxam** 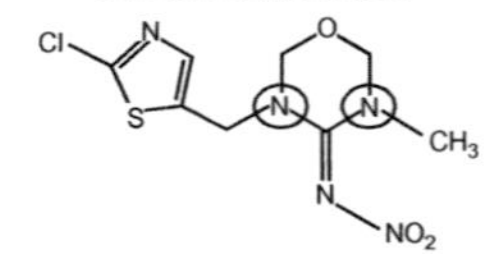 **Clothianidin** 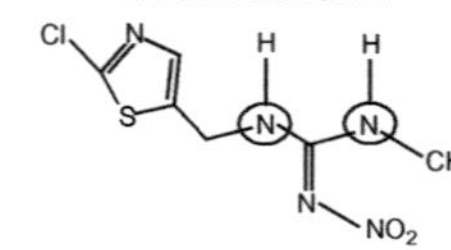	
Third-generation furanyls		**Dinotefuran** 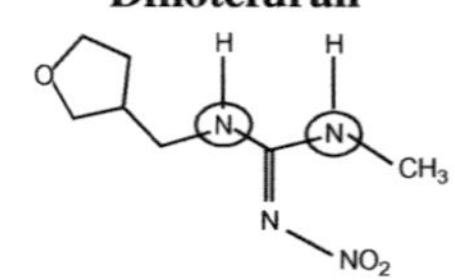	
Fourth-generation sulfoximines			**Sulfoxaflor** 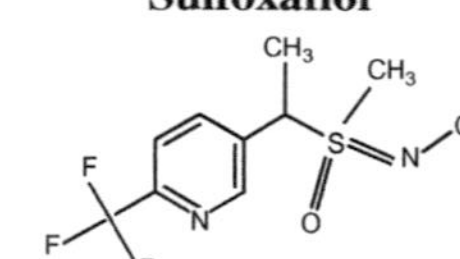

Neonicotinoids are synthetic derivatives of nicotine, acting as selective agonist activity at nicotinic acetylcholine receptors (nAChRs) in insects (Tomizawa and Casida, 2003). There is growing evidence that environmental chemicals can disturb the development of the central nervous system (CNS) and in this context, neonicotinoid have captivate considerable attention (Julvez et al., 2016). Neurotoxicity induced by insecticides, regardless of the sites of action (i.e., CNS, or peripheral nervous system, PNS), direct or indirect actions on the nervous system, as specificity of action on target sites, can be detected in terms of changes in the following areas, behavior, biochemistry, physiology and pathology.

2. Mechanism of action of neonicotinoids

Neonicotinoids act by interfering with neural transmission in the CNS. They bind to the nicotinic acetylcholine receptors (nAChR) in the postsynaptic neuron, acting as "false neurotransmitter" (agonist), which are ligand-gate ion channels that are present in both the PNS and CNS. While the nicotinoids are structurally similar to the neonicotinoids, they primarily differ by containing an ionizable basic amine or imine substituent (Tomizawa and Casida, 2005). This interference with acetylcholine (ACh) neurotransmitter signaling causes continuous activation of the receptor, leading to symptoms/signs of neurotoxicity. Typically, activation of brain nAChRs results in enhanced release of various key neurotransmitters, including dopamine, serotonin, glutamate and GABA (γ-aminobutyric acid) (Taly et al., 2009).

The neonicotinoids, are nicotinic agonists that interact with nicotinic acetylcholine receptors (AChRs) in a very different way than nicotine which confers selectivity, and thus bind more strongly to, insect than mammalian or other vertebrate receptors, so their toxicity to mammals, birds and fish are lower than it is to insects and the reversibility of intoxication higher. The neonicotinoids are not protonated but instead have an electronegative tip consisting of a nitro or cyano pharmacophore that imparts potency and selectivity, presumably by binding to a unique cationic subsite of the insect receptor. This is in marked contrast to the action of protonated nicotinoids, which require a cation-π interaction for binding to the vertebrate receptor. (Jeschke et al., 2011; Tomizawa and Casida, 2005).

The various assemblages of nicotinic acetylcholine receptor (nAChR) subtypes are broadly distributed in the brain. The subunit diversity of brain nAChRs (nine α-subunits and three β-subunits) poses a substantial challenge to pesticide discovery (Gotti et al., 2007). Certainly, the large number of nAChR homopentamers (α_7) and heteropentamers ($\alpha_4\beta_2$, $\alpha_6\alpha_4\beta_2\beta_3$, $\alpha6\beta2\beta3$) receptors exhibit distinct electrophysiological and kinetic properties and are differentially distributed throughout the nervous system (Taly et al., 2009) (see Table 3).

2.1 Mammals

It is important to note that the lower acute toxicity in mammals, birds and fish are due to differences in binding properties to the various nicotinic receptor subtypes in vertebrate species, as well as rapid elimination and poor penetration of the blood-brain barrier (BBB) (Sheets, 2001), but manifest some chronic toxicities in mammals. A decade ago, a male mouse study bring to light that four class of chloropyridinyl neonicotinoids (imidacloprid, nitenpyram, thiacloprid, acetamiprid) readily cross the BBB, and the concentrations are detectable in the brain at 60–70% of the serum within 15 min after a single intraperitoneal injection of 10 mg/kg bw

Table 3 Nicotinic binding sites and receptor subunits.

Brain Region	**Subunit of brain nAChRs**
Amygdala	$\alpha_4\beta_2$, α_7
Cerebellum	$\alpha_4\beta_2$, $\alpha_3\beta_4$, α_7, $\alpha_3\beta_2$
Cortex	$4\beta_2$, $A_4\beta_2$
Hippocampus	$\alpha_4\beta_2$, $\alpha_4\alpha_5\beta_2$, $\alpha_3\beta_4$, α_7
Interpeduncular nucleus	$\alpha_4\beta_2$, $\alpha_2\beta_2$, $\alpha_3\beta_3\beta_4$, α_7
Locus coeruleus	$\alpha_3\beta_4$, $\alpha_6\beta_2\beta_3$
Medial habenule	$\alpha_4\beta_2$, $\alpha_3\beta_4$, $\alpha_3\beta_3\beta_4$, α_7
Olfactory bulb	$\alpha_4\beta_2$, α_7
Pineal gland	α_7, $\alpha_3\beta_4$
Raphe nuclei	$\alpha_4\beta_2$
Spinal cord	$\alpha_4\beta_2$, $\alpha_3\beta_2$, α_7
Striatum	$\alpha_4\beta_2$, $\alpha_4\alpha_5\beta_2$, $\alpha_6\beta_2\beta_3$, $\alpha_6\alpha_4\beta_2\beta_3$
Substantia nigra ventral tegmental area	$\alpha_4\beta_2$, $\alpha_4\alpha_5\beta_2$, $\alpha_3\beta_4$, $\alpha_6\beta_2\beta_3$, α_7

(imidacloprid, thiacloprid and acetamiprid) or 20 mg/k bw (nitenpyram). The chemical analysis was performed also in liver and plasma at 15–240 min and 0–24 h in urine. The highest concentration of the parent compounds were observed in brain (11–16 μg/g for nitenpyram and thiacloprid, 6 μg/g for imidacloprid, and 3 μg/g for acetamiprid). Significant concentrations of acetamiprid achieved in the tissues were higher than the other neonicotinoids. These findings are generally consistent with earlier studies, receiving an oral administration of each individual neonicotinoids in other animal species. Referring to the urinary excretion of the parent compound was greatest after administration of nitenpyram and imidacloprid. Each of the compounds was cleaved to the same eight urinary metabolites derived from the chloropyridinylmethyl moiety (i.e., chloropyridinecarboxylic acid and its methylthio-, hydroxy-, and *N*-acetylcysteinyl derivatives and glycine, *O*-glucuronide, and sulfate conjugates thereof) (Ford and Casida, 2006).

A comparative pharmacokinetics study in the mice was reported using thiamethoxam, clothianidin, and the two desmethyl derivatives metabolites, and dinotefuran. These active substances were administered intraperitoneally to mice at 20 mg/kg bw for characterization of metabolites and pharmacokinetic analysis of brain, liver, plasma, and urine. The levels of these unchanged compounds and their desmethyl derivatives metabolites were compared in brain, liver and plasma after 15–240 min. The parent compound is present in larger amounts than the metabolites for at least the first 60 min with the single exception of clothianidin as a metabolite of clothianidin desmethyl derivative in brain. Each unchanged compound was recovered in the urine (19–55%) over the 24 h post administration period, and tissue residues are largely dissipated by 4 h. The highest concentration of the unchanged compounds for the three neonicotinoids were 11–18 μg/g in brain, 34–55 μg/g in liver, and 12–15 μg/g in plasma. The elimination half-life values relative to the highest concentration levels were similar (30–55 min) for the three tissues and three compounds except for an apparent largest permanence of thiamethoxam desmethyl derivative metabolite in liver (75 min) (Ford and Casida, 2006a). In most cases, neonicotinoids show their toxicity as a selective agonists at nAChRs in insects and mammals alike, it does induce the neurobehavioral and biochemical alterations in mature and developing mammals, and these adverse effects may be correlated with the affinity and distribution of their metabolites. The nAChR binding sites in the vertebrate nervous system are different from those in insects (i.e., they have less number of nicotinic receptors with high affinity), and in general they have lower numbers of nicotinic receptors. Rats that received high doses of thiamethoxam (50 or 100 mg/kg bw) for seven consecutive days showed an increase in anxiety behavior, may be correlated with a decrease in AChE activity (Rodrigues et al., 2010). Furthermore, in infant rat, after application of 24 mg/kg bw of clothianidin, was shown a significant deterioration of cognitive functions when compared to controls (Ozdemir et al., 2014). Using brain microdialysis in rats, studies showed that thiametoxam and clothianidin had an effect on striatal dopamine release in a dose-dependent manner, which was predominantly mediated by an exocytotic mechanism, Ca^{2+}, vesicular and tetrodotoxin-dependent (de Oliveira et al., 2010; Faro et al., 2012). However, oxidative stress may also be one toxicological mechanism for neonicotinoids, and occur in various species of animals (Wang et al., 2018). Biotransformations involve initial oxidation or reduction as both activation and detoxification mechanisms. The toxic effects of neonicotinoids might closely correlate with its metabolism process and oxidative stress. nAChRs facilitate neurotransmission in the CNS and PNS (Hassan-Puttaswamy et al., 2015). ROS, RNS, oxidative stress, Ca^{2+} level, and various signaling pathways also played critical roles in cellular apoptosis (Anadón, 2018). Though it has been documented that neonicotinoids activate the ERK cascade, which is triggered by primary action at the nAChRs with an involvement of intracellular Ca^{2+} mobilization, possibly mediated by inositol 1,4,5-trisphosphate (IP_3) (Tomizawa and Casida, 2002), the relationship between Ca^{2+} level, the oxidative stress, and the signal integration pathways induced by neonicotinoids are worthy of further investigation to effectively understand the toxic effects of neonicotinoids. Imidacloprid displayed sex-, tissue-, and duration-specific effects in its toxicity. Similarly, it was reported that the genotoxicity of imidacloprid is sex-related as hormones control the CYP450-mediated metabolism of Imidacloprid (Arslan et al., 2015).

Nicotinic acetylcholine receptors that play important roles in synaptic transmission in the CNS belong to a superfamily of ligand-gated ion channels (Dani and Biasi, 2001) There are three classes of nAChRs: (1) muscle subunits (α_1, β_1, γ, δ, ε); (2) neuronal $\alpha\beta$ combinations of subunits (α_2–α_6 and β_2–β_4), including $\alpha_4\beta_2$ (heteropentameric), which is the most abundant form of nAChR in the CNS; and (3) subunits (α_7–α_9) that are homomeric (and are inhibited by α-bungarotoxin) (α-BT), including α_7 cAChR (homopentameric), which is widely distributed in the CNS and is specifically involved in neuritic outgrowth (Girod et al., 1999).

The nAChRs of mammals and insects mediate the fast actions of ACh in the nervous system and in the case of mammals also at the neuromuscular junction. The nAChRs are members of the "cys-loop" superfamily of ionotropic neurotransmitter receptors, which also include ionotropic receptors for GABA, glycine and 5-HT (Raymond-Delpech et al., 2005). In the event that nAChRs, the integral ion channel becomes transiently permeable to cations following the binding of ACh (Unwin, 1993). The five polypeptide subunits share a common organization—a long extracellular N-terminal domain containing residues that contribute to the ACh binding site and four membrane spanning regions (M1-4) (Karlin, 2002). Two classes of subunits can be recognized: α subunits are defined by the presence of two adjacent cysteine residues in loop C, one of the several extracellular regions (loops A-F) that contribute to the ACh binding site (Karlin, 2002); non-α subunits do not possess this YxCC motif. An $(\alpha)2(\text{non-}\alpha)_3$ stoichiometry has been demonstrated for mammalian muscle and certain neuronal nAChRs. However, in the vertebrate nervous system other stoichiometries can occur (Lindstrom, 2000). The vertebrate nAChR subunit family is diverse with separate sets of genes encoding the five muscle (α1, β1, μ, δ, ε) and 12 neuronal (α2-10, β2-4) subunits. Functional diversity is evident in the mammalian nervous system where diverse expression patterns for the different subunits are observed, only some of which are overlapping. Also, nAChR subtypes differing in affinity for nicotinic ligands (including ACh), calcium permeability, compounds are widely classified as neurotoxicants if they can disrupt normal cholinergic signaling, such as the nicotinic cholinergic agonist nicotine (Crosby et al., 2015).

Neonicotinoids are regarded as neurotoxicants because they act as agonists against nAChRs in insects and mammals (Crosby et al., 2015). Compared to organophosphate pesticides (once widely used pesticides), neonicotinoids are thought to have reduced toxicity due to their presumed selectivity for insects over vertebrate nicotinic cholinergic receptors. Neonicotinoids selectively bind to insect nicotinic receptors with reduced action on the vertebrate nicotinic receptors (Tomizawa and Casida, 2003; Tomidawa and Casida, 2005). It is known that the favorable selective toxicity of imidacloprid to insects versus mammals is attributed to differences in their binding affinity or potency in the nAChR. Therefore, neonicotinoids might have lower neurotoxicity profiles for mammals, birds, fish, and other higher organisms. However, recent studies suggest that the neurotoxicity induced by neonicotinoids should be given more attention (Lin et al., 2013).

In spite of the original belief that neonicotinoids have low mammalian toxicity, there is increasing evidence that neonicotinoids could also cause a variety of toxic effects on animals and humans, such as neurotoxicity, immunotoxicity, hepatotoxicity, nephrotoxicity, and reproductive cytotoxic effects on vertebrates and invertebrates (Wang et al., 2018). Pesticides can exert their impact on vertebrates either directly, through their toxicity, or indirectly, for example, by reducing their food supply.

Toxicity correlates with agonist action and binding affinity at the vertebrate $\alpha_4\beta_2$ nAChR. Chronic exposure to neonicotinoid insecticides, and certain metabolites as well as nicotine, upregulates $\alpha_4\beta_2$ nAChR levels without altering the sensitivity of the binding site. The $\alpha_4\beta_2$ subtype consists of two α_4 and three β_2 subunits (heteropentameric) and the α_7 subtype is considered to be a homopentameric structure (Corringer et al., 2000). Recent articles showed that many diseases involving nAChR dysfunction result from perturbation of cholinergic nicotinic neurotransmission (Cotti and Clementi, 2004). Activation of α_7 nAChRs during development results in neuronal cell death.

The vertebrate nAChR is an agonist-gated ion channel responsible for rapid excitatory neurotransmission. It is a pentameric transmembrane complex in the superfamily of neurotransmitter-gated ion channels, including γ-aminobutyric acid ($GABA_A$ and $GABA_C$), glycine, and 5-HT_3 serotonin receptors. The nAChR consists of diverse subtypes assembled in combinations from 10 α, 4 β, γ, δ, and ε subunits (Tomizawa and Casida, 2005). In mammals, nicotinic receptors are made up of a number of subtypes, and located in many tissues, including autonomic ganglia, skeletal muscle (neuromuscular junctions), spinal cord, and a number of brain regions. In vertebrates, nAChRs are expressed at neuromuscular junctions, in the CNS and PNS (McGehee and Role, 1995) and on certain mechanosensory cells. However, the binding affinity of imidacloprid at the nicotinic receptors in mammals is much less than that of insect nicotinic receptors (Tomizawa and Casida, 1999).

Differences in binding properties to the various receptor subtypes contribute greatly to the much lower activity of neonicotinoids in vertebrate tissues, as compared to tissues from insects (Yamamoto et al., 1998). The relative specificity for the nicotinic receptor in insects is used to select compounds for trading development. The achievement of this strategy is reflected by very high margins of safety for these insecticides (Leicht, 1993).

Acute exposure to neurotoxicants usually are related to accident or intentional exposure. Clinical signs of acute neurotoxicity are obvious or are easier to be detected. However, slowly developing neurotoxicity by chronic exposure to low dose or subchronic doses of neurotoxicants is difficult to detect. Clinical signs induced by a neurotoxicant following acute or chronic exposure are not always identical. The acute toxicity (defined by lethal potency) of various neonicotinoid insecticides and related analogues in mammals is most closely related to potency at the α_7 nicotinic receptor subtype, with a activity at α_4, β_2, α_3, and α_1 nicotinic receptors having a decreasing effect on toxicity (Tomizawa and Casida, 1999). Toxicity in mammals involves complex interactions at multiple receptor sites with some of the receptor types even having a combination of agonist and antagonist effects on the synapse.

2.2 Insects

Neonicotinoids acts by interfering with the transmission of stimuli in the insect nervous system. They are mimics the action of Ach, but is not degraded by the enzyme AChE. They binds to the ACh receptor on the postsynaptic portion of nerve cells, resulting in persistent activation preventing transmission of impulse and leading to an accumulation of ACh. The ACh accumulation results in hyperexcitation, convulsions, paralysis, and death of the insect. Their high toxicities to insects has been attributed to selective binding affinity to insect nAChRs (Tomizawa and Casida, 2003; Tomizawa and Casida, 2005). It has been hypothesized that there are two binding sites with different affinities to imidacloprid higher at the agonist binding site and a lower affinity at the blocking site, thus imidacloprid has both agonistic and antagonistic effects on the nAChR channel (Nagata et al., 1998). This observation can be explained if it assumes that the two binding sites may have different affinities to imidacloprid, higher at the agonist binding site and a lower affinity at the blocking site.

Neonicotinoids act as selective agonists of insect nAChRs (Tomizawa and Casida, 2005), which belong to the "cys-loop" superfamily of ligand-gated ion channels (Ffrench-Constant et al., 2016). These chemical mimic the ACh neurotransmitter and are highly neurotoxic to insect. These nAChRs receptors are composed of five subunits (Jones et al., 2007), each subunit possesses four transmembrane domains (M1–M4), an extracellular amino-terminal domain involved in agonist binding and a large cytoplasmic loop between M3 and M4 containing several phosphorylation sites (Dupuis et al., 2012). Subunits were classified into two groups α and non α or β, depending on the presence or not of two adjacent cysteine residues in the extracellular domain, which play an important role for acetylcholine binding (Jones et al., 2007). In insect, these nAChR receptors are confined to CNS tissues (ganglia and brain).

In insects, several nAChR subunits have been cloned and the sequencing of the entire insect genome has revealed the existence of approximately 10 different nAChR subunit genes (Jones and Sattelle, 2010) suggesting a immense number of hypothetical nAChR subtypes. Combinations of nAChR subunits result in distinct receptors, with their own electrophysiological and pharmacological properties, which thereby influence sensitivity to neonicotinoids (Lansdell and Millar, 2000; Millar and Lansdell, 2010). Besides, previous studies have shown that neonicotinoid efficacy on nAChR subtypes depends on electro-pharmacological properties and many cellular and molecular factors such as conformational state, membrane potential, subunit composition and calcium-dependent phosphorylation/dephosphorylation process (Bodereau-Dubois et al., 2012). Neonicotinoid insecticides cause deep effects on nerve-impulse transmission through actions on insect postsynaptic nAChR (i.e., blocks postsynaptic nAChRs). Electrophysiological experiments in various neuronal preparations demonstrate that the response to treatment is biphasic consisting of an initial increase in the frequency of spontaneous discharge that is followed by a complete block to nerve-impulse propagation (Schroeder and Flattum, 1984). A variety of neurotoxic signs is evident in insects that have been treated with a neonicotinoid insecticide, including evidence of nicotinic stimulation. For example, the administration of imidacloprid to the American cockroach (*Periplaneta americana*) causes clinical signs of intoxication consisting in uncoordinated abdominal quivering, wing flexing, tremors (i.e., continuous repetitive twitching of skeletal muscle, usually palpable and visible), and violent whole body shaking, followed by prostration and death (Schroeder and Flattum, 1984).

Neonicotinoids are highly neurotoxic to honeybees and wild pollinators, their capacity to cross the ion-impermeable BBB surrounding the CNS and their strong binding to nAChR in the bee's CNS are responsible for a unique chronic and sublethal toxicity profile. Acting systemically, this new neonicotinoid class of neurotoxic insecticides following seed or soil application is taken up by plants, primarily through the roots, and translocates to all parts of the plant through xylemic and phloemic transport (Bromilow et al., 1990).

This mode of active of neonicotinoids allows control of the insects that attack the roots and the neck as well as feeding on the aerial part of the plant.

Acting on contact, the neonicotinoids are particularly suited for controlling many insects with biting and sucking mouth part specially if swallowed (Laurino et al., 2011).

3. Neurotoxicity

3.1 Acute neurotoxicity

A series of acute neurotoxicity studies were performed using imidacloprid, acetamiprid, thiamethoxam and thiacloprid. Table 4 lists the oral and dermal lethal doses (LD_{50}) and inhalatory lethal concentration (LC_{50}) values for the neonicotinoids acetamiprid, clothianidin, dinotefuran, imidacloprid, thiacloprid and thiamethoxam in rodents.

In rodents, the acute dietary reference dose for all populations of imidacloprid is 125 mg/kg bw/day. Acute toxicity of imidacloprid is characterized by nicotinic clinical signs (e.g., tremor) at relatively high levels of exposure (Sheets, 2010). To better understand physiological changes appearing after subchronic exposure to sublethal doses of imidacloprid insecticide using complementary approaches that include toxicology, electrophysiology, molecular biology and calcium imaging (Benzidane et al., 2017). It is known, cockroach neurosecretory cells identified as dorsal unpaired median (DUM) neurons, express two α-BT-insensitive nAChR subtypes, nAChR1 and nAChR2, which differ in their sensitivity to imidacloprid. α-BT is a selective α_7 nAChR antagonist. Although nAChR1 is sensitive to imidacloprid, nAChR2 is insensitive to this insecticide. In has been demonstrated that subchronic exposure to sublethal dose of imidacloprid differentially changes physiological and molecular properties of nAChR1 and nAChR2; this treatment decreased the sensitivity of nAChR1 to imidacloprid, reduced current density flowing through this nAChR subtype but did not affect its subunit composition (α_3, α_8 and β_1). Subchronic exposure to sublethal dose of imidacloprid also affected nAChR2 functions. However, these effects were different from those reported on nAChR1. The changes in nAChR2 conformational state could be related to a modification of the subunit composition (α_1, α_2 and β_1). Finally, the subchronic exposure affecting both nAChR1 and nAChR2 seemed to be linked to the elevation of the steady-state resting intracellular Ca^{2+} level (Benzidane et al., 2017). In conclusion, under subchronic exposure to sublethal dose of imidacloprid, cockroaches are capable of triggering adaptive mechanisms by reducing the participation of imidacloprid-sensitive nAChR1 and by optimizing functional properties of nAChR2, which is insensitive to this insecticide.

Table 4 LD_{50} and LC_{50} values for neonicotinoids.

Neonicotinoid	Animal specie	Route of exposure	LD50 (mg/kg)	LC50 (mg/L)	Classification[a]	References
Acetamiprid	Rat	Oral	195 (male)		Acute Tox cat. 3 H301 Toxic if swallowed	EC (2018)
		Inhalatory (4 h)		>1.15	No classification	EC (2018)
		Dermal	>2000		No classification	EC (2018)
Clothianidin	Rat	Oral	523 < LD50 < 1216 (female)		Acute Tox cat. 4 H302 Harmful if swallowed	EC (2014)
	Mice	Oral	389/465 (males/females)		Acute Tox cat. 4 H302 Harmful if swallowed	EC (2014)
	Rat	Inhalatory (4 h)		>6.14	No classification	EC (2014)
	Rat	Dermal	>2000		No classification	EC (2014)
Dinotefuram	Rat	Oral	2400		No classification	Sheets et al. (2016)
Imidacloprid	Mice	Oral	131/168 (males/females)		Acute Tox cat. 3 H301 Toxic if swallowed	ECHA (2018a)
	Rat	Inhalatory (4 h)		>5.32	No classification	EC (2015)
	Rat	Dermal	>5000		No classification	EC (2015)

Thiacloprid	Rat	Oral	177		Acute Tox cat. 3 H301 Toxic if swallowed	ECHA (2013)
	Rat	Inhalatory (4 h)		1.2 (females)	Acute Tox cat. 4 H332 Harmful if inhaled	ECHA (2015)
Thiapethoxam	Rat	Oral	1563		Acute Tox cat. 4 H302 Harmful if swallowed	Sheets et al. (2016)
	Mice	Oral	783		Acute Tox cat. 4 H302 Harmful if swallowed	ECHA (2018b)
	Rat	Inhalatory (4 h)		>3.72	No classification	ECHA (2018b)

[a]Classification based on CLP Regulation (EC, 2008)

It should also be noted that behavioral changes following acute and chronic exposure to neurotoxic agents are sensible and rapid indices of neurotoxicity. Usually, the methods designed for neurobehavioral testing are based on changes in motor function, sensory function, reactivity, learning, and natural occurring behaviors. The functional observation battery (FOB) for characterizing neurotoxic agents fall in three categories: (i) home cage and open field behaviors; (ii) manipulative response; and (iii) physiological measurements (Moser, 1989). For imidacloprid acute, subchronic, and developmental screening studies in rats, specific neurotoxicological parameters were investigated by FOB, automated motor activity measurements, and special neurohistopathology (EC, 2015). In the acute oral neurotoxicity screening study conducted in rats, imidacloprid produced clinical signs related to acute receptor-mediated cholinergic toxicity of this chloronicotinyl compound. Based on tremor and decreased motor/locomotor activity at 151 mg/kg bw/day, the NOAEL was 42 mg/kg bw/day (EC, 2015),

In a study reported, groups of rats were administered acetamiprid to evidence acute neurotoxicity and neuromorphological effects. The test substance was administered once orally at doses 10, 30 and 100 mg/kg bw. Clinical signs of behavioral effects were confined to the day of dosing with effect being observed at 30 and 100 mg/kg bw. Tremors and chewing movements were observed among males and females treated at 100 mg/kg bw. At this dose, males were difficult to handle, pupils were dilated in both sexes, animals felt cold to touch, and males showed increased urination. Gait and posture were also affected the day after exposure. At 30 mg/kg bw, there were a few animals observed with tremors and increased urination among males. The NOAEL for acute behavioral neurotoxicity was set at 10 mg/kg bw based on decreased locomotor activity observed at 30 mg/kg bw (EFSA, 2013).

Two additional studies were available (EC, 2018). In the first acute neurotoxicity study groups of rats were dosed with acetamiprid by oral gavage; the animals (3/sex/dose) were subjected to a FOB prior to treatment and at 6 h post dosing (time of peak effects) and on day 7 and 14 post dosing. Effects on the FOB and locomotor activity were confined to the day of dosing: (i) at 100 mg/kg bw, tremor and chewing (i.e, mastication is the act of chewing) movements of the mouth were seen. Pupils were dilated. Gait (i.e., the manner or style of locomotion) and posture were affected: hunched posture, walking on toes, slipping of hindlimbs. Forelimb grip strength was increased among males and landing foot splay was decreased among females. Locomotor activity was statistically significantly reduced in both sexes; (ii) at 30 mg/kgbw, there was evidence of tremor among males (one animal showed continuous tremor), one female showed tail tremor; (iii) No effects were observed at 10 mg/kgbw. In summary, the measured end-points included clinical signs (behavioral changes in both sexes) and reduced locomotor activities in both sexes at the highest dose and in males only at the medium dose were observed on the day of treatment. The NOAEL was considered to be 10 mg/kgbw.

In an acute oral neurotoxicity study adult Fischer 344 rats received clothianidin. Transient neurobehavioral effects were observed after acute oral administration of clothianidin, which were considered to be neurobehavioral evidence of systemic toxicity and/or signs of pharmacological overstimulation (route and dose levels were not indicated). Based on the following toxicological end-point: tremors, locomotor activity, hypothermia, the NOAEL was 60 mg/kgbw for males and 100 mg/kgbw for females (EC, 2014).

In the standard acute toxicity studies in rats, clinical signs associated with effects on nervous system were reported, including prostration, poor reflexes, decreased/reduced motility, apathy, spastic gait, spasmodic state, convulsions, and tremor. These effects occurred concurrently with other clinical signs of toxicity but they were not associated with any necropsy findings or changes in brain weight. Thiacloprid was also tested in several acute neurotoxicity studies performed in rats via oral route: (1) in a dose range-finding study (ECHA, 2015), tremors, decreased activity and repetitive chewing movements were found at lethal doses. Clinical signs that were slight and reversible within 24 h were reported also at sub-lethal doses. These consisted of slight repetitive chewing movements in males at 35 mg/kgbw, and of slight tremors and repetitive chewing movements in both sexes at 85 mg/kgbw; (2) in the main neurotoxicity study (ECHA, 2015), no lethality was observed up to the highest tested dose of 109 mg/kgbw. Clinical signs included tremors, decreased activity, ataxia, dilated pupils, ptosis, and reduced body temperature, which occurred on the day of treatment and was typically resolved by day 1, and by day 5 at the latest. Few animals showed these clinical signs at 22 mg/kgbw but toxicity was reported to be severe at the top dose of 109 mg/kgbw. Also reversibly decreased motor activity was reported in females at all doses (27, 41 and 71% at 22, 53 and 109 mg/kg bw, respectively, being statistically significant from 53 mg/kg), and at 109 mg/kg in males; (3) In an additional acute neurotoxicity study (ECHA, 2015), female rats were dosed with 3.1 and 11 mg/kg bw which were below the lowest dose of the above mentioned main study. Slight decreases in motor and locomotor activities were observed at the top dose. In conclusion, most of these effects are common in acute toxicity studies (e.g., tremors, decreased motility or reactivity, poor reflexes and prostration). The effects are considered as transient since clinical signs were reversible within 24 h and other effects resolved by day 7.

In two acute and one 13-week repeat dose toxicity study the potential neurotoxicity of thiacloprid has been investigated. Clinical signs of neurotoxicity, primarily decreased motor and locomotor activity, were noticed in rats after a single oral gavage but not in rats after dietary administration over 13-weeks. Based on effects seen in females at a dose of 11 mg/kg bw/day, the NOAEL for acute neurotoxicity was 3.1 mg/kg bw/day and the NOAEL for short-term toxicity was 2 mg/kg bw/day (ECHA, 2015).

3.2 Subacute neurotoxicity

In an oral subacute neurotoxicity study rats were exposed to imidacloprid (45 and 90 mg/kg bw) for the period of 28 days, produced significant decreased in spontaneous locomotor activity (SLA) and stimulates pain threshold along with pathological changes in the brain. The following neurotoxicity indicators (salivation, lachrymation, piloerection diarrhea, dyspnea, tremor, convulsion, paralysis) were taken as clinical signs of toxicity. AChE, ATPase and serum biochemical parameters such as creatine kinase, lactate dehydrogenase, sorbitol dehydrogenase and alkaline phosphatase levels were significantly decreased as result of imidacloprid exposure. Imidacloprid caused a significant decrease in antioxidant enzymes activity and non-enzymes level with increase in lipid peroxidation. The imidacloprid induced brain damage of biochemical markers and neurotoxicity in exposed rats (Lonare et al., 2014).

3.3 Subchronic neurotoxicity

In a 90-days oral toxicity study, female rats were administered imidacloprid at doses of 5, 10, 20 mg/kg bw/day. Morphological, biochemical, hematological and neuropathological evaluations were performed. Decrease in the body weight gain was observed at the high-dose and at necropsy the relative body weights of liver, kidney and adrenal was also significantly increased at this dose level. Clinical chemistry revealed at high-dose of imidacloprid significant elevation of serum GOT, GPT, glucose and BUN and resulted in 16–49% and 16–40% inhibition (dose-dependent) of AChE activity in brain and serum, respectively. The spontaneous locomotor activity was also decreased at high-dose exposure (i.e., a significant decrease in distance travel, ambulatory time, and stereotypic time) whereas there were no significant changes in hematology and urine parameters. The brain, liver and kidney of rats exposed with high-dose of imidacloprid had showed mild pathological changes. Imidacloprid produced necrosis of Purkinje cells with loss of dendrites and granules in the granular layer of the *cerebellum* in female rats as compared to the control (Bhardwaj et al., 2010).

In a subchronic (13-week) dietary neurotoxicity study, Fischer 344 rats were administered imidacloprid. No specific neurotoxic effects were observed at 196 mg/kg bw/day (highest dose level tested) (EC, 2015).

In a subchronic behavioral neurotoxicity study, rats were administered orally acetamiprid, dosed at 7.4, 14.8, 59.7 and 118 mg/kg bw/day for males and 8.5, 16.3, 67.6 and 134 mg/kg bw/day for females. While decreased body weight and food intake were observed in the high-dose level, no evidence of irreversible neurotoxic effects was observed in the dose range assayed. Based on non-neurotoxic endpoints, the respective subchronic NOAEL was set at 14.8 and 16.3 mg/kg bw/day for males and females, respectively. Altogether, neurotoxic effects occurred at approximately the same dose level as liver effects and general toxicity observed in the subchronic and reproduction toxicity studies (EFSA, 2013). In another 13-weeks toxicity study, rats were repeatedly given acetamiprid at a dose of 100, 200, 800 or 1600 ppm in the feed (equivalent to 7.4, 14.8, 59.7, 118 mg/kg bw/day for males and 8.5, 16.3, 67.6 and 134 mg/kg bw/day for females) (EC, 2018). All of the animals were subjected to FOB prior to treatment and after 4, 8, 13 weeks of treatment. Motor activity of each animal was quantitatively assessed at the same intervals. Clinical signs, body weight and food consumption was monitored. There were no mortalities and any clinical signs related with acetamiprid administration. Bodyweights of males and females treated at 800 and 1600 ppm were statistically significantly lower than controls. Cumulative food intake was also statistically significantly lower among males and females treated from 800 ppm. Feed conversion was impaired for males at 1600 ppm and for females from 800 ppm. Cumulative food consumption in rats treated with acetamiprid was statistically significant lower among males and females at 800 and 1600 ppm over 13 weeks. Statistically significant lower values for forelimb grip strength were recorded during week 4 among females from

800 ppm, during week 8 for all treated female groups (although there was no apparent dose relationship) and during week 13 for males at 1600 ppm. There were also records of statistically significant lower values for hindlimb grip strength among males from 800 ppm. There were no other behavioral changes observed which could be attributed to neurotoxicity. There was no evidence of neuropathology from the histopathological examination of the central and peripheral nerve system. In summary, based on food consumption and body weight effects the NOAEL for systemic toxicity was established at 200 ppm (equivalent to 14.8 and 16.3 mg/kg bw/day for males and females, respectively). There were no neuropathological or other findings to indicate changes in motor function and co-ordination except for the observation of lower grip strength values of the limbs (mainly for males at 1600 ppm). Taking into consideration all these factors, the lower grip strength values were not considered to be indicative of any irreversible neurotoxic effect.

In a subchronic (90 days) neurotoxicity screening study, Fischer 344 rats were administered clothianidin. No relevant treatment-related effects were seen in the FOB, motor activity assessments or histopathological examinations of nervous or muscle tissues. The NOAEL was 177 mg/kg bw/day in males, and 200 mg/kg bw/day in females (EC, 2014).

3.4 Developmental neurotoxicity in neonates

Their mode of insecticidal action, along with the neurodevelopmental toxicity of neonicotinoids, makes it crucial to clarify their neurotoxic potential. It is important to note that developmental neurotoxicity, addresses effects on the developing offspring, to detect gross neurological and behavioral abnormalities, so that the pregnant animals shall be administered test substance during gestation and early lactation. Usually is determining motor activity, response to auditory startle, assessment of learning, neuropathological evaluation, and brain weights. Gestational exposure to a single large, nonlethal dose of imidacloprid produced neurobehavioral deficits and an increased expression of glial fibrillary acidic protein (GFAP) in several brain regions of the offspring on postnatal day 30, corresponding to a human early adolescent age. Maternal exposure to a large sublethal dose of imidacloprid during gestation produced significant sensorimotor impairments that were reflected in beam walk time, inclined plane performance, and forepaw grip in the offspring on postnatal day 30, were associated with increased AChE activity in the midbrain, cortex and brainstem (125–145% increase) and in plasma (125% increase). These neurobehavioral deficits may reflect dysfunction at multiple anatomical areas in the CNS, PNS, or muscle. These changes were associated with increased AChE activity in the midbrain, cortex, and brainstem. Ligand binding densities for [^{3}H]cytosine for $\alpha_4\beta_2$ type nicotinic acetylcholine receptors did not show any significant change, whereas [^{3}H] AFDX 384, a ligand for m^2 muscarinic acetylcholine receptors, was significantly increased in the cortex. Histopathological evaluation of brain regions did not show any alteration in surviving neurons. On the other hand, there was an ascend in glial fibrillary acidic protein immunostaining in motor cortex layer III, CA1, CA3, and the dentate gyrus subfield of the hippocampus (Abou-Donia et al., 2008). These effects are mediated by a complex array of multiple pathways. For example, beam walking is an integrated form of behavior necessitating pertinent levels of consciousness, memory, and sensorimotor, cortical, subcortical, cerebellar, spinal cord, peripheral nervous system, neuromuscular junction, and muscular functions (Abou-Donia et al., 2001). Oral and intraperitoneal administration of imidacloprid, was absorbed from intestinal cell and over a period of time bio accumulated in brain whereby was responsible for tissue damage; imidacloprid and 7 analogues with LD_{50} values of 7–50 mg/kg (or intoxication clinical signs at 50 mg/kg bw) inhibited [^{3}H]nicotine binding by 50% (IC_{50}) at 12–800 nM whereas 21 other analogues that were not toxic at 50 mg/kg bw gave an IC_{50} value of >1000 nM, thereby correlating the toxicity with interaction at the [^{3}H]nicotine binding site (Chao and Casida, 1997).

The effects of long-term (14 days) and low dose (1 μM) exposure of neuron-enriched culture from neonatal rat *cerebellum* to nicotine and two neonicotinoids acetamiprid and imidacloprid. The cultured cerebellar neurons natively express the α_3, α_4 and α_7 nAChR subtipes in vitro (Didier et al., 1995; Kimura-Kuroda et al., 2016). In this study immunocytochemistry revealed no differences in the number or morphology of immature neurons or glial cells in any group compared to the control cultures, but nevertheless, a slight disturbance in Purkinje cell dendritic arborisation was observed in the exposed culture to neonicotinoids. Additionally, transcriptome analysis on total RNAs using microarrays and identifying significant differential expression between control and treated cultures in 34, 48, and 67 genes, respectively, was performed. Common to all exposure groups were 9 genes essential for neurodevelopment, suggesting that chronic neonicotinoid exposure alter the transcriptome of the developing mammalian brain in a similar way to nicotine exposure (Kimura-Kuroda et al., 2016)

A developmental neurotoxicity screening study was performed with imidacloprid. (EC, 2015). In this study, neurotoxic effects on the pups were noted at maternal dose levels of 80.4 mg/kg bw/day and higher. Based on the available data, the NOAEL was stabilized as 45.4 mg/kg bw/day. Behavioral changes in the acute experiment and, to a much lesser extent, in the developmental study were the only clinical signs that could be indicative of primary neurotoxic effects. However, there was no evidence of developmental neurotoxicity.

Sheets et al. (2016) reviewed the neonicotinoids for developmental neurotoxicity and found the decreased body weight of F1 animals during development (e.g., delayed sexual maturation, decreased brain weight, and morphometric measurements) and acute toxicity (decreased motor activity associated with peak exposure via the diet and milk).

For the guideline-based studies, treatment was administered via the diet or by gavage to primiparous female rats at three dose levels, plus a vehicle control ($\geq$20/dose level), from gestation day 0 or 6 to lactation day 21. F1 males and females were evaluated using measures of motor activity, acoustic startle response, cognition, brain morphometry, and neuropathology. The principal effects in F1 animals were associated with decreased body weight (delayed sexual maturation, decreased brain weight, and morphometric measurements) and acute toxicity (decreased activity during exposure) at high doses, without neuropathology or impaired cognition. No common effects were identified among the neonicotinoids that were consistent with developmental neurotoxicity or the neurodevelopmental effects associated with nicotine (a neurotoxin of brain development). Findings at high doses were associated with evidence of systemic toxicity, which indicates that these insecticides do not selectively affect the developing nervous system (Sheets et al., 2016).

The tendency of the damaged brain to activate two resident glial cell types, astroglia and microglia, can be exploited for detection and quantification of neurotoxicity. The response of microglia to toxic exposures of the brain also represents a sensitive index of brain damage. Microglia are the major immune cell type in the CNS and play critical roles in embryonic and early postnatal development and maintenance of the CNS environment. In some contexts, microglial activation in response to CNS injury can cause or exacerbate neuropathological changes (Rivest, 2009 Falta); whereas activated microglia can serve a neuro-supportive role in the healthy brain by producing neuroprotective agents and scavenging cellular debris. The effects of neonicotinoid insult on microglia in the developing CNS are unknown.

However, two studies have been performed. In the first neurodevelopmental toxicity study acetamiprid was given to ICR mice (8 weeks of age) and acclimated for 2 weeks before use. Acetamiprid was suspended in 0.5% carboxymethylcellulose sodium and administered to pregnant mice by oral gavage (5.0 mL/kg bw) from embryonic day to embryonic day 13 or from embryonic day 6 to embryonic day 18, with the period from embryonic day 6 to embryonic day 13 corresponding to mouse neocortex organogenesis (Kagawa and Nagao, 2018). The effects of repeated prenatal acetamiprid exposure (5 mg/kg bw, oral doses administered to pregnant dams) on neurogenesis, neuronal distribution and microglial activation in the dorsal telencephalon on embryonic day 14 and in the neocortex on postnatal day 14 was evaluated. Immunohistochemical and morphological analyses on embryonic day 14 revealed hypoplasia of the cortical plate and decreased neurogenesis in mice exposed to acetamiprid from embryonic day 6 to 13, whereas newborn acetamiprid-exposed mice showed an abnormal neuronal distribution in the neocortex. Furthermore, acetamiprid-exposed ICR mice showed increased numbers of Iba1-immunoreactive and amoeboid-type microglia as well as an increased M1/M2 microglial ratio. These findings suggest that prenatal acetamiprid-exposure induces neurodevelopmental toxicity and increases microglial activation in the developing brain. In conclusion, the outcomes of this study indicate that prenatal exposure to acetamiprid produces significant neurodevelopmental toxicity and microglial activation in the mouse neocortex. In a second study, ICR mice were exposed to acetamiprid, imidacloprid (both 5 mg/kg bw/day) or nicotine (0.5 mg/kg bw/day) from postnatal day 12–26 by oral gavage. On postnatal day 27, brains were removed, and neurogenesis and microglial activation in the hippocampal dentate gyrus were examined via immunohistochemistry; it was found a reduction in neurogenesis in the hippocampal dentate gyrus (i.e., this part of the brain region is thought to contribute to the formation of new episodic memories) of neonates following the treatment with the three substances. To sum up, acetamiprid and imidacloprid impairs neurogenesis and alters microglial profiles in the developing hippocampal dentate gyrus.

Besides, has been observed that neonicotinoid-exposed newborns showed an increase in the number of amoeboid-type and activated M1-type microglia. These findings indicate that neonatal neonicotinoid exposure might have affected neuronal cells in the hippocampal dentate gyrus, possibly resulting from the promotion of neuronal cell death in the subgranular zone and granule cell layer of the hippocampal dentate gyrus.

Overall, it was suggested that neonicotinoids induce neurodevelopmental toxicity associated with abnormal microglial profiles in the hippocampus of mouse neonates (Nakayama et al., 2019)

As regards the developmental neurotoxicity study with acetamiprid (EC, 2018), pregnant female Sprague-Dawley rats (25/group) were exposed by gavage from gestation day 6 through lactation day 21 daily exposed to 0, 2.5, 10 or 45 mg acetamiprid/kg bw/day. F0 maternal toxicity was expressed at the dose level of 45 mg/kg/day by a single mortality and reductions in body weight gain and food consumption. F1 developmental toxicity was expressed at the dose level of 45 mg/kg/day by early postnatal mortality and reduced post-weaning body weights. Deficits in auditory startle response occurred in the 45 mg/kg/day group F1 males and females without concomitant effects in other FOB) neuropathology or brain morphometry. Based on the results of this study, the NOAEL (no-observed-adverse-effect level) for maternal toxicity, developmental toxicity and developmental neurotoxicity was considered to be 10 mg/kg/day.

As refers to clothianidin administered orally via the diet to Crl:CD BR VAF/Plus presumed pregnant rats was performed (EC, 2014). There was some indication for developmental neurotoxicity in rats at doses which also induced reductions of maternal and offspring pre-weaning body weight gain and a slight decrease of offspring viability after weaning Female offspring exhibited reduced motor activity on post-natal day 62, which, in the absence of exposure to the test substance at the time of testing, could indicate residual neurodevelopmental changes. The NOAEL for developmental neurotoxicity 42.9 mg/kg bw/day.

In a study designed specifically to investigate developmental neurotoxicity of thiacloprid (ECHA, 2015), there was a decrease in pup weight and a delay in sexual maturation, but no signs of developmental neurotoxicity.

3.5 Neurotoxicity in target species

Due to their high neurotoxicity for insects, neonicotinoids are extremely effective in controlling certain pests. Imidacloprid acts on several types of post-synaptic nicotinic acetylcholine receptors located only within the CNS of insects. Following binding to the nicotinic receptor, nerve impulses are spontaneously discharged at first, followed by failure of the neuron to propagate any sign (Sheets, 2001). Sustained activation of the receptor results from the inability of AChE to break down the insecticide whereby this binding process is irreversible (Matsuda and Sattelle, 2005).

3.6 Neurotoxicity in non-target species

The impact of neonicotinoid insecticides on insect pollinators is highly controversial. Surrounded by the extensive diversity of pollinating species bees are the most important. When neonicotinoids were first used, beekeepers started describing different disorders and clinical signs such as bees not returning to the hive, disoriented bees, bees gathered close together in small groups on the ground, abnormal foraging behavior, massive bee losses in spring, and queen losses. Even if the poisoned honeybees were able to return to the colony, their memory and communication abilities could be impaired (Decourtye and Devillers, 2010). Sublethal concentrations of neonicotinoids alter the behavior of social bees and reduce survival of entire colonies (Decourtye and Devillers, 2010; Whitehorn et al., 2012) and impair the olfactory memory and learning capacity of honeybees and the orientation and foraging activity. However, critics argue that the reported negative effects only arise from neonicotinoid concentrations that are greater than those found in the nectar and pollen of pesticide-treated plants. Furthermore, it has been suggested that bees could choose to forage on other available flowers and hence avoid or dilute exposure (Godfray et al., 2014).

Using a two-choice feeding assay, was showed that the honeybee (*Apis mellifera*), and the buff-tailed bumblebee (*Bombus terrestris*), do not avoid nectar-relevant concentrations of three of the most commonly used neonicotinoids, imidacloprid, thiamethoxam, and clothianidin, in food. Moreover, bees of both species prefer to eat more of sucrose solutions laced with imidacloprid, or thiamethoxam than sucrose alone. Stimulation with imidacloprid, thiamethoxam and clothianidin neither elicited spiking responses from gustatory neurons in the bees' mouthparts, nor inhibited the responses of sucrose-sensitive neurons. Our data indicate that bees cannot taste neonicotinoids and are not repelled by them. Instead, bees preferred solutions containing imidacloprid or thiamethoxam, even though the consumption of these pesticides caused them to eat less food overall. This work shows that bees cannot control their exposure to neonicotinoids in food and implies that treating flowering crops with imidacloprid and thiamethoxam presents a sizeable hazard to foraging bees (Kessler et al., 2015). In January 2013, the European Food Safety Authority (EFSA) published three scientific opinion on the risks posed by the three neonicotinoids (i.e., clothianidin, thiamethoxam, imidacloprid) on request of European Commission to perform an updated risk assessment as regards the risk to bees from the uses of the three neonicotinoid pesticides as seed treatments and granules (EFSA, 2013, 2013a, 2013b, EFSA 2013c). EFSA examined lethal as well as sub-lethal effects on honeybees. It concluded that the insecticides pose "*high acute*" risks for bees. In particular, EFSA identified highly acute risks to honeybees from exposure via dust, from consumption of residues in contaminated pollen and nectar and from exposure via guttation fluid (in the case of maize).

The Commission Implementing Regulation (EU) No 485/2013 (EU, 2013) amended the conditions for approval of the active substances (clothianidin, thiamethoxam or imidacloprid), for use in plant protection products, all belonging to the group of neonicotinoids. The specific provisions of the approval were amended to restrict the uses of clothianidin, thiamethoxam and imidacloprid, in order to provide for specific risk mitigation measures for the protection of bees and to limit the use of the plant protection products containing these active substances to professional users. In particular the uses as seed treatment and soil treatment of plant protection products containing clothianidin, thiamethoxam or imidacloprid should be prohibited for crops attractive to bees and for cereals, except for uses in greenhouses and for winter cereals. Among others, the regulation forbids their use as seed treatment, soil treatment or foliar application for the following crops: maize, oilseed rape, soya, barley, millet, oats, rice, rye, sorghum and wheat. The entire list, which also includes ornamental plants and fruits, can be found in Annex I of the regulation. As of December 1, 2013, the marketing of seeds treated with the three neonicotinoids will also be banned. The Regulation also lists a number of exceptions. For instance, the ban does not apply to uses in closed systems such as greenhouses, nor to crops which are considered non-attractive to bees, such as cereals sown in winter. Another exception is that foliar treatments are allowed once the crop's flowering period is over. Non-professional use of the three insecticides is also prohibited. The regulation stipulates that, within 2 years of its entry into force, the European Commission will initiate a review of new scientific information on the pesticides. The Commission will then decide whether it is most appropriate to lift the ban, prolong it temporarily, or make it permanent.

Pesticides can produce four types of effects on honeybees: lethal effects and sublethal effects from acute or chronic exposure. Lethal effects from chronic exposure refer to honeybee mortality that occurs after prolonged exposure (Van der Sluijs et al., 2013). Neonicotinoids are high toxic to honeybees and other bee species (*Bombus species*, *Osmia lignaria*, *Megachile rotundata*) when are administered orally or by contact. Recently imidacloprid has raised concern because of its ability to cause honeybee colony collapse disorder. It is known that bee health remains of paramount importance since it concerns biodiversity, food production and the environment. At field realistic doses, neonicotinoids, which are nerve agents, cause a wide range of adverse sublethal effects in honeybee and bumblebee (*Bombus terrestris*) colonies, affecting colony performance through impairment of foraging success, brood and larval development, memory and learning, damage to the central nervous system, susceptibility to diseases, hive hygiene, among others (Van der Sluijs et al., 2013). Neonicotinoids cause a wide range of harm to individual bees such as damaging memory and reducing queen numbers. Honeybees' exposure to neonicotinoids can occur through ingestion, contact and inhalation (aerosols). Many possible exposure pathways can exist: (i) intake of food that contain residues; (ii) nesting material (resin, wax, etc.); (iii) direct contact with spray drift and dust drift during application; (iv) contact with contaminated plants, soil, water; (v) use of cooling water in the hive; and (vi) inhalation of contaminated air (EFSA, 2012). For bumble bees and other wild bees that nest in soil, contact with contaminated soil is an additional pathway of concern.

The neonicotinoids modified the foraging ability of bees (Fischer et al., 2014). Foraging activity and orientation in bees is coordinated by ritualized modes of communication, including the waggle dance (Feltham et al., 2014).

Non-target effects of pesticides use are an important global issue. Neurotoxic pesticides, such as neonicotinoids, negatively affect the cognitive capacity and fitness of non-target species, and could also modify interspecific interactions (Barbieri et al., 2013). There is increasing evidence that pesticide use at lethal and sublethal concentrations is contributing towards pollinator declines and affecting behavioral responses of non-target organisms (Desneux et al., 2007). Sublethal contamination with imidacloprid also affected the brain development and motility of callow stingless bee workers (*Melipona quadrifasciata anthidioides*) (Tome et al., 2012). Pesticides such as the neonicotinoids are widely used and are effective in the control of many insect pests (Tomizawa and Casida., 2005). These chemicals interact with acetylcholine receptors, and directly affect the central and peripheral nervous system of insects (Brown et al., 2006). Neonicotinoids can impair the cognitive function of insects to such an extent that their ability to interpret external signals and learn is reduced, or even lost, owing to their neurotoxic action (Williamson and Wright, 2013). Exposure of pollinators such as bees to pesticides has demonstrated a range of physiological and behavioral changes (Decourtye et al., 2004). Bumble-bees exposed to sublethal concentrations of a neonicotinoid had longer foraging trips, decreased food collection and produced fewer workers, and hives had higher worker mortality and loss while foraging (Fischer et al., 2014). Another study showed that small doses of two neonicotinoids (imidacloprid and clothianidin) inhibited the neuronal responses in the brain of honeybees (Palmer et al., 2013), and therefore demonstrated that neonicotinoids act in zones of insect brain responsible for cognition, learning and behavior.

The role of behavior in determining the success of species and shaping communities is well established. The dose-dependent impacts of neonicotinoids in the neuronal activity of insect brains could impair cognition and learning of neurobehavioral tasks (Palmer et al., 2013).

As refers to clothianidin and thiamethoxam are highly toxic both via ingestion and indirect contact. By using the indirect contact test thiamethoxam was lethal to a concentration 20 times lower than the field one, showing a degree of danger long after administration (Laurino et al., 2011). The acute ingestion toxicity value (LD_{50}) was 1.8–3.8 ng/honey bee at 24 h for clothianidin and 5.0 ng/honey bee at 48 h for thiamethoxam (Tomlin, 2003). Hazard quotient (HQ) [calculated using in field application rate (g/ha) lowest and highest, maximum application rates/(laboratory oral LD_{50} values (μg/bee)] of both active ingredient are exceedingly high when compared with those of other insecticides and their order of magnitude is similar to the imidacloprid hazard quotient regarding bumble bees (Marletto et al., 2003, Laurino et al., 2011).

There has been considerable controversy in recent years over the effects of neonicotinoids on honey bees, culminating in a European Commission regulation in 2013 restricting certain uses of these pesticides on flowering crops. Most uses of neonicotinoid pesticides represent a risk to wild bees and honeybees, which are critical to sustainable agriculture (Anadón, 2018). EFSA in 2018 has updated its risk assessments of three neonicotinoids (clothianidin, imidacloprid and thiamethoxam) that are currently subject to restrictions in the EU because of the threat they pose to bees when are applied as seed treatments and granules. Bees can be exposed to neonicotinoids in multiple ways, depending on the use of the pesticide. The assessments indicated that in many cases bees foraging on the treated crop in the field as well as in its vicinity are likely to be exposed to harmful levels of the neonicotinoid pesticides (EFSA, 2018). This is because pollen and nectar of the treated crop contain pesticide residues, and plants in the vicinity can also be contaminated by dust drifting away from the field.

In addition, the soil where the crop is planted can become contaminated with the pesticide. In some situations, the pesticide may persist and accumulate in the soil. These residues end up in the pollen and nectar of the newly grown plants. Therefore, in some cases, bees might still be exposed to harmful level of neonicotinoids pesticides through this route.

4. Poisoning of neonicotinoids

4.1 Animal poisoning

There is limited published information detailing adverse effects of imidacloprid in dogs and cats. Because the drug is bitter tasting, oral contact may cause excessive salivation. Signs of poisoning from imidacloprid are similar to nicotinic signs, including lethargy, vomiting, diarrhea, hypersalivation, initial tremors, muscle weakness, and ataxia. Acute oral ingestions can cause salivation and vomiting. At very high oral imidacroprid exposures may lead to lethargy, vomiting, diarrhea, salivation, muscle weakness, and ataxia, which are all indicative of nicotinic signs (Wismer, 2004). Other signs of exposure at high doses are tremors, uncoordinated gait, and reduced motility/activity, apathy onset of signs in minutes, fast reversible (Thyssen and Machemer, 1999).

4.2 Human poisoning

Published studies regarding human exposure to neonicotinoids and toxicity is quite rare in spite of its widespread use. One case reports of attempted suicide (22-year-old male with clinical toxicity due to intentional ingestion of 30 mL of imidacloprid at a concentration of 17.8%, described signs of toxicity including fever, drowsiness, dizziness, vomiting, disorientation, an palpitation. Although the initial tachycardia could be attributed to nicotinic effects of the compound, it is unclear if the bradycardia was due to toxic effects or metabolic disturbances. The vomiting again could have been secondary to gastric irritation; however, this was not evaluated with an upper gastrointestinal scopy (Deepu et al., 2007; Shadnia, 2007; Wu et al., 2001). In two of these cases reports, the authors concluded that the other ingredients in the formulated product ingested by the victims were more likely to account for many of the observed clinical signs since acute poisoning was reported following ingestion of its formulation containing 10% imidacloprid. In one case report of human poisoning, the toxicant, a combination of 9.7% imidacloprid, less than 2% surfactant, and *N*-methyl pyrrolidone (solvent), produced drowsiness, disorientation, dizziness, as well as oral and gastroesophageal erosions and hemorrhagic gastritis. The patient developed a productive cough, fever, leukocytosis, and hyperglycemia and ultimately made an uneventful recovery in 4 days (Wu et al., 2001). In a second case, 69-year-old woman ingested a formulated product containing 9.6% imidacloprid in *N*-methyl pyrrolide solution. Clinical symptoms of toxicity soon after the ingestion included disorientation, sweating, vomiting, and increased heart and respiratory rates, but woman suffered severe cardiac toxicity and death appeared 12 h after the exposure (Huang et al., 2006).

A case report, 24-year-old farmer who accidentally inhaled a pesticide containing 17.8% imidacloprid while spraying on his farm. Patient had apparently become unconscious after accidentally inhaling the insecticide spray, further accentuating the exposure. He was disoriented, agitated, incoherent, sweating and breathless following the exposure and appeared severe neuropsychiatric manifestation and rhabdomyolysis (Agarwal and Srinivas, 2008). Rhabdomyolysis might be related to direct nicotinic end-plate stimulation or fever seen in the index patient. In a previous report ventricular fibrillation observed was attributed to either imidacloprid, its diluent (imidacloprid "inerts"), or coexisting coronary artery, disease, hypoxia, and hypercarbia (Huang et al., 2006). In addition, given the prominent neurological manifetations, central nicotinic stimulation seems to be the dominant pathogenic cause (Agarwal and Srinivas, 2008).

Lastly a case report, includes a previously healthy 35-year-old male farmer (body weight, 85 kg) was found by his family to be drowsy with severe nausea and copious vomiting. The patient reported that approximately 350 mL of imidacloprid was ingested in a suicide attempt about 30 min previously. Disorientation, drowsiness, dizziness, and palpitations had developed by the time of arrival at a local hospital (1 h after ingestion) and copious vomiting, which progress to coma, tachycardia, hypertension, apnostic respiration, mydriasis with sluggish reaction to light, fever, leukocytosis, hypokalemia, hypernatremia, and finally bradicardia and cardiopulmonary arrest (Shadnia, 2008).

The fast-growing use of neonicotinoids in recent years has seen a concurrent dramatic increase in the number of acute neonicotinoid poisoning cases reported worldwide, such as in Taiwan (Phua et al., 2009). Alarmingly, reports of severe human toxicity attributed to neonicotinoids are emerging. Thiacloprid, and imidacloprid poisonings resulted from deliberate ingestion in humans, manifesting with neurotoxicities, such as status epilepticus, convulsions, and hypotension (Imamura et al., 2010; Lin et al., 2013; Vinod et al., 2015; Yeh et al., 2010). The absence of an effective antidote raises concern in this regard (Vinod et al., 2015).

Latterly, it was reported the occurrence of subacute neonicotinoid intoxication through daily consumption of contaminated food in Japan, country were chlorinated neonicotinoids are used for a wide variety of crops, and the maximum residual limit is set higher than in other countries, especially for fruits and tea leaves (Marfo et al., 2015). These authors detected high levels of neonicotinoid metabolites in the urine of patients who showed typical clinical symptoms of neonicotinoid intoxication such as headache, general fatigue, palpitation/chest pain, abdominal pain, muscle pain/weakness/spasm, cough, finger tremor, and short-term memory loss. The detection of *N*-desmethyl-acetamiprid was the most frequent and highest in typical symptomatic group, followed by in atypical symptomatic group and in non-symptomatic group (6.0%, 2.2 μg/mL). Thiamethoxam was detected in typical and atypical symptomatic groups. Nitenpyram and clothianidin were detected in typical symptomatic and in non-symptomatic groups and thiacloprid was detected in atypical symptomatic group. The cases in typical symptomatic group with detection of *N*-desmethyl-acetamiprid and thiamethoxam were aged 5 to 62 years and 13 to 62 years, respectively and its detection was associated with increased prevalence of the symptoms.

5. Remark conclusions

There have been scarce studies of neonicotinoid-induced toxicity in the nervous systems of vertebrates, and these studies were conducted with only a few of neonicotinoids. A series of acute neurotoxicity studies were performed with imidacloprid, acetamiprid, thiamethoxam and thiacloprid, and showed common clinical signs such as tremors, decreased motility or reactivity, poor reflexes and prostration. In oral subacute neurotoxicity studies exposed to imidacloprid produced significant decreased in spontaneous locomotor activity and stimulates pain threshold along with pathological changes in the brain; AChE, ATPase and serum biochemical parameters such as creatine kinase, lactate dehydrogenase, sorbitol dehydrogenase and alkaline phosphatase levels were significantly decreased as result of imidacloprid exposure. In a 90-days oral toxicity study, female rats were administered imidacloprid the clinical chemistry revealed at high-dose of imidacloprid significant elevation of serum GOT, GPT, glucose and BUN and an inhibition (dose-dependent) of AChE activity in brain and serum respectively. The spontaneous locomotor activity was also decreased at high-dose exposure. The brain, liver and kidney of rats exposed with high-dose of imidacloprid had showed mild pathological changes. Moreover, imidacloprid produced necrosis of Purkinje cells with loss of dendrites and granules in the granular layer of the *cerebellum* in female rats. Maternal exposure to a large sublethal doses of imidacloprid during gestation produced significant sensorimotor impairments in the offspring on postnatal day 30, associated with increased AChE activity in the midbrain, cortex and brainstem and in plasma. There was an increase in glial fibrillary acidic protein immunostaining in motor cortex layer III, CA1, CA3, and the dentate gyrus subfield of the hippocampus. The exposure of neuron-enriched culture from neonatal rat *cerebellum* to nicotine and two neonicotinoids acetamiprid and imidacloprid produced a slight disturbance in Purkinje cell dendritic arborisation. Chronic neonicotinoid exposure alter the analysis of total transcriptome of the developing mammalian brain in a similar way to nicotine exposure. A developmental neurotoxicity screening study evidentiate that imidacloprid produced neurotoxic effects on the pups at maternal dose levels and higher. The prenatal exposure to acetamiprid produces significant neurodevelopmental toxicity and microglial activation in the mouse neocortex. Exposition to acetamiprid, imidacloprid or nicotine from postnatal day 12 to 26 impairs neurogenesis and alters microglial profiles in the developing hippocampal dentate gyrus in mice. Besides, has been observed that neonicotinoid-exposed newborns might have affected neuronal cells in the hippocampal dentate gyrus, possibly resulting from the promotion of neuronal cell death in the subgranular zone and granule cell layer of the hippocampal dentate gyrus. Acetamiprid rats exposed from gestation day 6 through lactation day 21 daily was observed at the higher dose level maternal toxicity and deficits in auditory startle response in neonatal animals (both sexes). With clothianidin administered orally there was some indication for developmental neurotoxicity in rats, after weaning female offspring exhibited reduced motor activity on post-natal day 62.

Due to their high neurotoxicity for insects (target species), neonicotinoids are extremely effective in controlling certain pests. Imidacloprid acts on several types of post-synaptic nicotinic ACh receptors located only within the CNS of insects. The impact of neonicotinoid insecticides on insect pollinators (non-target species) is highly controversial. When neonicotinoids were initiated to be used, beekeepers started describing different disorders and clinical signs (i.e., bees not returning to the hive, disoriented bees, bees gathered close together in small groups on the ground, abnormal foraging behavior, massive bee losses in spring, and queen losses). Sublethal concentrations of neonicotinoids alter the behavior of social bees and reduce survival of entire colonies and impair the olfactory memory and learning capacity of honeybees and the orientation and foraging activity. In January 2013, the EFSA published three scientific opinion on the risks posed by the three neonicotinoids (clothianidin, thiamethoxam, and imidacloprid), examined lethal as well as sub-lethal effects on honeybees; it concluded that the insecticides pose *"high acute"* risks for bees. The same year, the Commission Implementing Regulation (EU) No 485/2013 forbids the use of clothianidin, thiamethoxam and imidacloprid on crops attractive to honeybees. The regulation forbids their use as seed treatment, soil treatment or foliar application for the following crops: maize, oilseed rape, soya, barley, millet, oats, rice, rye, sorghum and wheat and stipulates that, within 2 years of its entry into force, the European Commission will initiate a review of new scientific information on the pesticides. The Commission will then decide whether it is most appropriate to lift the ban, prolong it temporarily, or make it permanent. Recently, has been published, the Commission Implementing Regulation (EU) 2018/783 of 29 May 2018 amending Implementing Regulation (EU) No 540/2011 (EU, 2018) as regards the conditions of approval of the active substance imidacloprid. This regulation established that seeds treated with plant protection products containing imidacloprid shall not be placed on the market or used, except where: (a) the seeds are intended to be used only in permanent greenhouses; and (b) the resulting crop stays within a permanent greenhouse during its entire life cycle.

Acknowledgments

This work was supported by Project S2013/ABI-2728 (ALIBIRD-CM Program) from Comunidad de Madrid, and by Project Ref. RTA2015-00010-C03-03 from Ministerio de Economía, Industria y Competitividad, Spain.

References

Abou-Donia, M.B., Goldstein, L.B., Jones, K.H., Abdel-Rehman, A.A., Damodran, T.V., Dechkovskaia, A.M., Bullman, S.L., Amir, B.E., Khan, W.A., 2001. Effect of daily dermal application of DEET and permethrin, alone and in combination, on sensorimotor performance, blood brain barrier, and blood-tests barrier in rat. J. Toxicol. Environ. Health A 62, 101–119.

Abou-Donia, M.B., Goldstein, L.B., Bullman, S., Tu, T., Khan, W.A., Dechkovskaia, A.M., et al., 2008. Imidacloprid induces neurobehavioral deficits and increases expression of glial fibrillary acidic protein in the motor cortex and hippocampus in offspring rats following in utero exposure. J. Toxicol. Environ. Health 71, 119–130.

Agarwal, R., Srinivas, R., 2008. Severe neuropsychiatric manifestations and rhabdomyolysis in a patient with imidacloprid poisoning. Am. J. Emerg. Med. 25, 844–856.

Anadón, A., 2018. Neonicotinoid toxicity: overview including societal issues. J. Vet. Pharmacol. Ther. 41 (Suppl. 1), 35–36.

Arslan, M., Sevgiler, Y., Buyukleyla, M., Yardimci, M., Yilmaz, M., Rencuzogullari, E., 2015. Sex-related effects of imidacloprid modulated by piperonyl butoxide and menadione in rats. Part II: genotoxic and cytotoxic potential. Drug Chem. Toxicol. 39 (1), 81–86.

Barbieri, R.F., Lester, P.J., Miller, A.S., Ryan, K.G., 2013. A neurotoxic pesticide changes the outcome of aggressive interactions between native and invasive ants. Proc. R. Soc. B Biol. Sci. 280 (1772), 20132157.

Benzidane, Y., Goven, D., Abd-Ella, A.A., Deshayes, C., Lapied, B., Raymond, V., 2017. Subchronic exposure to sublethal dose of imidacloprid changes electrophysiological properties and expression pattern of nicotinic acetylcholine receptor subtypes in insect neurosecretory cells. Neurotoxicology 62, 239–247.

Bhardwaj, S., Srivastava, M.K., Kapoor, U., Srivastava, L.P., 2010. A 90 days oral toxicity of imidacloprid in female rats: morphological, biochemical and histopathological evaluations. Food Chem. Toxicol. 48, 1185–1190.

Bodereau-Dubois, B., List, O., Calas-List, D., Marques, O., Communal, P.Y., Thany, S.H., Lapied, B., 2012. Transmembrane potential polarization, calcium influx, and receptor conformational state modulate the sensitivity of the imidacloprid-insensitive neuronal insect nicotinic acetylcholine receptor to neonicotinoid insecticides. J. Pharmacol. Exp. Ther. 341, 326–339.

Bromilow, R.H., Chamberlain, K., Evans, A.A., 1990. Physicochemical aspects of phloem translocation of herbicides. Weed Sci 38, 305–314.

Brown, L.A., Ihara, M., Buckingham, S.D., Matsuda, K., Sattelle, D.B., 2006. Neonicotinoid insecticides display partial and super agonist actions on native insect nicotinic acetylcholine receptors. J. Neurochem. 99, 608–615.

Chao, L.S., Casida, J.E., 1997. Interaction of imidacloprid metabolites and analogs with the nicotinic acetylcholine receptor of mouse brain in relation to toxicity. Pestic. Biochem. Physiol. 58, 77–88.

Corringer, J.-P., Le Novère, N., Changeux, J.-P., 2000. Nicotinic receptors at the amino acid level. Annu. Rev. Pharmacol. Toxicol. 40, 431–458.

Cotti, C., Clementi, F., 2004. Neuronal nicotinic receptors: from structure to pathology. Prog. Neurobiol. 74, 363–396.

Crosby, E.B., Bailey, J.M., Oliveri, A.N., Levin, E.D., 2015. Neurobehavioral impairments caused by developmental imidacloprid exposure in zebrafish. Neurotoxicol. Teratol. 49, 81–90.

Dani, J.A., Biasi, M., 2001. Cellular mechanisms of nicotine addiction. Pharmacol. Biochem. Behav. 70, 439–446.

de Oliveira, I.M., Nunes, B.V., Barbosa, D.R., Pallares, A.M., Faro, L.R., 2010. Effects of the neonicotinoids thiametoxam and clothianidin on in vivo dopamine release in rat striatum. Toxicol. Lett. 192, 294–297.

Decourtye, A., Devillers, J., 2010. Ecotoxicity of neonicotinoid insecticides to bees. Adv. Exp. Med. Biol. 683, 85–95.

Decourtye, A., Devillers, J., Cluzeau, S., Charreton, M., Pham-Delegue, M.H., 2004. Effects of imidacloprid and deltamethrin on associative learning in honeybees under semi-field and laboratory conditions. Ecotoxicol. Environ. Saf. 57, 410–419.

Deepu, D., George, I.A., Peter, J.V., 2007. Toxicology of the newer neonicotinoid insecticides: imidacloprid poisoning in a human. Clin. Toxicol. 45, 485–486.

Desneux, N., Decourtye, A., Delpuech, J.M., 2007. The sublethal effects of pesticides on beneficial arthropods. Annu. Rev. Entomol. 52, 81–106.

Didier, M., Berman, S.A., Lindstrom, J., Bursztajn, S., 1995. Characterization of nicotinic acetylcholine receptors expressed in primary cultures of cerebellar granule cells. Brain Res. Mol. Brain Res. 30, 17–28.

Dupuis, J., Louis, T., Gauthier, M., Raymond, V., 2012. Insights from honeybee (Apis mellifera) and fly (Drosophila melanogaster) nicotinic acetylcholine receptors: from genes to behavioral functions. Neurosci. Biobehav. Rev. 36, 1553–1564.

EC, 2015. Competent authority report of active substance imidacloprid evaluated by Germany concerning the making available on the market and use of biocidal product. https://echa.europa.eu/es/information-on-chemicals/biocidal-active-substances/-/disas/factsheet/37/PT18.

EC, 2018. Competent authority report of active substance acetamiprid evaluated by Belgium concerning the making available on the market and use of biocidal product. https://echa.europa.eu/es/information-on-chemicals/biocidal-active-substances/-/disas/factsheet/1235/PT18.

EC, 2008. Regulation (EC) No 1272/2008 of the European Parliament and of the Council of 16 December 2008 on Classification, Labelling and Packaging of Substances and Mixtures, Amending and Repealing Directives 67/548/EEC and 1999/45/EC, and Amending Regulation (EC) No 1907/2006.

EC, 2014. Regulation (EU) No 528/2012 concerning the making available on the market and use of biocidal products. Evaluation of active substances. Assessment Report. Clothianidin Product-type 18 (Insecticides, Acaricides and Products to control other Arthropods).

ECHA, 2013. CLH Report for Thiacloprid. Proposal for harmonised classification and labelling based on Regulation (EC) No 1272/2008 (CLP Regulation) by UK. https://echa.europa.eu/documents/10162/4cb7927b-3ce8-94de-dc29-a1e70578d4af.

ECHA, 2015. Committee for Risk Assessment. Opinion proposing harmonised classification and labelling at EU level of Thiacloprid (2015). https://echa.europa.eu/documents/10162/52bd518e-4cda-0115-cbbc-038f1e462d5c.

ECHA, 2018a. CLH Report for Imidacloprid. Proposal for harmonised classification and labelling based on Regulation (EC) No 1272/2008 (CLP Regulation) by Germany. https://echa.europa.eu/documents/10162/fee1fda7-1b3b-c226-19de-e87ed5f6711f.

ECHA, 2018b. CLH Report for Thiamethoxam, Proposal for harmonised classification and labelling based on Regulation (EC) No 1272/2008 (CLP Regulation) by France and Spain. https://echa.europa.eu/documents/10162/b3750dbe-7250-26f0-a94c-9becb622a07c.

EFSA, 2012. European food safety authority: scientific opinion on the science behind the development of a risk assessment of plant protection products on bees (*Apis mellifera*, *Bombus* spp. and solitary bees). EFSA J. 10, 2668. https://doi.org/10.2903/j.efsa.2012.2668.

EFSA, 2013. Scientific opinion on the developmental neurotoxicity potential of acetamiprid and imidacloprid. EFSA J. 11 (12), 3471.

EFSA, 2013a. Conclusion on the peer review of the pesticide risk assessment for bees for the active substance thiamethoxam. EFSA J. 11 (1), 3067. 2013.

EFSA, 2013b. Conclusion on the peer review of the pesticide risk assessment for bees for the active substance imidacloprid. EFSA J. 11 (1), 3068. 2013.

EFSA, 2013c. Conclusion on the peer review of the pesticide risk assessment for bees for the active substance clothianidin. EFSA J. 11 (1), 3066. 2013.

EFSA, 2018. Evaluation of the Data on Clothianidin, Imidacloprid and Thiamethoxam for the Updated Risk Assessment to Bees for Seed Treatments and Granules in the EU. Technical report. APPROVED: 01 February 2018.

Elbert, A., Haas, M., Springer, B., Thielert, W., Nauen, R., 2008. Applied aspects of neonicotinoid uses in crop protection. Pest Manag. Sci. 64, 1099–1105.

EU, 2013. Commission Implementing Regulation (EU) No 485/201 of 24 May 2013 amending Implementing Regulation (EU) No 540/2011, as regards the conditions of approval of the active substances clothianidin, thiamethoxam and imidacloprid, and prohibiting the use and sale of seeds treated with plant protection products containing those active substances (Official Journal of the European Union L 139/12, 25.5.2013).

EU, 2018. Commission Implementing Regulation (EU) 2018/783 of 29 May 2018 Amending Implementing Regulation (EU) No 540/2011 as Regards the Conditions of Approval of the Active Substance Imidacloprid (Official Journal of the European Union L 132/31, 30.5.2018).

Faro, L.R., Oliveira, I.M., Duran, R., Alfonso, M., 2012. In vivo neurochemical characterization of clothianidin induced striatal dopamine release. Toxicology 302, 197–202.

Feltham, H., Park, K., Goulson, D., 2014. Field realistic doses of pesticide imidacloprid reduce bumblebee pollen foraging efficiency. Ecotoxicology 23, 317–323.

Ffrench-Constant, R.H., Williamson, M.S., Davies, T.G., Bass, C., 2016. Ion channels as insecticide targets. J. Neurogenet. 30, 163–177.

Fischer, J., Müller, T., Spatz, A.-K., Greggers, U., Grünewald, B., Menzel, R., 2014. Neonicotinoids interfere with specific components of navigation in honeybees. PLoS One 9, e91364.

Ford, K.A., Casida, J.E., 2006. Chloropyridinyl neonicotinoid insecticides: diverse molecular substituents contribute to facile metabolism in mice. Chem. Res. Toxicol. 19, 944–951.

Ford, K.A., Casida, J.E., 2006a. Unique and common metabolites of thiamethoxam, clothianidin, and dinotefuran in mice. Chem. Res. Toxicol. 19, 1549–1556.

Gibbons, D., Morrissey, C., Mineau, P., 2015. A review of the direct and indirect effects of neonicotinoids and fipronil on vertebrate wildlife. Environ. Sci. Pollut. Res. 22 (1), 103–118.

Girod, R., Crabtree, G., Ernstrom, G., Ramirez-Latorre, J., McGehee, D., Turner, J., Role, L., 1999. Heteromeric complexes of alpha 5 and/or alpha 7 subunits. Effects of calcium and potential role in nicotine-induced persynaptic facilitation. Ann. N. Y. Acad. Sci. 868, 578–590.

Godfray, H.C., Blacquière, T., Field, L.M., Hails, R.S., Petrokofsky, G., Potts, S.G., Raine, N.E., Vanbergen, A.J., McLean, A.R., 2014. A restatement of the natural science evidence base concerning neonicotinoid insecticides and insect pollinators. Proc. Biol. Sci. 281, 20140558.

Gotti, C., et al., 2007. Heterogeneity and complexity of native brain nicotinic receptors. Biochem. Pharmacol. 74, 1102–1111.

Hassan-Puttaswamy, V., Adams, D.J., Kini, R.M., 2015. A distinct functional site in omega-neurotoxins: novel antagonists of nicotinic acetylcholine receptors from snake venom. ACS Chem. Biol. 10 (12), 2805–2815.

Huang, N., Lin, S., Chou, C., Hung, Y., Chung, H., Huang, S., 2006. Fatal ventricular fibrillation in a patient with acute imidacloprid poisoning. Am. J. Emerg. Med. 24 (7), 883–885.

Imamura, T., Yanagawa, Y., Nishikawa, K., Matsumoto, N., Sakamoto, T., 2010. Two cases of acute poisoning with acetamiprid in humans. Clin. Toxicol. 48, 851–853.

Jeschke, P., Nauen, R., Schindler, M., Elbert, A., 2011. Overview of the status and global strategy for neonicotinoids. J. Agric. Food Chem. 59, 2897–2908.

Jones, A.K., Sattelle, D.B., 2010. Diversity of insect nicotinic acetylcholine receptor subunits. Adv. Exp. Med. Biol. 683, 25–43.

Jones, A.K., Brown, L.A., Sattelle, D.B., 2007. Insect nicotinic acetylcholine receptor gene families: from genetic model organism to vector, pest and beneficial species. Invert. Neurosci. 7, 67–73.

Julvez, J., Méndez, M., Fernandez-Barres, S., Romaguera, D., Vioque, J., Llop, S., Ibarluzea, J., Guxens, M., Avella-Garcia, C., Tardón, A., Riaño, I., Andiarena, A.,

Robinson, O., Arija, V., Esnaola, M., Ballester, F., Sunyer, J., 2016. Maternal consumption of seafood in pregnancy and child neuropsychological development: a longitudinal study based on a population with high consumption levels. Am. J. Epidemiol. 183 (3), 169–182.

Kagawa, N., Nagao, T., 2018. Neurodevelopmental toxicity in the mouse neocortex following prenatal exposure to acetamiprid. J. Appl. Toxicol. 38, 1521–1528.

Karlin, A., 2002. Emerging structure of the nicotinic acetylcholine receptors. Nat. Rev. Neurosci. 3, 102–114.

Kessler, S.C., Tiedeken, E.J., Simcock, K.L., Derveau, S., Mitchell, J., Softley, S., Stout, J.C., Wright, G.A., 2015. Bees prefer foods containing neonicotinoid pesticides. Nature 521 (7550), 74–76.

Kimura-Kuroda, J., Nishito, Y., Yanagisawa, H., Kuroda, Y., Komuta, Y., Kawano, H., Hayashi, M., 2016. Neonicotinoid insecticides alter the gene expression profile of neuron-enriched cultures from neonatal rat cerebellum. Int. J. Environ. Res. Public Health 13, 987.

Lansdell, S.J., Millar, N.S., 2000. The influence of nicotinic receptor subunit composition upon agonist, alpha-bungarotoxin and insecticide (imidacloprid) binding affinity. Neuropharmacology 39, 671–679.

Laurino, D., Porporato, M., Patetta, A., Manino, A., 2011. Toxicity of neonicotinoid insecticides to honey bees: laboratory tests. Bull. Insectol. 64 (1), 107–113.

Leicht, W., 1993. Imidacloprid: a chloronicotinyl insecticide. Pestic. Outlook 4 (3), 17–21.

Lin, P.C., Lin, H.J., Liao, Y.Y., Guo, H.R., Chen, K.T., 2013. Acute poisoning with neonicotinoid insecticides: a case report and literature review. Basic Clin. Pharmacol. Toxicol. 112, 282–286.

Lindstrom, J., 2000. The structure of neuronal nicotinic receptors. In: Clementi, F., Fornasari, D., Gotti, C. (Eds.), Neuronal nicotinic receptors. Handbook of experimental pharmacology. In: vol. 144. Springer, Berlin Heidelberg New York, pp. 101–162.

Lonare, M., Kumar, M., Raut, S., Badgujar, P., Doltade, S., Telang, A., 2014. Evaluation of imidacloprid-induced neurotoxicity in male rats: a protective effect of curcumin. Neurochem. Int. 78, 122–129.

Marfo, J.T., Fujioka, K., Ikenaka, Y., Nakayama, S.M., Mizukawa, H., Aoyama, Y., Ishizuka, M., Taira, K., 2015. Relationship between urinary N-Desmethyl-Acetamiprid and typical symptoms including neurological findings: a prevalence case-control study. PLoS One 10, e0142172.

Marletto, F., Patetta, A., Manino, A., 2003. Laboratory assessment of pesticide toxicity to bumblebees. Bull. Insectol. 56 (1), 155–158.

Matsuda, K., Sattelle, D.B., 2005. New discoveries in agrochemicals. In: ACS Symposium Series. 892, p.172.

McGehee, D.S., Role, L.W., 1995. Physiological diversity of nicotinic acetylcholine receptors expressed by vertebrate neurons. Annu. Rev. Physiol. 57, 521–546.

Millar, N.S., Lansdell, S.J., 2010. Characterisation of insect nicotinic acetylcholine receptors by heterologous expression. Adv. Exp. Med. Biol. 683, 65–73.

Moser, V.C., 1989. Screening approaches to neurotoxicity: a functional observation battery. J. Am. Coll. Toxicol. 8, 85–93.

Nagata, K., Song, J.H., Shono, T., Narahashi, T., 1998. Modulation of the neural nicotinic acetycholine receptor channel by the nitromethylene heterocycle imidacloprid. J. Pharmacol. Exp. Ther. 285, 731–738.

Nakayama, A., Yoshida, M., Kagawa, N., Nagao, T., 2019. The neonicotinoids acetamiprid and imidacloprid impair neurogenesis and alter the microglial profile in the hippocamal dentete gyrus of mouse neonates. J. Appl. Toxicol. 39, 877–887.

Nauen, R., Tietjen, K., Wagner, K., Elbert, A., 1998. Efficacy of plant metabolites of imidacloprid against *Myzus persicae* and *Aphis gossypii* (Homoptera: Aphididae). Pestic. Sci. 52, 53–57.

Ozdemir, H.H., Kara, M., Yumrutas, O., Uckardes, F., Eraslan, E., Demir, C.F., Bal, R., 2014. Determination of the effects on learning and memory performance and related gene expressions of clothianidin in rat models. Cogn. Neurodyn. 8, 411–416.

Palmer, M.J., Moffat, C., Saranzewa, N., Harvey, J., Wright, G.A., Connolly, C.N., 2013. Cholinergic pesticides cause mushroom body neuronal inactivation in honeybees. Nat. Commun. 4, 1634.

Phua, D., Lin, C., Wu, M., Deng, J.F., Yang, C.C., 2009. Neonicotinoid insecticides: an emerging cause of acute pesticide poisoning. Clin. Toxicol. 47, 336–341.

Raymond-Delpech, V., Matsuda, K., Sattelle, B.M., Rauh, J.J., Sattelle, D.B., 2005. Ion channels: molecular targets of neuroactive insecticides. Invertebr. Neurosci. 5, 119–133 (2005).

Rivest, S., 2009. Regulation of innate immune responses in the brain. Nat. Rev. Immunol. 9 (6), 429.

Rodrigues, K.J., Santana, M.B., Do Nascimento, J.L., Picanco-Diniz, D.L., Maues, L.A., Santos, S.N., Ferreira, V.M., Alfonso, M., Duran, R., Faro, L.R., 2010. Behavioural and biochemical effects of neonicotinoid thiamethoxam on the cholinergic system in rats. Ecotoxicol. Environ. Saf. 73, 101–107.

Schroeder, M.E., Flattum, R.F., 1984. The mode of action and neurotoxic properties of the nitromethylene heterocycle insecticides. Pestic. Biochem. Physiol. 22, 148–160.

Shadnia, S., 2007. Fatal intoxication with imidacloprid insecticide. Am. J. Emerg. Med. 26, 634.e1–634.e4.

Sheets, L.P., 2001. The neonicotinoids insecticides. In: Massaro, E.J. (Ed.), Handbook of Neurotoxicology. In: vol. 1. Human Press Inc, Totowa, NJ, pp. 79–87.

Sheets, L.P., 2010. Imidacloprid: a neonicotinoid insecticide. In: Hayes' Handbook of Pesticide Toxicology. Elsevier Inc., pp. 2055–2064. Chapter 95.

Sheets, L.P., Li, A.A., Minnema, D.J., Collier, R.H., Creek, M.R., Peffer, R.C., 2016. A critical review of neonicotinoid insecticides for developmental neurotoxicity. Crit. Rev. Toxicol. 46 (2), 153–190.

Taly, A., Corringer, P.-J., Guedin, D., Lestage, P., Changeux, J.P., 2009. Nicotinic receptors: allosteric transitions and therapeutic targets in the nervous system. Nat. Rev. Drug Discov. 8, 733–750.

Thyssen, J., Machemer, L., 1999. Imidacloprid: toxicology and metabolism. In: Yamamoto, I., Casida, J.E. (Eds.), Nicotinoid Insecticides and the Nicotinic Acetylcholine Receptor. Springer-Verlag, Tokyo, pp. 213–222. Chapter 9.

Tome, H.V.V., Martins, G.F., Lima, M.A.P., Campos, L.A.O., Guedes, R.N.C., 2012. Imidacloprid-induced impairment of mushroom bodies and behavior of the native stingless bee Melipona quadrifasciata anthidioides. PLoS One 7, e38406.

Tomizawa, M., Casida, J.E., 1999. Minor structural changes in nicotinoid insecticides confer differential subtype selectivity for mammalian nicotinic acetylcholine receptors. Br. J. Pharmacol. 127, 115–122.

Tomizawa, M., Casida, J.E., 2002. Desnitro-imidacloprid activates the extracellular signal-regulated kinase cascade via the nicotinic receptor and intracellular calcium mobilization in N1E-115 cells. Toxicol. Appl. Pharmacol. 184, 180–186.

Tomizawa, M., Casida, J.E., 2003. Selective toxicity of neonicotinoids attributable to specificity of insect and mammalian nicotinic receptors. Annu. Rev. Entomol. 48, 339–364.

Tomizawa, M., Casida, J.E., 2005. Neonicotinoid insecticide toxicology: mechanisms of selective action. Annu. Rev. Pharmacol. Toxicol. 45, 247–268.

Tomlin, C., 2003. The Pesticide Manual, 13th ed. Br. Crop Protection Council, Alton, Hampshire, UK, p. 1344.

Unwin, N., 1993. Nicotinic acetylcholine receptor at 9 A resolution. J. Mol. Biol. 299, 1101–1124.

van der Sluijs, J.P., Simon-Delso, N., Goulson, D., Maxim, L., Bonmatin, J.-M., Belzunces, L.P., 2013. Neonicotinoids, bee disorders and the sustainability of pollinator services. Curr. Opin. Environ. Sustain. 5, 293–305.

Vinod, K.V., Srikant, S., Thiruvikramaprakash, G., Dutta, T.K., 2015. A fatal case of thiacloprid poisoning. Am. J. Emerg. Med. 33, 310–316.

Wang, X., Anadón, A., Wu, Q., Qiao, F., Ares, I., Martínez-Larrañaga, M.R., Yuan, Z., Martínez, M.A., 2018. Mechanism of neonicotinoid toxicity: impact on oxidative stress and metabolism. Annu. Rev. Pharmacol. Toxicol. 58, 471–507.

Whitehorn, P.R., O'Connor, S., Wackers, F.L., Goulson, D., 2012. Neonicotinoid pesticide reduces bumble bee colony growth and queen production. Science 336, 351–352.

Williamson, S.M., Wright, G.A., 2013. Exposure to multiple cholinergic pesticides impairs olfactory learning and memory in honeybees. J. Exp. Biol. 216, 1799–1807.

Wismer, T., 2004. Novel insecticides. In: Plumlee, K.H. (Ed.), Clinical Veterinary Toxicology. Mosby, St. Louis, MO, pp. 184–185.

Wu, I., Lin, J., Cheng, E., 2001. Acute poisoning with the neonicotinoid insecticide imidacloprid in N-methyl pyrrolidone. Clin. Toxicol. 39 (6), 617–621.

Yamamoto, I., Tomizawa, M., Saito, T., Miyamoto, T., Walcott, E.C., Sumikawa, K., 1998. Structural factors contributing to insecticidal and selective actions of neonicotinoids. Arch. Insect Biochem. Physiol. 37, 24–32.

Yeh, I.J., Lin, T.J., Hwang, D.Y., 2010. Acute multiple organ failure with imidacloprid and alcohol ingestion. Am. J. Emerg. Med. 28 (255), e1–e3.

Further reading

EU, 2014. Competent authority report of active substance clothianidin evaluated by Germany concerning the making available on the market and use of biocidal product. https://echa.europa.eu/es/information-on-chemicals/biocidal-active-substances/-/disas/factsheet/15/PT18.

Sano, K., Isobe, T., Yang, J., Win-Shwe, T.T., Yoshikane, M., Nakayama, S.F., Kawashima, T., Suzuki, G., Hashimoto, S., Nohara, K., Tohyama, C., Maekawa, F., 2016. *In utero* and lactational exposure to acetamiprid induces abnormalities in socio-sexual and anxiety-related behaviors of male mice. Front. Neurosci. 10, 228.

Shadnia, S., Moghaddam, H.H., 2008. Fatal intoxication with imidacloprid insecticide. Am. J. Emerg. Med. 26 (5), 634.e1–634.e4.

CHAPTER FIVE

Rotenone neurotoxicity: Relevance to Parkinson's disease

Vivek Lawana[a,b], Jason R. Cannon[a,b,*]
[a]School of Health Sciences, Purdue University, West Lafayette, IN, United States
[b]Purdue Institute for Integrative Neuroscience, Purdue University, West Lafayette, IN, United States
*Corresponding author: e-mail address: cannonjr@purdue.edu

Contents

Advances in Neurotoxicology, Volume 4
ISSN 2468-7480
https://doi.org/10.1016/bs.ant.2019.11.004

1. Background

1.1 Historical use

Specific plants produce rotenone, an isoflavone, as a naturally evolved phenomenon to protect against insects. It is specifically produced in the leaves, seeds and stem on Mexican turnip (*Pachyrhizus erosus*) plants, also known as Jicama vine plants; and roots on Leguminosae family of plants belonging to genus *Derris*. The first reported discovery of rotenone's insecticide effects dates back to 1840's, where it was shown that caterpillars eating rotenone-producing plant leaf shown immediate lethal effects (Metcalf, 1948). A Science Services (1933) (Washington, DC) report noted that these plants have been used for centuries as fish poison (piscicide) by natives of tropical countries of East Indies to harvest fish. In addition, rotenone is also present in South American plant called "Cubé" (*Lonchocarpus utilis*) and a North American plant, *Nolina lindheimeriana*, commonly known as "Devil's shoestring." In 1895, a French botanist Emmanuel Geoffroy successfully extracted the toxic chemical from these *Lonchocarpus nicou* plant and named it nicouline (Turner, 1932). Later in 1902, a Japanese botanist, isolated similar compound from *Derris elliptica* plant, crystalized it's pure form and named it rotenone based on plants Japanese name *roten* (Turner, 1932).

1.2 Commercial uses

The WHO categorizes rotenone as a moderately safe pesticide (WHO-IPCS, 1992). Rotenone was one of the most widely used pesticide, insecticide and piscicide around the world. Rotenone was first registered in 1947 in the USA. In 1952, US approved the use of rotenone to control the population of fish in rivers and lakes. In the USA, Prentiss Incorporated, Foreign Domestic Chemicals Corporation and Tifa International LLC produces and sells rotenone for large-scale use as a pesticide and fish toxin. Following several concerning reports indicating potential neurological effects of rotenone, the widely used rotenone as an agriculture pesticide was dropped since 1990 and was voluntarily stopped by all three manufacturers in year 2006, when rotenone was included in Restricted Use Pesticide (RUP) category by the USEPA. However, it is still used to control fish population in the USA. Being a non-specific fish toxin, rotenone is used only in lakes and river where there are no known endangered species. A USEPA (2007) report determined that the maximum rotenone concentrations up to 50 ppb in river and 250 ppb in lakes, reservoirs and ponds are allowed. According to these guidelines, rotenone must be sold only to Certified Applicators and should only be used in private or public bodies with clear notice to common public.

1.3 Chemical properties

Rotenone (CAS # 83-79-4, molecular formula $C_{23}H_{22}O_6$, molecular weight 394.4 g/mol), chemically termed as (2R,6aS,12aS)-1,2,6,6a,12,12a-hexahydro-2-isopropenyl-8,9-dimethoxychromeno[3,4-*b*]furo(2,3-h)chromen-6-one, belongs to a larger class of compounds called rotenoids and is a colorless, odorless chemical with a melting point of 166 °C. In plants, rotenone is degraded by photodecomposition resulting in relatively nontoxic metabolites including rotenone, dehydrorotenone, dimethoxy salicylic acid, rissic acid and tubaic acid (Cheng et al., 1972), while in soil, rotenone is photodecomposed to hydroxyrotenone with 50% disappearance time of 5–7 h (Cavoski et al., 2007). In natural water systems, rotenone is decomposed mainly by hydrolysis, but also by oxidation and photodecomposition, to less toxic compounds. The rate of decomposition in water varies depending on temperature, pH and hardness of water (Finlayson et al., 2001). For instance, in a natural pond, rotenone half-life is calculated to be of ~12 h at water temperature of about 24 °C and dramatically increases to 84 h at 0 °C (Dawson et al., 1991). Since the rotenone half-life in water (especially at higher altitudes and lower temperatures) is much longer than needed to kill fish population, excess rotenone is oxidized by potassium permanganate to lesser toxic compounds (Finlayson et al., 2001).

The log *P* value rotenone is reported to be ~4.10 indicating high lipophilicity contributing to efficacy in passing biological barriers (Hansch, 1995). Although lipophilic, bioaccumulation through food-chain is limited due to half-lives of ~<1 day (Robertson and Smith-Vaniz, 2008). Even more, reports suggest that humans consuming fish exposed to rotenone have no notable symptoms of toxicity, which might be due to lower bioaccumulation and high degradation during high temperature cooking (Robertson and Smith-Vaniz, 2008).

The structure of rotenone, as shown in Fig. 1, was first presented by Laforge et al. (1933). The structure shows that rotenone contains a central dihydro-γ-pyrone ring with dihydrobenzopyran system on left and dihydrobenzofuran system on right. The central hexagonal rings contribute to rotenone-mediated mitochondrial toxicity.

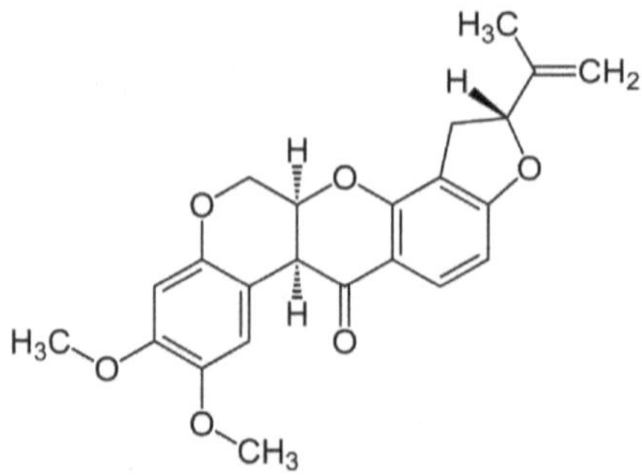

Fig. 1 Chemical structure of rotenone showing three-ring center. The structure was drawn using an online ChemDraw chemical drawing software by PerkinElmer.

1.4 Rotenone toxicity: Summary and introduction to selective neurotoxicity

While previously treated as an organic treatment, more recently, rotenone use has been regulated. Commercial rotenone products are classified as RUP due to known acute inhalation, acute oral and severe aquatic toxicity, resulting in classification as a category IV toxicant (USEPA, 2007). Rotenone has less than 10% absorption through skin, hence, to exert severe, systemic toxicity in humans, it must be supplied directly into the body (oral, inhalation, subcutaneous or intravenous routes). The US environmental protection agency (EPA) reports 40 ppb levels of rotenone in drinking water to be a level of concern for susceptible populations (infants, children, elderly and pregnant females). The most common symptoms of rotenone toxicity are eye and dermal irritation, throat irritation, nausea and coughing. Rotenone induced reproductive defects may occur at dose 8–10-times higher than NOAEL (0.5 mg/kg/day, 1–30 days). However, it is unlikely for humans to unintentionally get exposed to such high levels. Since it has very poor bioaccumulation, a human would need to consume ~274,000 perch from rotenone-treated lakes to develop acute toxicity (USEPA, 2007).

Rotenone exerts toxicity in humans and other animals by inhibiting mitochondrial respiration. In this context, Bougria et al. (1995) showed that mitochondrial IC_{50} of rotenone is $2.45 \pm 0.8 \times 10^{-10}$ M for mitochondrial extracted from rat striatum, $3.5 \pm 1.6 \times 10^{-10}$ M for cortical mitochondria and $2.3 \pm 1.3 \times 10^{-10}$ M for hepatic mitochondria in vitro. Due to high lipophilicity, rotenone can rapidly cross the blood brain barrier (Ravenstijn et al., 2008). However, early in vivo study with systemic rotenone exposure of a high dose (10–18 mg/kg/day) failed to develop of PD pathology but caused non-specific lesions in brain and systematically inhibited mitochondrial complex I activity (Ferrante et al., 1997). In contrast, an earlier study in rats showed that single exposure of very high dose (5 mM local concentration, ~500,000-fold higher than the complex I IC_{50}) directly to parenchyma depleted striatal levels of dopamine and its metabolites (Heikkila et al., 1985). While earlier studies had failed to provide a convincing link to PD relevant pathology, a groundbreaking report in rats at far lower doses (below that which inhibit ATP production), showed that systemic rotenone treatment produced selective loss of dopamine neurons in the Substantia nigra (SN) and aggregation of α-synuclein, similar to Lewy bodies; both of which are hallmarks of clinical PD (Betarbet et al., 2000). The models has been since supported by numerous publications from the original lab as well as many others (Aksoy et al., 2017; Cannon et al., 2009; Kaur et al., 2011; Sherer et al., 2003a; Xiong et al., 2013; Zhang et al., 2017).

2. Risk for Parkinson's disease: Intersection between epidemiology and laboratory studies

First registered in the USA in 1947, rotenone was widely used both in agriculture and in households until 2006, when its registration was withdrawn (USEPA, 2007). Due to its broad-spectrum use under various brand names, the epidemiological assessment of its disease risk is challenging. A study performed by Dhillon et al. (2008) in eastern Texas revealed that the use of rotenone for household gardening purposes increased the risk of PD (OR = 10.9; 95% confidence interval (CI) 2.9–34.3; $P < 0.05$). A case-control study of 84,000 agriculture participants from North Carolina and Iowa showed a strong correlation (OR = 2.5; 95% CI 1.3–4.7; $P < 0.05$) between rotenone exposure and PD (Tanner et al., 2011). To the contrary, a different case-control survey from Iowa or North Carolina showed marginal risk of rotenone associated to PD (OD = 1.7; 95% CI 0.6–4.7) (Kamel et al., 2007).

Interestingly, a recent case-control study reported a striking difference between males and female cases of PD exposed to rotenone. Following an in vivo rat study that clearly identified increased susceptibility in males (De Miranda et al., 2019), Goldman and Tanner (2019) showed that men were at higher risk of rotenone-induced PD (OR 2.8; 95% CI 1.4–5.7) compared to females (OR 1.1; 95% CI 0.3–4.3). Future studies are needed to better understand the different mechanisms that varyingly alter the dopaminergic toxicity in males versus in females, though it is important to point out that the recent study in rats suggest that differences in inflammation, protein aggregation and iron transport may influence sex-specific sensitivity (De Miranda et al., 2019). Taken together, epidemiology studies in general support a link between rotenone exposure and PD. Such studies have been both prompted by and supported by a strong in vivo model literature.

3. Experimental PD models and mechanisms of neurotoxicity

3.1 General overview

Rotenone is now a widely used toxin in a variety of animal models to produce PD-relevant pathology and behavioral phenotypes phenotype (Cannon and Greenamyre, 2010). The primary mechanism by which rotenone produces selective dopaminergic neurotoxicity is through complex I inhibition followed by oxidative stress generation and resultant oxidative damage (Betarbet et al., 2000; Sanders and Greenamyre, 2013). While multiple complex I inhibitors, notably MPTP, have been used to model PD, there are some key differences, especially with respect to cellular uptake. Due to its lipophilic structure, rotenone readily penetrates in to the brain (Greenamyre et al., 2010) where it binds to and inhibits the mitochondrial complex I of the ETC (Cabeza-Arvelaiz and Schiestl, 2012; Choi et al., 2011; Sanders and Greenamyre, 2013). Other models such as MPTP selectively accumulate in dopamine neurons through the dopamine transporter, inhibiting complex I only in dopaminergic neurons (Cannon and Greenamyre, 2010, 2011). This is an important distinction because the rotenone model tests the hypothesis that dopamine neurons exhibit heightened sensitivity to complex I inhibition relative to both non-neuronal and non-dopaminergic neurons that also experience complex I inhibition. Importantly, this hypothesis was informed by human data showing that complex I inhibition occurs systemically in human patients (Parker et al., 1989; Schapira et al., 1989).

Rotenone specifically blocks electron transfer from complex I to ubiquinone, halting the production of ATP via NADH (Sanders and Greenamyre, 2013; Sherer et al., 2003b). In both cell culture and animal models, rotenone's interruption of the ETC significantly increases free radicals, resulting in the accumulation of ROS and oxidative stress (Marella et al., 2008; Sanders and Greenamyre, 2013; Sherer et al., 2003b). Rotenone treatment also depletes cellular antioxidant defense and increases lipid peroxidation in dopaminergic neurons (Chinta and Andersen, 2006; Xiong et al., 2012a).

Treatment with natural antioxidants has been shown to reduce the dopaminergic neurodegeneration induced by rotenone (Karuppagounder et al., 2013; Xiong et al., 2012a). Based on data from rotenone models and other complex I inhibitors an adverse outcome pathway has been developed, where toxin binding and inhibition are the primary initiators of selective dopaminergic neurotoxicity (Bal-Price et al., 2018).

3.2 Development of rotenone rat model

In 2000, researchers first reported the in vivo effects on rotenone in two different strains of rats (Betarbet et al., 2000). In this study, 2-month-old rats were exposed to various doses of rotenone via a subcutaneous mini-pump with direct cannulation to jugular vein, providing continuous slow release of rotenone. Higher doses (>7 mg/kg/day) demonstrated cardiovascular and systemic toxicity with mortality, while, 2–3 mg/kg/day dose for 1–5 weeks found to be optimal for inducing selective dopaminergic neurotoxicity. The study concluded that estimated tissue rotenone concentration of ~30 nM inhibited mitochondrial complex I in liver but not in brain in an acute exposure study. Acute rotenone treatment was shown to increase dopamine release in the presence of high glutamate without causing notable neurodegeneration (Leng et al., 2003). In contrast, rats exposed to 2.5–2.75 mg/kg/day rotenone showed progressive loss of nigral dopaminergic neurons and striatal terminals from day 7 to day 36 (Betarbet et al., 2000). Further, systemic rotenone treatment resulted in aberrant accumulation of cytoplasmic α-synuclein positive inclusions in nigral dopamine neurons, a key pathological hallmark of PD (Betarbet et al., 2000; Sherer et al., 2003c; Spillantini et al., 1997). Notably, no adverse effect on postsynaptic striatal neurons, indicating rotenone's selectivity for dopaminergic neurons in nigrostriatal system (Betarbet et al., 2000; Sherer et al., 2003c).

In rats, rotenone was found to be metabolized by CYP450 enzymes, especially by CYP3A4, to 12aβ-hydroxy analogues (Caboni et al., 2004). Chronic subcutaneous infusion of rotenone leads to brain accumulation of about 0.11 ppm after 14 days and doubles at 28 days, indicating that prolonged low dose exposure to the piscicide may lead to the increasing brain concentration. The study also reported similar trend for some of the toxic metabolites of rotenone, though levels of metabolites were much lower than rotenone itself. Interestingly, the rotenone concentration in brain directly correlated with extent of lesion and neurodegeneration. (Caboni et al., 2004).

Rats exposed to rotenone (1.5–2.5 mg/kg/day for 60 days) exhibit significant loss of dopaminergic catecholamine neurotransmitters, namely dopamine (DA) and its metabolites [dihydroxyphenylacetic acid (DOPAC) and homovanillic acid (HVA)], in striatum, with loss of tyrosine hydroxylase (TH), a marker for dopaminergic neurons, in SN lysates (Alam and Schmidt, 2002), which was ameliorated by co-treatment with tetrahydroisoquinoline treatment (Antkiewicz-Michaluk et al., 2003). Furthermore, rotenone inhibited the uptake of dopamine up on exposure to rat striatal synaptosomes (Maragos et al., 2002). In addition to neurodegeneration and cytoplasmic accumulation, some of the earlier studies reported time-dependent progressive locomotor behavioral deficits in rats treated with rotenone (Alam and Schmidt, 2002; Hoglinger et al., 2003). In addition, selective accumulation of inflammatory glial cells in SN region post rotenone exposure in rats has also been observed similar to human PD (Sherer et al., 2003a; Tansey and Goldberg, 2010).

In 2009, another key study demonstrated similar trend of rotenone-mediated dopaminergic toxicity using different age groups including young adults (3-month-old), adult (7-month-old) and middle-aged (12–14-month-old). In this study, investigators injected rotenone (2.75–3.00 mg/kg/day) via intraperitoneal route until the animals showed behavioral PD phenotype (limited movement, feeding, and/or grooming) (Cannon et al., 2009). The data showed that all age groups exposed to rotenone had moderate weight loss of 10–15% on the 7 days of exposure. In addition, all rotenone-treated animals showed time-dependent development of severe PD-like behaviors including bradykinesia, postural instability, rearing behavior, debilitating rigidity and some mortality, which were dopamine-dependent (apomorphine-responsive) (Cannon et al., 2009). Similar to previous findings, Cannon et al. (2009) reported chronic rotenone-induced striatal lesions in all age groups, which were specific to dopaminergic terminals as cell bodies of postsynaptic neurons appeared unaffected. The loss of dopamine terminals was further supported by reduced dopamine levels in striatal sections and loss of tyrosine hydroxylase (TH)-immunoreactive (IR) neurons in SN (Cannon et al., 2009). In addition, a recent report showed that chronic rotenone rat model exerts key electrophysiological characteristics of PD along with disease pathophysiological and behavioral impairments (von Wrangel et al., 2015). Taken together, the literature strongly supports that specific rat rotenone models reproduce the key biochemical, pathological, and behavioral aspects of clinical PD.

3.2.1 Controversies of rat rotenone model

It is worth noting that some studies have challenged the selectivity of the rat rotenone model and utility to model PD (Hoglinger et al., 2003). In contradiction to several studies from Greenamyre group and others (Betarbet et al., 2000; Cannon et al., 2009; Gao et al., 2003; Marella et al., 2008;

Mastroberardino et al., 2009; Monti et al., 2010; Saravanan et al., 2006; Sherer et al., 2003c; Tapias et al., 2010), Ravenstijn et al. (2008) claimed that chronic subcutaneous infusion of rotenone only exerts a minor nigral lesion and the nigrostriatal injury is not directly mediated by rotenone, but as a consequence of peripheral toxicity. The authors showed that unlike systemic treatment, intracerebral injection of rotenone causes more specific dopaminergic toxicity in nigrostriatal region without causing systemic toxicity, including weight loss. Intriguingly, in another study, the chronic rotenone treatment (2.5 mg/kg/day, 30 days) via intranasal route to mice or rats did not cause PD pathology or phenotype (Rojo et al., 2007). Contrary to proposed selectivity of rotenone as dopaminergic toxicant in rats (Betarbet et al., 2000; Cannon et al., 2009; Sherer et al., 2003c), a report showed that rotenone's toxic effect are 80% systemic and only 20% in CNS. The systemic toxicity included thinning of gastrointestinal walls, hepatotoxicity and nephrotic remodeling, while CNS toxicity was not limited to dopaminergic neurons (Lapointe et al., 2004). Hoglinger et al. (2005) also showed that chronic rotenone exposure caused neurotoxicity in cerebral region with Alzheimer's-like tauopathy, indicating non-dopaminergic effects of rotenone in CNS.

Studies challenging the utility and importance of the rotenone model as a tool to understand the pathogenesis of PD are in the minority with respect to the literature as a whole. Further, issues related to variability (both magnitude of effect, and number of animals with a PD phenotype) and mortality have largely been addressed through more recent iterations of the model utilizing a refined delivery vehicle and dosing regimen (Alam and Schmidt, 2004; Cannon and Greenamyre, 2010; Cannon et al., 2009).

3.3 Mouse models of rotenone exposure

In general, mouse rotenone models have been far less utilized, likely due to inconsistency. Significant species differences are known to occur in PD models. For example, rats are generally insensitive to MPTP, along with many strains of mice (Cannon and Greenamyre, 2011). While several studies show no effect of rotenone on dopaminergic neurons, others show neurotoxic effects of rotenone similar to rat models. Thiffault et al. (2000) first showed that mice chronically treated with rotenone (5–15 mg/kg/day, <21 days, s.c.) did not exhibit dopaminergic neurodegeneration or altered striatal dopamine levels, even at the highest exposed dose. Additionally, Richter et al. (2007) reported no neuropathological characteristics of PD in rotenone-treated (2.5–5.0 mg/kg/day for 30–45 days) mice even after

12 months of exposure. The study, however, reports partial deficits of locomotor behaviors in rotenone treated animals, which could be attributed to systemic toxicity.

Contrary to these studies, Takeuchi et al. (2009) used ~10-times higher doses (30 mg/kg/day) versus typical rat studies for 4 weeks to induce dopaminergic neurodegeneration. In another study, Inden et al. (2007) showed that oral administration of rotenone (10–30 mg/kg/day for 28 days) caused marked loss of TH-IR neurons in SN and striatum and depleted striatal levels of dopamine and DOPAC, without affecting other neurons. Researchers also reported accumulation of activated microglia and astrocytes in midbrain sections with intracellular accumulation of α-synuclein aggregates in dopaminergic neurons.

Other studies have used oral delivery in mice. A study by Inden et al. (2011), showed that high dose rotenone exposure (100 mg/kg/day, p.o.) for 28 days showed inconsistent dopaminergic neurodegeneration in mice, having complete loss of TH-IR in few animals and no effects in others. In contrast, a study showed that 30 mg/kg/day rotenone (orally) for 28 or 56 days caused time-dependent loss of dopaminergic neurons from mice SN. In addition, this study also showed that chronic rotenone treatment (30 mg/kg/day, 28 or 56 days) caused time-dependent increase in α-synuclein protein expression and accumulation of microglia and astrocytes in in the substantia nigra. Most significant, a chronic rotenone (5 mg/kg) study via intragastric route caused neurochemical, histopathological alterations (including α-synuclein aggregates) and behavioral deterioration (Pan-Montojo et al., 2010).

A recent study exposed rotenone to mice via "environment-contact" to evaluate the toxic effects (Liu et al., 2015). In this study, researcher sprayed rotenone (5–100 mg/kg) solution to mice cages (without bedding) and mice were left in these cages for 2 h daily for up to 14 day. Rotenone-induced animal mortality was dose-dependent and highest exposed dose of 100 mg/day was reported to be extremely toxic and mice died within 24 h of exposure and 30 mg/kg killed all animals by day 8, hence rest studied used lowest concentrations (5 mg/kg) for biochemical, histological and behavioral analysis. The researchers showed that 5 mg/kg rotenone caused progressive loss of dopaminergic neurons in SN and striatum, depleted striatal dopamine levels, induced α-synuclein aggregation and impaired locomotor behavior in a time-dependent manner (Liu et al., 2015). The study proposes more relevant route of exposure model for rotenone, however, there are some obvious and major concerns relative to the delivery regimen, particularly whether absorption is through skin, inhalation, or ingestion, along with the strength of the data supporting the pathology findings.

In summary, while some researchers have reported PD-relevant mouse models. Efficacy, and the magnitude of PD-relevant phenotypes appears to be highly variable in the literature.

3.4 Non-mammalian in vivo rotenone models

Originally used as fish-poison, rotenone's toxicity in zebrafish (*Danio rerio*) model has been investigated. Earlier investigation by Bretaud et al. (2004) revealed that only at doses higher than 30 μg/L rotenone exert toxicity in postfertilization embryos, which effects lasted up to 4 days causing PD phenotype leading to mortality. However, concentrations higher than 75 μg/L caused immediate fatality in fish. Interestingly, like other models, use of lower dose (2 μg/L) for longer period (4 weeks) lead to reduction in TH levels and depleted neurotransmitter levels of DA and DOPAC, but no change in α-synuclein expression. In addition, study also showed that chronic low dose rotenone treatment generated motor and non-motor symptoms of PD in zebrafish (Wang et al., 2017). Given the high sensitivity of fish to rotenone, selective neurotoxicity is generally unlikely to be observed.

In contrast to fish, rotenone exposure in *Drosophila melanogaster* (fruit fly) has been used to model PD. Here, a PD-relevant *Drosophila* model was generated by exposing flies to lower doses of rotenone (<500 μM) for 7 days or longer. This study reported loss of dopaminergic neurons accompanied with compromised motor activity (Coulom and Birman, 2004). Hosamani and Muralidhara (2009) showed the key role of oxidative stress in rotenone-induced dopaminergic neurotoxicity in *Drosophila*, which was ameliorated by an antioxidant extract of *Bacopa monnieri*. It is worth noting that several more recent studies have utilized the rotenone fly model as reliable non-mammalian model for PD (Li et al., 2010; Pandareesh et al., 2016; St Laurent et al., 2013; Stephano et al., 2018).

Caenorhabditis elegans (*C. elegans*), nematodes, are another commonly used model for PD. Besides being transparent, this model is of particular advantage of having only eight dopaminergic neurons that can be readily assessed for pathological changes in response to neurotoxic exposures (Sammi et al., 2018). One to six days treatment nematodes with rotenone (1.5–50 μM) caused time- and dose-dependent mitochondrial complex I inhibition, oxidative damage, α-synuclein aggregation leading to selective dopaminergic neurodegeneration and hypokinetic phenotype (Mocko et al., 2010; Ruan et al., 2010; Ved et al., 2005). This model is also used as PD model for neuroprotective and genetic studies (Chikka et al., 2016; Maulika et al., 2019).

Additionally, rotenone appears to be a dopaminergic neurotoxin in a wide variety of non-traditional animal models. For example, rotenone-treated snails (*Lymnaea stagnalis*) and frogs (*Lithobates pipiens*) have also been proposed as PD models, however, due to better characterization and ease of use of other models, these models have not been widely used in PD research (Xiong et al., 2012b).

3.5 In vitro rotenone models

Primary rodent mesencephalic neuron/glia cultures have been extensively used and widely accepted models for studying PD (Xiong et al., 2012a). Based on a rat study, 25–30 nM brain rotenone concentrations can effectively block mitochondrial complex I and decrease mitochondrial membrane potential (Greenamyre et al., 1999). Studies in primary midbrain cultures show that lower concentrations of (20–30 nM for 11–24 h) induce selective dopaminergic neurotoxicity without affecting other mesencephalic cells (Ahmadi et al., 2003; Hajieva et al., 2009; Moon et al., 2005). Gao et al. (2002) showed that rotenone concentration as low as 1 nM can induce selective dopaminergic toxicity in neuron/glia primary rat culture after 8 days of exposure. These data coincide with the in vivo data and show that rotenone selectively targets dopaminergic neurons. As expected, selectivity is dose-dependent, where concentrations higher than 40 nM have been shown to promote non-specific neurotoxicity in primary cortical neuronal culture (Chen et al., 2006; Mullett and Hinkle, 2009).

The immortalized cell lines, especially human dopaminergic SH-SH5Y neuroblastoma cells, rat N27 dopaminergic cells and rat adrenal pheochromocytoma PC-12 cells, have been regularly used in neurotoxicological and PD research (Gao et al., 2012; Westerink and Ewing, 2008; Xicoy et al., 2017). Unlike primary culture, neuronal cell lines are more resistant to rotenone toxicity. Thus, higher doses of up to 100 μM and above have been used to induce cell death (Hsuan et al., 2006; Pan et al., 2005; Ruan et al., 2010). Interestingly, Borland et al. (2008) showed that chronic (21 days) exposure of low concentration (50 nM) rotenone in human SH-SY5Y neuroblastoma cells induced cytotoxicity by promoting oxidative stress, mitochondrial dysfunction and UPS impairments, similar to rat in vivo model. Similar results were reported in murine PC-12 and N27 dopaminergic neuronal cells (Sai et al., 2008; Shinomol et al., 2012). The varying sensitivity of neuronal cells can be attributed to variance in cell handling, for example, higher passages of SH-SY5Y cells were shown to be resistant to rotenone-mediated toxic effects (Xiong et al., 2012b).

The Lund human mesencephalic (LUHMES) cell line were originated from a healthy human embryonic mesencephalon (8 weeks old) and were genetically engineered by tetracycline-regulated v-myc-vector transfection to develop immortalized cells. Unlike primary mesencephalic cultures and previously mentioned cell lines, the LUHMES cell differentiation generates homogenous population of mature dopaminergic neurons (Harischandra et al., 2019). Smirnova et al. (2016) recently showed that rotenone exposure to a 3D culture of LUHMES cells causes time- and dose-dependent cytotoxicity, beginning at 100 nM for 12 h exposure leading to complete cell loss at 10 μM rotenone for 48 h. Further studies are needed to determine the molecular toxicological evidences of rotenone-mediated LUHMES cell death.

Lastly, in recent advancements of PD models, human induced pluripotent stem cells (hiPSCs)-derived dopaminergic neurons are becoming highly acceptable in vitro models (Li et al., 2018). In this context, a recent study investigated the effects of two well-known dopaminergic neurotoxicants, manganese and rotenone, on hiPSCs-derived dopaminergic neurons. The study found that although both manganese (>200 μM) and rotenone (>10 μM) induced oxidative stress by depleting reduced glutathione, a cellular antioxidant, levels and promoting lipid peroxidation from 2.5 to 24 h posttreatment, only rotenone was able to cause neurite degeneration (Neely et al., 2017). Interestingly, this study highlights the key mechanistic differences between rodent dopaminergic neuronal models and postmitotic human dopaminergic neurons. For instance, acute (24 h) 10 μM rotenone exposure failed to show any oxidative damage and cell death in hiPSC-derived dopaminergic neurons, unlike murine PC-12 cells. Also, contrary to rodent dopaminergic neuronal cells, rotenone exposure to hiPSC-derived dopaminergic neurons caused neurite degeneration at lower concentrations without causing notable oxidative stress, suggesting possible interaction of rotenone directly to tubulin (Neely et al., 2017). It is important to note such mechanistic differences between cellular model systems when extrapolating to in vivo models.

Further, in a subchronic study, Pistollato et al. (2017) showed that rotenone (0.1–500 nM for 14 days) activates the Nrf2/ARE signaling pathway in response to oxidative stress in hiPSCs-derived dopaminergic neurons. The study also showed that astrocytes derived from hiPSCs were also activated by the subchronic rotenone treatment. Further characterization of hiPSC-derived neurons is an evolving area of research and future studies can further demonstrate the standard rotenone-exposure model, but, present data demonstrate that hiPSCs-derived dopaminergic neurons are susceptible to rotenone treatment and show promise for an alternative reliable in vitro model system for PD.

3.6 Non-motor symptoms of rotenone exposure

In addition to the well-recognized motor deficits, studies show that most PD patients develop one or more of non-motor symptoms much earlier than inception of motor symptoms (Pfeiffer, 2016). Among these, GI disturbances such as constipation, bloating, and dysphagia have been reported in pre-Parkinsonian and Parkinsonian patients (Lionnet et al., 2018; Natale et al., 2008; Pfeiffer, 2018). Virga et al. (2018) reported that rotenone markedly depletes mitochondrial velocity and ATP levels in primary enteric neurons, leading to apoptogenic signaling and neurodegeneration, which could be reverted by overexpression of an antiapoptotic gene (Bcl_{xl}). Likewise, chronic rotenone exposure (3 mg/kg/day, s.c. minipump) showed delayed gastric emptying, reduced defecation and impaired enteric neuronal function after 3–4 weeks (Greene et al., 2009). Even further, rotenone treatment (2 mg/kg/day, i.p. for 28 days) to rats resulted in accumulation of α-synuclein positive intracellular aggregates in enteric myenteric plexus neurons 6 months after last rotenone exposure that were similar to enteric Lewy bodies found in idiopathic PD patients (Drolet et al., 2009). In addition, mice treated rotenone (50 mg/kg/day) exhibited loss of dopaminergic neurons and increased astrocytic infiltration in the enteric myenteric plexus. A recent study further evaluated the effect of rotenone (2.75 mg/kg/day, i.p.) in rat gut and showed that systemic rotenone treatment for 4 weeks severely slowed the gastric emptying time (gastroparesis), altered small intestinal microbiota and thickened the mucosal walls (Johnson et al., 2018), showing similarity to clinical PD cases (Pfeiffer, 2018).

Tasselli et al. (2013) utilized an oral gavage delivery of rotenone (30 mg/kg/day, 28 days) in mice. The study reported that mice developed partial GI symptoms potentially relevant to clinical PD, including lower gastric emptying volume and accumulation of α-synuclein in jejunum. Hence, according to suggestion by Martinez and Greenamyre (2012), although rotenone model does not completely mimic PD-like ENS pathology and

GI symptoms, it is highly suited as a preclinical model to study GI pathology related to PD. In fact, a recent study recapitulated the GI effects of chronic rotenone in mice and further showed that mice treated with unpredictable restraint stress for 12 weeks had exaggerated response to rotenone both in GI as well as in brain (Dodiya et al., 2018). This study suggests that a reliable rotenone model with regards to enteric pathology can be achieved by chronic restraint stress-mediated pro-inflammatory signals leading to PD in presence of the mitochondrial complex I inhibitor.

Accumulation of α-synuclein Lewy aggregates in the olfactory bulb (OB) is believed to contribute to PD-associated hyposmia (reviewed in Fullard et al., 2016). In this context, Betarbet et al. (2006) investigated the accumulation of α-synuclein aggregates in OB of rotenone-treated rats, but did not identify the overt aggregation. A detailed study of rotenone olfactory effects in rats was first investigated by Rodrigues et al. (2014). Similar to clinical PD, chronic rotenone treatment resulted in hyposmia symptoms in >50% of treated rats and increased TH-IR, but reduced dopamine levels in OB (Johnson et al., 2015; Rodrigues et al., 2014). The authors suggest that nigrostriatal dopaminergic neurodegeneration in PD may trigger compensatory dopaminergic neurogenesis in OB, however, the immature dopaminergic neurons fail to generate required dopamine release, contributing to pre-disease symptoms.

In addition, the rotenone rat model also reported depletion of dopamine in retina (Biehlmaier et al., 2007). Investigation of retina of rats exposed to rotenone showed selective degeneration of retinal dopaminergic neurons further causing toxicity and photoreceptor cell death and functional impairments in retina (Esteve-Rudd et al., 2011).

Further, most PD patients suffer from mild to severe depression, emotional deprivation and sleep irregularities (Pfeiffer, 2016). Likewise, rotenone treatment progressively lowered the sleep quality and deteriorated sleep-wake behavior in rats by decreasing the frequency of rapid eye movement (REM) sleep, increasing the number of awakenings and enhancing rapid eye movement sleep during dark phase, while increased day-time sleeping, similar to PD patients (Garcia-Garcia et al., 2005; Lima, 2013; Yi et al., 2007).

In rats, chronic rotenone induced similar depression-like behaviors after >3 weeks of exposure, which was, interestingly, linked with brain ratio of dopamine and serotonin (Santiago et al., 2010). Beside CNS dopaminergic toxicity, rat study showed that chronic rotenone treatment also causes extranigral motoneuron degeneration in cervical and lumbar regions of spinal cord, which further contributes to symptoms related to Parkinsonism (Samantaray et al., 2007). In addition, rotenone treatment has also reported to weaken memory and deteriorate cognitive senses (Dos Santos et al., 2013; Jia et al., 2014; Kaur et al., 2011).

4. Mechanism of rotenone toxicity: Mechanism of action and key advances in understanding PD pathogenesis

The majority of neurotoxicity data suggest that rotenone causes selective dopaminergic neurodegeneration by inhibiting electron transport from complex I (NADH-quinone oxidoreductase) to complex II of the ETC on mitochondrial inner membrane, inhibiting mitochondrial respiration (Sherer et al., 2003b). In this context, rotenone (3 mg/kg/day, s.c. continuous infusion) treatment for 7 days in rats caused up to 40% reduction in state 3/3 U respiration, without altering state 4 respiration in brain mitochondria. Interestingly, liver mitochondrial respiration was not affected, further illustrating selective sensitivity of neuronal mitochondria. Partial inhibition of complex I and II can lead to aberrant generation of reactive oxygen species. More specifically, the superoxide production site is believed to be a protein-associated ubiquinone species between N2 cluster of complex I and ubiquinone, a specific site of action for rotenone (Ohnishi et al., 2005). This inhibition has been found to increase the mitochondrial permeability transition (mPT) and forming mitochondrial membrane permeability transition pores on mitochondrial outer membrane. This can cascade to higher intracellular oxidative stress and activation of apoptogenic pathways.

Intriguingly, adeno-associated virus (AAV)-mediated delivery of an "alternative" NADH oxidoreductase (Ndi1) in rat striatum protected against rotenone (Marella et al., 2008). The study demonstrated that unilaterally injected Ndi1 AAV reduced oxidative damage, dopaminergic neuronal loss, and behavioral impairments induced by chronic rotenone exposure. Ndi1, a complex I of *Saccharomyces cerevisiae* mitochondria, does not interact with mammalian complex I inhibitors. Thus, this finding further illustrates the selectivity of rotenone-induced dopaminergic neurotoxicity for mitochondrial complex I.

The following segments further details the several mechanisms altered by rotenone in different preclinical models. Graphical illustration in Fig. 2 encapsulates the key mechanisms affected by rotenone exposure that leads to selective degeneration of dopaminergic neurons, responsible PD-like

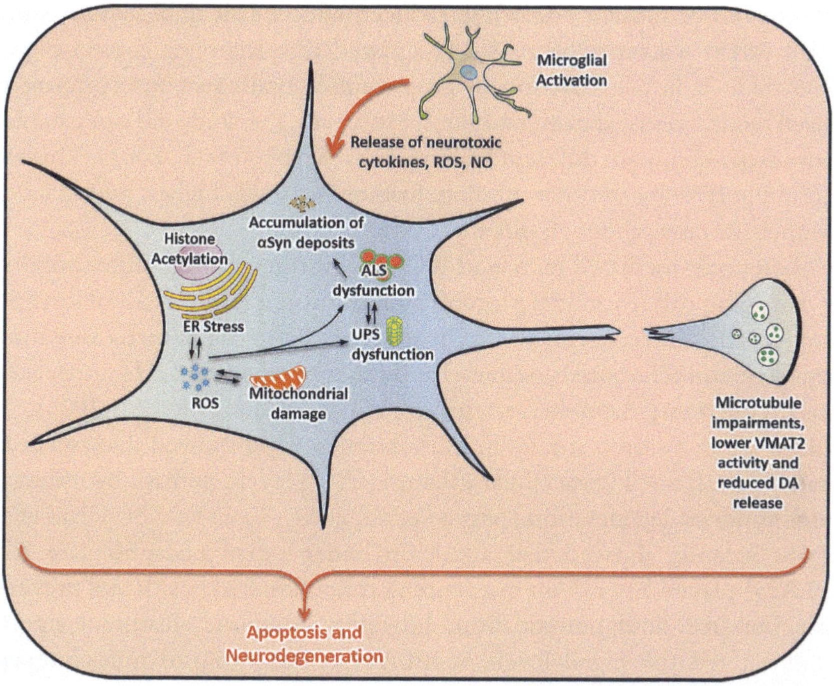

Fig. 2 Schematic presentation of proposed or established molecular mechanisms and targets of rotenone responsible for promoting dopaminergic neurotoxicity. αSyn—α-synuclein; ALS—autophagolysosomal system; DA—dopamine; ER—endoplasmic reticulum; NO—nitric oxide; ROS—reactive oxygen species; UPS—ubiquitin-proteasome system; VMAT2—Vesicular monoamine transporter 2. The figure was drawn using Biomedical Powerpoint® tools and exported using Adobe® Photoshop®.

phenotype. Targeting one or more of these toxicity mechanisms may mitigate the toxicity of this mitochondrial complex I inhibitor and may reduce dopaminergic neuronal loss.

4.1 Oxidative stress, mitochondrial dysfunction, and apoptosis

Oxidative stress has been repeatedly implicated in PD pathogenesis (Gotz et al., 1990; Hwang, 2013). Sherer et al. (2003b) showed that rats chronically treated with rotenone (s.c.) exhibited elevated oxidative stress particularly in midbrain and olfactory bulb. Likewise, 40-day rotenone treatment (4 mg/kg/day, i.p.) markedly increased nitric oxide synthase levels and 3-nitrotyrosinase levels in SN and striatum of rat, indicating rotenone-mediated increased nitrosative stress (He et al., 2003). In addition, 1 year old rats treated with rotenone (1.5 mg/kg/day, s.c.) showed increased the nigral levels of nitric oxide while reduced antioxidant thiol levels (Xiong et al., 2015). Interestingly, a study showed that rotenone caused higher superoxide radical generation in hippocampal brain sections compare to striatal section derived from rat brains. However, this study did not examine neurodegeneration in different brain regions (Saybasili et al., 2001). This was also supported by another finding further showing higher peroxidation products in rat striatum (Bashkatova et al., 2004).

Rotenone-mediated increased ROS can further damage mitochondria, by damaging mitochondrial proteins and disturbing mitochondrial integrity (Sanders and Greenamyre, 2013). In addition, ROS can in turn cause more damage to mitochondrial complex I and other complexes of ETC, especially those containing iron-sulfur clusters (Garcia-Garcia et al., 2005). In addition, patch-clamp studies in the substantia nigra showed that rotenone treatment increased intracellular calcium (Ca^{2+}) levels, perhaps by inducing mitochondrial degeneration (Nistico et al., 2011). Even further, Chao et al. (2018) recently showed that a calcium-independent phospholipase A2γ (iPLA2γ) plays a key role in maintaining mitochondrial health and prevents mitochondrial lipid peroxidation, however, rotenone treatment significantly altered the nigral levels of iPLA2γ contributing to mitochondrial degeneration and neurotoxicity.

With regards to cell death, rotenone treatment has been found to induce the upregulation of apoptotic pathways. For example, chronic rotenone treatment increased nigral expression of apoptogenic p53 and Bax protein expressions (Wu et al., 2013a). Numerous other studies from in vivo and in vitro models show that rotenone exposure may influence apoptosis, where it leads to cytoplasmic release of cytochrome C, activation of caspase-9, caspase-3 and bax, and depletion of anti-apoptotic Bcl-2 levels (Abdelkader et al., 2016; Ethell and Fei, 2009; Luo et al., 2017; Watabe and Nakaki, 2004).

Rotenone treatment also altered the mitochondrial protein transport mechanisms. A recent study in rats showed that α-synuclein bind to the translocase of the outer membrane 20 (TOM20) protein on the outer layer of mitochondria, effectively blocking the protein transportation of mitochondria, which ultimately led to impaired mitochondrial respiration, increases mitochondrial superoxide generation and loss of mitochondrial membrane potential, contributing to neurodegeneration. Even further, partial knock down of α-synuclein in vivo reverted the mitochondrial protein transportation and altered neuronal loss. Also, over expression of TOM20 in SH-SY5Y cells reduced rotenone-mediated toxicity, further confirming the molecular mechanisms of rotenone-mediated mitochondrial toxicity in dopaminergic neurons (Di Maio et al., 2016). Importantly, using the rotenone model, this study identified a potentially novel biochemical pathway of significant importance to the pathogenesis of PD.

Parkin knockout mice showed high sensitivity to rotenone treatment (Casarejos et al., 2006). Similar studies from in vivo and in vitro models show that rotenone exposure depletes the levels of parkin, which in turn reduces the ability of neurons to remove impaired mitochondria, further contributing to neurotoxicity (George et al., 2010; Koch et al., 2009; Norris et al., 2015; Sonia Angeline et al., 2012; Wu et al., 2014). Peng et al. (2019) recently showed a pivotal role of parkin in maintaining mitochondrial health by controlling mitochondrial biogenesis, fission/fusion, and mitophagy in the presence of rotenone. While in the presence of potent antioxidant, *N*-acetyl cysteine, rotenone-mediated parkin loss was altered in rat nigrostriatal system resulting in neuroprotection (Rahimmi et al., 2015).

Additionally, rotenone exposure alters both mitochondrial dynamics and biogenesis as reported by Peng et al. (2017). In their in vitro experiments, rotenone downregulated the cellular level of peroxisome proliferator-activated receptor gamma coactivator 1-alpha (PGC1α), a mitochondrial biogenesis marker required for regulating mitochondrial fission/fusion. In contrast, overexpressing PGC1α, inhibiting mitochondrial fission or stimulating mitochondrial fusion protected the cells against rotenone-mediated neurotoxicity. In another study, Peng et al. (2016) demonstrate that resveratrol, a natural antioxidant, reversed rotenone-induced cell death by recovering the loss of Dynamin-related protein (Drp1), fission 1 (Fis1), optic atrophy-1 (Opa-1) and mitofusin-2 (Mfn-2) (regulators of mitochondrial fission/fusion) and PGC1α/mitochondrial transcription factor A (TFAM) (regulators of mitochondrial biogenesis) in the rat SN and PC-12 dopaminergic cells. Arnold et al. (2011) demonstrate that rotenone changes mitochondrial morphology by first inducing fusion as an early compensatory response, followed by damage-inducing fission. More recently, the inhibition of fission was shown to reduce rotenone-induced neurotoxicity in primary neurons (Imbeault et al., 2014; Peng et al., 2016).

4.2 Microtubule deformation

It is important to note that alternate primary mechanisms of action have also been explored. For example, some studies suggest non-mitochondrial cellular targets for rotenone, especially microtubules. Ren et al. (2005) and

Feng (2006) showed that rotenone disturbed the tubulin-associated microtubule polymerization selectively in primary rat midbrain dopaminergic neurons in vitro. This further disrupts vesicular transport leading to increased cytosolic dopamine and oxidative stress. Even further, rotenone-mediated neurotoxicity was reversed by addition of nerve growth factor and was potentiated by blocking the microtubule-associated protein kinase (MEK) kinase (Jiang et al., 2006).

Rotenone has been also shown to modulate several key protein kinases linked to dopaminergic neurotoxicity. Activation of c-Jun-N-terminal protein kinase (JNK) and its translocation to mitochondria occurs most commonly post-rotenone treatment (Chen et al., 2008a; Jang et al., 2015; Klintworth et al., 2007; Newhouse et al., 2004). Rotenone-induced activation of JNK inhibits Vesicular monoamine transporter 2 (VMAT2) activity blocking the synaptic release of dopamine, which increases cytosolic dopamine and ROS levels (Choi et al., 2015). Activation of MAP kinases, such as p38 and extracellular signal-regulated kinase ½ (ERK1/2), has also been demonstrated in rotenone models. Rifampicin inhibition of glycogen synthase kinase 3 beta (GSK3β) phosphorylation was shown by Wu et al. (2018) to alleviate rotenone-mediated neurotoxicity.

While other primary mechanisms of neurotoxic action have been explored, the bulk of the literature supports complex I inhibition as key pathogenetic event in mediating dopaminergic neuronal cell death.

4.3 Dysfunctional proteostasis and α-synuclein aggregates

Impaired protein clearance has been linked with neurotoxicity in PD, resulting in accumulation of α-synuclein aggregates (Spillantini et al., 1997). Rotenone exposure has also been linked to accumulation of cytoplasmic aggregates containing α-synuclein (Betarbet et al., 2006; Sherer et al., 2003c). A mechanistic study also showed that rotenone treatment impaired the ubiquitin-proteasome system (UPS) activity, specifically in ventral midbrain, contributing to lower protein clearance and increased accumulation (Betarbet et al., 2006). In addition, daily intraperitoneal injection of rotenone also resulted in accumulation of α-synuclein (co-localized with poly-ubiquitin) indicating dysfunctional UPS (Cannon et al., 2009).

Rotenone treatment has also been shown to modulate other protein clearance mechanisms. In this context, rotenone-based models have been checked for the regulation of autophagic flux (Xiong et al., 2013). A recent rat study showed that an 8-week treatment with 2 mg/kg/day rotenone generated PD like behavioral deficits and the loss of nigrostriatal

TH neurons, which was attributed to impaired autophagic flux (Wu et al., 2015). Using the PC-12 dopaminergic in vitro model, Wu et al. (2015) further showed that rotenone treatment leads to the accumulation of microtubule-associated proteins 1A/1B light chain 3B (LC3B)-tagged autophagic vacuoles, the upregulation of p62, beclin 1 and LC3B II, and the loss of lysosomal associated membrane protein 2 (LAMP2) with reduced nuclear translocation of Transcription Factor EB (TFEB, a lysosomal biogenesis marker). Even further, semiquantitative assessment by Filomeni et al. (2012) revealed that rotenone treatment increases the accumulation of autophagosome at early stage and not at the mature or late stages. Notably, rapamycin or kaempferol pretreatment can protect against rotenone-induced neurotoxicity by elevating autophagic flux (Filomeni et al., 2012; Pan et al., 2009a). Few studies have focused on the neuropathological assessment of in vivo alterations in autophagy. Recent examples, include where chronic rotenone treatment in rats disrupted autophagy by increasing the levels of LC3B and cargo-carrier protein, optineurin (OPTN), both in nigral dopaminergic neurons and extra-nigral non-dopaminergic PD-relevant brain nuclei (Wise and Cannon, 2016; Wise et al., 2018). Furthermore, autophagy inducers were shown to reduce oxidative stress and mitochondrial dysfunction in rotenone-treated neuronal cells expressing the A53T mutant α-synuclein (Choubey et al., 2011; Filomeni et al., 2012). Pan et al. (2009b) reported that knocking down the ATG5 gene blocked rapamycin-mediated protection against rotenone. These findings collectively indicate that rotenone impairs the autophagic flux in cells, which contributes to increased oxidative stress and reduced protein clearance, inducing cytotoxic effects. In addition, heat shock proteins (HSPs) have been shown to degrade the α-synuclein aggregation (Thakur and Nehru, 2014). Interestingly, rotenone treatment altered the protein levels of several key HSPs in rats and in in vitro cultures, further contributing to impairments in proteostasis mechanisms (Kilpatrick et al., 2013; Thakur and Nehru, 2014). Even further, an antiulcer agent, carbenoxolone, have been shown to reduce rotenone toxicity by upregulating HSP-dependent antioxidant and cell surviving mechanisms in rats (Thakur and Nehru, 2015).

Even further, rotenone-induced disruption of protein clearance mechanisms also extends to microglia. A recent report showed that lipopolysaccharide (LPS)-primed microglia accumulate LC3B-positive puncta and show upregulation of LC3B II and p62 levels, with reduced TFEB expression, indicating dysfunctional autophagolysosomal system (ALS)

(Lawana et al., 2017). In this study, the LPS/rotenone-mediated ALS malfunction in microglia was believed to contribute to the activation of NACHT, LRR and PYD domains-containing protein 3 (NLRP3) inflammasomes and the microglial release of inflammatory markers.

Besides interfering with protein clearance, rotenone exposure also reduces the neuronal expression of parkin and DJ-1, which are responsible for mitochondrial clearance through mitophagy (Gao et al., 2012; Mili et al., 2012). Evidence from electron microscopic analysis of rotenone-treated rat brain slices confirmed impaired mitochondrial clearance (Filomeni et al., 2012). Interestingly, Chu et al. (2013) showed that a low dose of rotenone can induce mitophagy by externalization of cardiolipin from the mitochondrial inner membrane. Using a different approach to attenuate rotenone's dose effect, di Rita et al. (2018) recently showed that overexpressing the mitochondrial fusion mediator AMBRA1^{ActA} in rotenone-treated SH-SY5Y neuronal cells upregulated mitophagy and reduced oxidative stress. These findings suggest that rotenone's effect on mitophagy, either stimulatory or inhibitory, depends on its dose and duration of exposure.

4.4 Impaired iron homeostasis

Hyperaccumulation of iron in SN have been reported almost a century ago and role of iron is since investigated for dopaminergic neuronal health and PD pathogenesis (reviewed in Iversen and Iversen, 2007). In fact, the healthy SN has higher concentration of iron compare to liver (Gotz et al., 2004). In addition, genetic mutation of iron transporters have been linked with PD (Guerreiro et al., 2006), while treatment with iron chelator have been shown protective (Kaur et al., 2003). However, the role of iron in PD pathogenesis has long been controversial. Importantly, chronic rotenone treatment also caused aberrant increase in labile iron pool (Mastroberardino et al., 2009). This study also reported increased levels of transferrin/transferrin receptor 2 (Tf/Tfr2), required for iron transport into mitochondria. This study highlights another example where a new potentially critical pathogenic pathway was discovered in the rotenone model. The above mechanism was identified in rotenone-treated rats and confirmed in PD patient tissue, in the same report.

4.5 ER stress

Rotenone have been shown to modulate endoplasmic reticulum (ER) stress in PD models (Goswami et al., 2016a; Tong et al., 2016). Wu et al. (2013b) demonstrated that treating Lewis rats with rotenone increased the expression of glucose-regulated protein (GRP78), activating transcription factor 4 (ATF4) and C/EBP homologous protein (CHOP), well-established markers for ER stress, in the ventral midbrain, while inhibiting ER stress in the rotenone-treated rats with an angiotensin II receptor blocker reduced dopaminergic neuronal loss in the SN. Early experiments in HeLa cells showed that rotenone-induced cell death depends on ER stress-mediated ROS generation and calcium deregulation causing the activation of nuclear factor κB (NF-κB) (Pahl and Baeuerle, 1996). Another study demonstrated that rotenone-mediated ER stress in human neuroblastoma SK-N-MC cells induces phosphorylation of the unfolded protein response (UPR) marker PRKR-like endoplasmic reticulum kinase (PERK), which then phosphorylates the eukaryotic initiation factor 2α (eIF2α) transcription factor, resulting in neuronal death (Chen et al., 2008b). An inhibitor of eIF2α-mediated ER stress, salubrinal, can dramatically reduce rotenone-induced cell death by reducing the activation of caspase-12 and -3, as demonstrated by Goswami et al. (2016a,b), both in vitro in neuro-2A cells and in vivo in TH-positive neurons of the rat midbrain. When rotenone-treated mice are co-treated with 4-phenylbutyrate, a partial ER stress inhibitor and chemical chaperon that reverses protein aggregation, not only are aggregated α-synuclein deposits reduced, but so are GRP78 upregulation and caspase-12 activation (Inden et al., 2007). In another study, administering a peroxisome proliferator-activated receptor (PPAR)-β/δ agonist also reduced ER stress and dopaminergic neurodegeneration in the rotenone rat model (Tong et al., 2016). Collectively, studies from different models suggest that rotenone-induced mitochondrial dysfunction and oxidative stress can contribute to ER stress and subsequent neuronal degeneration.

4.6 Rotenone and nueroinflammation

Chronic rotenone neurotoxicity also activates microglia (Fleming et al., 2004). A recent report suggests that rotenone treatment induced oxidative stress and mitochondrial dysfunction in rat SN occurs in non-dopaminergic neurons as well, which ultimately deter dopaminergic neurons (Elgayar et al., 2018). Numerous studies show that chronic low dose exposure in rats leads to excessive microgliosis in nigrostriatal regions (Emmrich et al., 2013; Gao et al., 2002; Liang et al., 2015; Sherer et al., 2003a; Zhang et al., 2014; Zhou et al., 2007). Similarly, intranigral injection also readily accumulated microglia in rat SN 14 days after exposure (Norazit et al., 2010). Rotenone also activates the microglial NADPH oxidase, gp91phox (phagocyte oxidase), releasing excessive neurotoxic oxidative species (Gao et al., 2003; Uversky, 2004; Zhou et al., 2012). Using a neuron-glia mixed culture, Emmirch et al. (2013) showed that while low levels of rotenone failed to change mitochondrial integrity in microglia, it did increase microglial phagocytosis of dopaminergic neurons. A recent study showed that microglia pre-exposed to LPS are extremely sensitive to rotenone exposure, which leads to a dramatic release of inflammatory mediators [nitric oxide (NO), interleukin (IL)-1β, IL-18, tumor necrosis factor α (TNFα), NLRP3, caspase 1, etc.] via c-Abl kinase-dependent activation of the NLRP3 inflammasome complex (Lawana et al., 2017).

In addition, rotenone exposure also activates other inflammatory cells in SN (Tansey and Goldberg, 2010). Chronic rotenone exposure in rats increased the expression of connexin 43, a gap junction marker on astrocytes, in SN and globus pallidus region containing dopaminergic neurons (Kawasaki et al., 2009). Further, Abdelsalam and Safar (2015) showed that low-dose rotenone exposure for 21 days (1.5 mg/k/day, s.c.) activated receptor for advanced glycation end product (RAGE) signaling in rat brain. The rotenone-mediated RAGE induction further activated transcription factor nuclear factor-κB (NF-κB) in brain, which the further lead to transcription of proinflammatory genes, including tumor necrosis factor-α (TNFα), inducible nitric oxide synthase (iNOS) and IL-1β.

Activation of microglia and astrocytes leads to increased levels of proinflammatory cytokines, which then bind to cytokine receptors trigger neurotoxic intracellular signaling including production of iNOS and activation of NADP oxidases causing neuronal oxidative stress (Bailey et al., 2006; Garcia-Dominguez et al., 2018; Sarkar et al., 2017). Accordingly, Li et al. (2012) showed that rats chronically treated with rotenone resulted in time-dependent increase in microglia, along with IL-1β and TNFα. Even further, use of a natural anti-inflammatory drug resulted in reduced nigral inflammation and lowered the loss of dopaminergic neuron.

4.7 Rotenone-gene interaction models

Gene-environment interactions have repeatedly been postulated to underlie the pathogenesis of PD (Fleming, 2017). To date, the rotenone model has been instrumental in identifying mechanisms of such interactions. Here, several investigators have used known genetic models of PD and exposed them to rotenone to evaluate effects on dopaminergic neurotoxicity. As an example, some familial cases of PD involve a recessive mutation in DJ-1 (*PARK7*) (Ariga et al., 2013), which is also linked with increased oxidative stress in dopaminergic neurons (Lotharius and Brundin, 2002). Rats chronically infused with rotenone showed oxidative modification of DJ-1 selectively in dopaminergic neurons of SN (Betarbet et al., 2006). In mice and SH-SY5Y neuroblastoma cells, rotenone increased the protein expression of DJ-1 (Inden et al., 2007). In addition, *C. elegans* and *Drosophila* DJ-1 knockout models showed increased sensitivity to rotenone by promoting oxidative stress (Meulener et al., 2005; Ved et al., 2005). Further, Gao et al. (2012) demonstrated that overexpression of DJ-1 ameliorated rotenone toxicity by promoting autophagic flux and clearing damage mitochondria though mitophagy. Importantly, De Miranda et al. (2018) recently reported that astrocyte-specific overexpression of human DJ-1 protected dopaminergic neurons from rotenone toxicity in rats. The study showed that astrocytes overexpressing DJ-1 reduced rotenone-mediated neuroinflammation and lowered the neuronal accumulation of α-synuclein by alleviating functional autophagic flux.

The physiological role of α-synuclein in dopaminergic neurons and PD pathogenesis is still unclear. However, in both specific experimental models and PD patients (both sporadic and inherited) show abnormal accumulation of intracellular α-synuclein positive aggregates (Stefanis, 2012). Several studies evaluated the crosstalk between rotenone and α-synuclein in promoting dopaminergic neurotoxicity. For instance, Mulcahy et al. (2012) first showed that AAV-induced overexpression of α-synuclein in rat SN dramatically synergized rotenone-mediated dopaminergic neurodegeneration. In a follow up study, Mulcahy et al. (2013) showed that nigral α-synuclein overexpression exacerbated behavioral impairments in rotenone treated rats. Accordingly, shRNA-mediated knockdown of α-synuclein from rat SN markedly reduced the loss of dopaminergic neurons after rotenone exposure and alleviated rotenone-induced motor deficits (Zharikov et al., 2015). To the contrary, an in vitro study performed in human SH-SY5Y neuroblastoma cells showed that ectopic expression of wild-type (WT) α-synuclein in these neuroblastoma cells marginally protected against the rotenone-mediated toxicity by rescuing mitochondrial membrane potential (Choong and Say, 2011). Interestingly, the study reported that mutant related to fPD (A30P, A53T and E46K) exaggerated rotenone-mediated dopaminergic toxicity.

The E46K mutation in human α-synuclein is PD pathogenic. Cannon et al. (2013) investigated the effect of rotenone in human E46K α-synuclein expressing rat brain. Here, the researchers created transgenic (Tg) rats that expressed human α-synuclein mutant through all brain regions affected in

PD; and showed high nitrosative stress and depletion of dopamine levels in striatum without significant dopaminergic neurodegeneration. However, chronic rotenone exposure to adult E46K Tg rats markedly developed PD-like hypokinetic behaviors and induced dopaminergic neurodegeneration (Cannon et al., 2013). Even further, unilateral intranigral rotenone treatment to the E46K Tg rats showed dopaminergic α-synuclein aggregate formations (Rcom-H'cheo-Gauthier et al., 2016). Similar effects were reported with A30P and A53T α-synuclein mutant models exposed to rotenone (George et al., 2010; Ved et al., 2005; Yy et al., 2011). Dadakhujaev et al. (2010) reported that mutant α-synuclein overexpressing SH-SY5Y cells have dysfunctional autophagy.

Besides *SNCA*, *LRRK2* is another autosomal recessive gene associated with fPD. *LRRK2* gene is responsible for transcribing leucine-rich repeat kinase 2 (LRRK2), however, its exact role in neurons is not still unclear (Bardien et al., 2011). Two studies in *C. elegans* first studied the effect of rotenone in LRRK2 mutants (G2019S and R1441C) (Saha et al., 2009; Wolozin et al., 2008). The studies suggested that overexpression of wild-type human LRRK2 or kinase disabled R1441C mutants protected against rotenone-mediated neurotoxicity, while the G2019S mutant exaggerated the effects of rotenone. Similar results were reported in fly PD models exposed to rotenone (25 μM) (Ng et al., 2009; Wang et al., 2008). Venderova et al. (2009) further reported that LRRK2 mutants, I1122V, Y1699C and I2020T, expressing *Drosophila* demonstrated increased toxicity with rotenone (100 μM) exposure. Importantly, a recent study also found that LRRK2 kinase activity was selectively enhanced in substantia nigra dopamine neurons in brain tissue from PD patients. Similarly, this finding was shown in the rat rotenone model in the same study, where the effect was found to occur through an oxidative mechanism that enhanced phosphorylation of the LRRK2 substrate Rab10 (Di Maio et al., 2018).

4.7.1 Rotenone and neuroprotectant evaluation

L-DOPA is a precursor for dopamine and is commonly used for to treat motor symptoms in humans. In animal models, it is regularly used as well, especially to determine if behavioral deficits are dopamine-dependent. For example, 10 mg/kg L-DOPA treatment (i.p. plus peripheral decarboxylase inhibitor) reversed the rotenone's parkinsonian effects in rats (Alam and Schmidt, 2002, 2004), further validating the rotenone rat model of PD. Furthermore, an inhibitor of monoamine oxidase-B (MAO-B), an enzyme for dopamine metabolism, deprenyl, has been used as antidepressant and antiparkinsonian agent. Interestingly, deprenyl co-treatment was shown to reduce the oxidative stress and loss of TH-positive neurons mediated by rotenone in rat nigrostriatal regions and improved animal's locomotion (Saravanan et al., 2006). Likewise, another antiparkinsonian drug and dopamine receptor agonist, pramipexole, protected against rotenone-mediated dopaminergic neurotoxicity in mice (Inden et al., 2009). Inhibitors of adenosine A_{2A} receptors (ZM241385) but not A1 receptor antagonists (8-cyclopentyl-1, 3-dipropylxanthine) demonstrated partial protection from rotenone-mediated dopaminergic neurotoxicity in rats and improved motor behaviors, suggesting an important role of pre- and post-synaptic adenosine receptors in striatum (Fathalla et al., 2016). The neuroprotectant, cilostazol, prevented rotenone toxicity by promoting autophagy, which lead to activation of Nuclear receptor related 1 (Nurr1) receptor, responsible for maintaining dopaminergic neuronal health and inhibiting microglial activation (Hedya et al., 2018).

Dietary administration of antioxidants have been shown to be protective against PD in a population-based studies (de Rijk et al., 1997; Yang et al., 2017). Several studies have used antioxidants, such as thymoquinone, centrophenoxine, quercetin, edaravone, Dl-3-n-butylphthalide, α-lipoic acid, to abolish rotenone-mediated neurotoxicity and improved behavioral deficits in different experimental models of PD (Karuppagounder et al., 2013; Nehru et al., 2008; Radad et al., 2009; Xiong et al., 2011, 2012a; Zaitone et al., 2012b). Likewise, sesamol and naringenin, natural antioxidant flavonoids, reduced oxidative stress and prevented rotenone-mediated dopaminergic neurodegeneration in rats by inhibiting apoptotic activation of caspases and by alleviating UPS function (Sonia Angeline et al., 2013). Hydrogen sulfide (H_2S), an endogenous gas, have been shown to be neuroprotective due to its antioxidant properties (Lu et al., 2008; Yang et al., 2018). According to a report by Hu et al. (2010), chronic rotenone treatment markedly depleted H_2S levels in SN. Interestingly, the study showed that pretreatment with sodium hydrosulfide (1.68 mg/kg/day, i.p. for 3 days), a precursor for H_2S, reduced oxidative stress and ameliorated the dopaminergic neuronal loss in SN induced by rotenone. Further, SH-SY5Y cells pre-exposed to the redox cofactor of ETC, pyrroloquinoline quinone, reduced rotenone-induced neurotoxicity by maintaining mitochondrial respiration (Zhang et al., 2016).

Antioxidants have been widely investigated as potential neuroprotective treatments in PD. While promising in model systems, such treatments have not resulted in effective therapies. At least in some cases, the rotenone model confirms that antioxidants may not be protective or even have deleterious effects. As one example, Tapias et al. (2014) showed that co-administration of commonly believed neuroprotective antioxidant, pomegranate juice (Braidy et al., 2013; Essa et al., 2013; Hartman et al., 2006), exacerbated rotenone-mediated dopaminergic neurodegeneration. The investigators in this study showed that orally administered pomegranate juice heightened the loss of dopaminergic terminals, increased nigrostriatal inflammation and elevated the levels of nigral pro-apoptotic caspases in rotenone-treated rats. Increased nitrosative and oxidative stress suggested possible pro-oxidant effects of pomegranate juice in rat brain. Although not conclusive, the researchers proposed that presence of polyphenolic phytochemical mixture in pomegranate juice can act as pro-oxidants, unlike a single flavonoid used in high concentration which can act an antioxidant.

Next, anti-apoptotic drugs such as vanillin, montelukast and ursodeoxycholic acid, markedly reduced rotenone-mediated dopaminergic neurodegeneration in rats (Abdelkader et al., 2016; Dhanalakshmi et al., 2016; Mansour et al., 2018). Along the same line of research, rats treated with NOS inhibitor, *N*-ω-nitro-l-arginine methyl ester, also protected against rotenone-mediated TH loss in SN and striatum. Likewise, the free radical scavenger, melatonin, reduced rotenone-induced oxidative stress in rat SN (Bassani et al., 2014; Carriere et al., 2016; Lin et al., 2008; Pahl and Baeuerle, 1996; Saravanan et al., 2007). However, these studies was contradicted by Tapias et al. (2010) showing potentiating effects of melatonin on rotenone-mediated dopaminergic toxicity in rats. Importantly, that study examined neurochemical and pathological changes, including rigorous quantitative analysis of cell loss.

Parkinsonian effects in rats chronically treated with rotenone were alleviated by valproic acid's dietary administration (Monti et al., 2010). Valproic acid is a common anti-epileptic agent and have demonstrated neuroprotective properties (Eleuteri et al., 2009; Morland et al., 2004). Nuclear translocation of α-synuclein have been shown to inhibit acetylation of histones at various lysine sites, which can attribute to neurotoxicity (Kontopoulos et al., 2006; Leng and Chuang, 2006). Interestingly, valproic acid undermined the rotenone-induced α-synuclein nuclear translocation and inhibited the hyperacetylation of histone H3, perhaps blocking the further production of cytotoxic α-synuclein and reducing rotenone's neurotoxicity (Monti et al., 2010). Even more, valproic acid also ameliorated rotenone-dependent behavioral deficits in rats (Carriere et al., 2014).

Cigarette smoking have been linked with reduced risk of PD (reviewed in Ma et al., 2017). To evaluate in context of rotenone, Takeuchi et al. (2009) showed reduced loss of nigral TH-IR in mice co-treated with nicotine systemically. Iptakalim, a brain permeable potassium channel opener, was shown to block rotenone-induced dopaminergic neuronal loss by markedly reducing the activation of microglia and reducing neuroinflammation in SN (Zhou et al., 2007). Similarly, caffeine exposure also have been shown to be beneficial against rotenone-mediated dopaminergic toxicity in rats (Khadrawy et al., 2017).

Anti-inflammatory agents have been believed to reduce neuronal loss during several progressive neurodegenerative diseases including PD (Bornebroek et al., 2007; Gagne and Power, 2010; Ren et al., 2005). In like manner, drugs like, piracetam and vinpocetine, protected against rotenone-induced toxicity in rats by abolishing microglial activation (Zaitone et al., 2012a). In addition, an inhibitor of histamine-4 receptor (microglial phenotype regulating receptor), JNJ7777120, reduced rotenone-mediated pro-inflammatory cascade in SN and kept microglia in resting stage, which in turn rescued dopaminergic neurons and altered behavioral deficits (Zhou et al., 2019).

Lawana et al. (2017) showed that dasatinib, an inhibitor of c-Abl tyrosine kinase, blocked rotenone-activated signaling cascade required for gene expression and maturation of pro-inflammatory cytokines (IL-1β and IL-18) in microglia, which in turn abrogated the loss of TH neurons in primary neuron/glia mixed culture. Likewise, indirect inhibitor of proinflammatory transcription regulator NF-κB, vildagliptin (a selective inhibitor of dipeptidyl peptidase-4), abolished rotenone toxicity by downregulating genes associated with inflammation, oxidative stress and apoptosis in rat brain (Abdelsalam and Safar, 2015). Recently, Zaminelli et al. (2014) reported that chronic administration of a non-steroidal anti-inflammatory drug, ibuprofen, markedly reduced the nigral levels of cyclooxygenase (COX), alleviated oxidative stress and significantly reduced dopaminergic neurodegeneration in rotenone-treated rats. Similar protective effects were reported with salicylic acid against rotenone in rats (Madathil et al., 2013).

PD has been "cured" in animal models many times, with poor translation laboratory successes to clinical efficacy. The rotenone model has been utilized to identify both potential neuroprotective compounds. Notably, it has also identified treatments that increase toxicity. In some cases, these treatments were found to be neuroprotective in other models. While rotenone model may potentially have higher translation potential relative to other models, this will need to be empirically determined.

5. Conclusions

Evidence shows that rotenone triggers systemic mitochondrial ETC disruptions and nigral ROS accumulation, which precedes the deregulation of mitochondrial kinetics, elevated ER stress, redox-sensitive kinase activation, impaired proteostasis, accumulation of α-synuclein deposits and other pro-apoptotic and cytotoxic factors, causing aberrant loss of dopaminergic neurons. Additionally, rotenone was shown both in vivo and in vitro to activate microglia, which also magnifies neuron loss. Indeed, rotenone models revealed key features that are identical to clinical PD. Successful characterization of rotenone-based rat model led to development of mice model, followed by genetically manipulative non-mammalian *Drosophila* and *C. elegans* models and numerous in vitro models. Thus, in the field of PD preclinical research, rotenone is playing a major role in model development and predicting the PD pathology in brain and systematically. Use of known genetic mutations with even lower rotenone exposure may declutter the complexity of PD pathogenesis. Collectively, based on its suite of pathological markers and behavioral phenotypes, the rotenone model of PD will be regarded as one of the more highly reproducible and translatable experimental toxin-based PD models. Given the discoveries already made in the past ~20 years the rat rotenone model has been in existence, many more advances are expected with continually evolving iterations of rotenone exposure models as a valuable tool to study dopaminergic neurotoxicity.

References

Abdelkader, N.F., Safar, M.M., Salem, H.A., 2016. Ursodeoxycholic acid ameliorates apoptotic cascade in the rotenone model of Parkinson's disease: modulation of mitochondrial perturbations. Mol. Neurobiol. 53, 810–817.

Abdelsalam, R.M., Safar, M.M., 2015. Neuroprotective effects of vildagliptin in rat rotenone Parkinson's disease model: role of RAGE-NFkappaB and Nrf2-antioxidant signaling pathways. J. Neurochem. 133, 700–707.

Ahmadi, F.A., Linseman, D.A., Grammatopoulos, T.N., Jones, S.M., Bouchard, R.J., Freed, C.R., Heidenreich, K.A., Zawada, W.M., 2003. The pesticide rotenone induces caspase-3-mediated apoptosis in ventral mesencephalic dopaminergic neurons. J. Neurochem. 87, 914–921.

Aksoy, D., Solmaz, V., Cavusoglu, T., Meral, A., Ates, U., Erbas, O., 2017. Neuroprotective effects of exenatide in a rotenone-induced rat model of Parkinson's disease. Am. J. Med. Sci. 354, 319–324.

Alam, M., Schmidt, W.J., 2002. Rotenone destroys dopaminergic neurons and induces parkinsonian symptoms in rats. Behav. Brain Res. 136, 317–324.

Alam, M., Schmidt, W.J., 2004. L-DOPA reverses the hypokinetic behaviour and rigidity in rotenone-treated rats. Behav. Brain Res. 153, 439–446.

Antkiewicz-Michaluk, L., Karolewicz, B., Romanska, I., Michaluk, J., Bojarski, A.J., Vetulani, J., 2003. 1-methyl-1,2,3,4-tetrahydroisoquinoline protects against rotenone-induced mortality and biochemical changes in rat brain. Eur. J. Pharmacol. 466, 263–269.

Ariga, H., Takahashi-Niki, K., Kato, I., Maita, H., Niki, T., Iguchi-Ariga, S.M., 2013. Neuroprotective function of DJ-1 in Parkinson's disease. Oxid. Med. Cell. Longev. 2013, 683920.

Arnold, B., Cassady, S.J., Vanlaar, V.S., Berman, S.B., 2011. Integrating multiple aspects of mitochondrial dynamics in neurons: age-related differences and dynamic changes in a chronic rotenone model. Neurobiol. Dis. 41, 189–200.

Bailey, S.L., Carpentier, P.A., McMahon, E.J., Begolka, W.S., Miller, S.D., 2006. Innate and adaptive immune responses of the central nervous system. Crit. Rev. Immunol. 26, 149–188.

Bal-Price, A., Leist, M., Schildknecht, S., Tschudi-Monnet, F., Paini, A., Terron, A., 2018. Adverse Outcome Pathway on Inhibition of the Mitochondrial Complex I of Nigro-Striatal Neurons Leading to Parkinsonian Motor Deficits. OECD iLibrary.

Bardien, S., Lesage, S., Brice, A., Carr, J., 2011. Genetic characteristics of leucine-rich repeat kinase 2 (LRRK2) associated Parkinson's disease. Parkinsonism Relat. Disord. 17, 501–508.

Bashkatova, V., Alam, M., Vanin, A., Schmidt, W.J., 2004. Chronic administration of rotenone increases levels of nitric oxide and lipid peroxidation products in rat brain. Exp. Neurol. 186, 235–241.

Bassani, T.B., Gradowski, R.W., Zaminelli, T., Barbiero, J.K., Santiago, R.M., Boschen, S.L., da Cunha, C., Lima, M.M., Andreatini, R., Vital, M.A., 2014. Neuroprotective and antidepressant-like effects of melatonin in a rotenone-induced Parkinson's disease model in rats. Brain Res. 1593, 95–105.

Betarbet, R., Sherer, T.B., Mackenzie, G., Garcia-Osuna, M., Panov, A.V., Greenamyre, J.T., 2000. Chronic systemic pesticide exposure reproduces features of Parkinson's disease. Nat. Neurosci. 3, 1301–1306.

Betarbet, R., Canet-Aviles, R.M., Sherer, T.B., Mastroberardino, P.G., McLendon, C., Kim, J.H., Lund, S., Na, H.M., Taylor, G., Bence, N.F., Kopito, R., Seo, B.B., Yagi, T., Yagi, A., Klinefelter, G., Cookson, M.R., Greenamyre, J.T., 2006. Intersecting pathways to neurodegeneration in Parkinson's disease: effects of the pesticide rotenone on DJ-1, alpha-synuclein, and the ubiquitin-proteasome system. Neurobiol. Dis. 22, 404–420.

Biehlmaier, O., Alam, M., Schmidt, W.J., 2007. A rat model of Parkinsonism shows depletion of dopamine in the retina. Neurochem. Int. 50, 189–195.

Borland, M.K., Trimmer, P.A., Rubinstein, J.D., Keeney, P.M., Mohanakumar, K.P., LIU, L., Bennett, J.P., 2008. Chronic, low-dose rotenone reproduces Lewy neurites found in early stages of Parkinson's disease, reduces mitochondrial movement and slowly kills differentiated SH-SY5Y neural cells. Mol. Neurodegener. 3, 1–12.

Bornebroek, M., De Lau, L.M., Haag, M.D., Koudstaal, P.J., Hofman, A., Stricker, B.H., Breteler, M.M., 2007. Nonsteroidal anti-inflammatory drugs and the risk of Parkinson disease. Neuroepidemiology 28, 193–196.

Bougria, M., Vitorica, J., Cano, J., Machado, A., 1995. Implication of dopamine transporter system on 1-methyl-4-phenylpyridinium and rotenone effect in striatal synaptosomes. Eur. J. Pharmacol., Mol. Pharmacol. Sect. 291, 407–415.

Braidy, N., Selvaraju, S., Essa, M.M., Vaishnav, R., Al-Adawi, S., Al-Asmi, A., Al-Senawi, H., Abd Alrahman Alobaidy, A., Lakhtakia, R., Guillemin, G.J., 2013. Neuroprotective effects of a variety of pomegranate juice extracts against MPTP-induced cytotoxicity and oxidative stress in human primary neurons. Oxid. Med. Cell. Longev. 2013, 685909.

Bretaud, S., Lee, S., Guo, S., 2004. Sensitivity of zebrafish to environmental toxins implicated in Parkinson's disease. Neurotoxicol. Teratol. 26, 857–864.

Cabeza-Arvelaiz, Y., Schiestl, R.H., 2012. Transcriptome analysis of a rotenone model of parkinsonism reveals complex I-tied and -untied toxicity mechanisms common to neurodegenerative diseases. PLoS One 7, e44700.

Caboni, P., Sherer, T.B., Zhang, N.J., Taylor, G., Na, H.M., Greenamyre, J.T., Casida, J.E., 2004. Rotenone, deguelin, their metabolites, and the rat model of Parkinson's disease. Chem. Res. Toxicol. 17, 1540–1548.

Cannon, J.R., Greenamyre, J.T., 2010. Chapter 2—Neurotoxic in vivo models of Parkinson's disease: recent advances. In: Björklund, A., Cenci, M.A. (Eds.), Progress in Brain Research. Elsevier.

Cannon, J.R., Greenamyre, J.T., 2011. The role of environmental exposures in neurodegeneration and neurodegenerative diseases. Toxicol. Sci. 124, 225–250.

Cannon, J.R., Tapias, V., Na, H.M., Honick, A.S., Drolet, R.E., Greenamyre, J.T., 2009. A highly reproducible rotenone model of Parkinson's disease. Neurobiol. Dis. 34, 279–290.

Cannon, J.R., Geghman, K.D., Tapias, V., Sew, T., Dail, M.K., Li, C., Greenamyre, J.T., 2013. Expression of human E46K-mutated alpha-synuclein in BAC-transgenic rats replicates early-stage Parkinson's disease features and enhances vulnerability to mitochondrial impairment. Exp. Neurol. 240, 44–56.

Carriere, C.H., Kang, N.H., Niles, L.P., 2014. Neuroprotection by valproic acid in an intrastriatal rotenone model of Parkinson's disease. Neuroscience 267, 114–121.

Carriere, C.H., Kang, N.H., Niles, L.P., 2016. Chronic low-dose melatonin treatment maintains nigrostriatal integrity in an intrastriatal rotenone model of Parkinson's disease. Brain Res. 1633, 115–125.

Casarejos, M.J., Menendez, J., Solano, R.M., Rodriguez-Navarro, J.A., De Garcia Yebenes, J., Mena, M.A., 2006. Susceptibility to rotenone is increased in neurons from parkin null mice and is reduced by minocycline. J. Neurochem. 97, 934–946.

Cavoski, I., Caboni, P., Sarais, G., Cabras, P., Miano, T., 2007. Photodegradation of rotenone in soils under environmental conditions. J. Agric. Food Chem. 55, 7069–7074.

Chao, H., Liu, Y., Fu, X., Xu, X., Bao, Z., Lin, C., Li, Z., Liu, Y., Wang, X., You, Y., Liu, N., Ji, J., 2018. Lowered iPLA2gamma activity causes increased mitochondrial lipid peroxidation and mitochondrial dysfunction in a rotenone-induced model of Parkinson's disease. Exp. Neurol. 300, 74–86.

Chen, M.J., Yap, Y.W., Choy, M.S., Koh, C.H., Seet, S.J., Duan, W., Whiteman, M., Cheung, N.S., 2006. Early induction of calpains in rotenone-mediated neuronal apoptosis. Neurosci. Lett. 397, 69–73.

Chen, S., Zhang, X.J., Yang, D.H., Du, Y.L., Li, L., Li, X.P., Ming, M., Le, W.D., 2008a. D2/D3 receptor agonist ropinirole protects dopaminergic cell line against rotenone-induced apoptosis through inhibition of caspase- and JNK-dependent pathways. FEBS Lett. 582, 603–610.

Chen, Y.Y., Chen, G., Fan, Z., Luo, J., Ke, Z.J., 2008b. GSK3beta and endoplasmic reticulum stress mediate rotenone-induced death of SK-N-MC neuroblastoma cells. Biochem. Pharmacol. 76, 128–138.

Cheng, H.M., Casida, J.E., Yamamoto, I., 1972. Rotenone photodecomposition. J. Agric. Food Chem. 20, 850–856.

Chikka, M.R., Anbalagan, C., Dvorak, K., Dombeck, K., Prahlad, V., 2016. The mitochondria-regulated immune pathway activated in the C-elegans intestine is neuroprotective. Cell Rep. 16, 2399–2414.

Chinta, S.J., Andersen, J.K., 2006. Reversible inhibition of mitochondrial complex I activity following chronic dopaminergic glutathione depletion in vitro: implications for Parkinson's disease. Free Radic. Biol. Med. 41, 1442–1448.

Choi, W.S., Palmiter, R.D., Xia, Z., 2011. Loss of mitochondrial complex I activity potentiates dopamine neuron death induced by microtubule dysfunction in a Parkinson's disease model. J. Cell Biol. 192, 873–882.

Choi, W.S., Kim, H.W., Xia, Z.G., 2015. JNK inhibition of VMAT2 contributes to rotenone-induced oxidative stress and dopamine neuron death. Toxicology 328, 75–81.

Choong, C.J., Say, Y.H., 2011. Neuroprotection of alpha-synuclein under acute and chronic rotenone and maneb treatment is abolished by its familial Parkinson's disease mutations A30P, A53T and E46K. Neurotoxicology 32, 857–863.

Choubey, V., Safiulina, D., Vaarmann, A., Cagalinec, M., Wareski, P., Kuum, M., Zharkovsky, A., Kaasik, A., 2011. Mutant A53T alpha-synuclein induces neuronal death by increasing mitochondrial autophagy. J. Biol. Chem. 286, 10814–10824.

Chu, C.T., Ji, J., Dagda, R.K., Jiang, J.F., Tyurina, Y.Y., Kapralov, A.A., Tyurin, V.A., Yanamala, N., Shrivastava, I.H., Mohammadyani, D., Wang, K.Z.Q., Zhu, J.H., Klein-Seetharaman, J., Balasubramanian, K., Amoscato, A.A., Borisenko, G., Huang, Z.T., Gusdon, A.M., Cheikhi, A., Steer, E.K., Wang, R., Baty, C., Watkins, S., Bahar, I., Bayir, H., Kagan, V.E., 2013. Cardiolipin externalization to the outer mitochondrial membrane acts as an elimination signal for mitophagy in neuronal cells. Nat. Cell Biol. 15, 1197–U168.

Coulom, H., Birman, S., 2004. Chronic exposure to rotenone models sporadic Parkinson's disease in Drosophila melanogaster. J. Neurosci. 24, 10993–10998.

Dadakhujaev, S., Noh, H.S., Jung, E.J., Cha, J.Y., Baek, S.M., Ha, J.H., Kim, D.R., 2010. Autophagy protects the rotenone-induced cell death in alpha-synuclein overexpressing SH-SY5Y cells. Neurosci. Lett. 472, 47–52.

Dawson, V.K., Gingerich, W.H., Davis, R.A., Gilderhus, P.A., 1991. Rotenone persistence in freshwater ponds: effects of temperature and sediment adsorption. N. Am. J. Fish Manag. 11, 226–231.

De Miranda, B.R., Rocha, E.M., Bai, Q., El Ayadi, A., Hinkle, D., Burton, E.A., Greenamyre, J.T., 2018. Astrocyte-specific DJ-1 overexpression protects against rotenone-induced neurotoxicity in a rat model of Parkinson's disease. Neurobiol. Dis. 115, 101–114.

De Miranda, B.R., Fazzari, M., Rocha, E.M., Castro, S., Greenamyre, J.T., 2019. Sex differences in rotenone sensitivity reflect the male-to-female ratio in human Parkinson's disease incidence. Toxicol. Sci. 170, 133–143.

De Rijk, M.C., Breteler, M.M., den Breeijen, J.H., Launer, L.J., Grobbee, D.E., van der MECHE, F.G., Hofman, A., 1997. Dietary antioxidants and Parkinson disease. The Rotterdam study. Arch. Neurol. 54, 762–765.

Dhanalakshmi, C., Janakiraman, U., Manivasagam, T., Justin Thenmozhi, A., Essa, M.M., Kalandar, A., Khan, M.A., Guillemin, G.J., 2016. Vanillin attenuated behavioural impairments, neurochemical deficts, oxidative stress and apoptosis against rotenone induced rat model of Parkinson's disease. Neurochem. Res. 41, 1899–1910.

Dhillon, A.S., Tarbutton, G.L., Levin, J.L., Plotkin, G.M., Lowry, L.K., Nalbone, J.T., Shepherd, S., 2008. Pesticide/environmental exposures and Parkinson's disease in East Texas. J. Agromedicine 13, 37–48.

di Maio, R., Barrett, P.J., Hoffman, E.K., Barrett, C.W., Zharikov, A., Borah, A., Hu, X.P., Mccoy, J., Chu, C.T., Burton, E.A., Hastings, T.G., Greenamyre, J.T., 2016. Alpha-synuclein binds to TOM20 and inhibits mitochondrial protein import in Parkinson's disease. Sci. Transl. Med. 8, 1–14. 342ra78.

Di Maio, R., Hoffman, E.K., Rocha, E.M., Keeney, M.T., Sanders, L.H., de Miranda, B.R., Zharikov, A., van Laar, A., Stepan, A.F., Lanz, T.A., Kofler, J.K., Burton, E.A., Alessi, D.R., Hastings, T.G., Greenamyre, J.T., 2018. LRRK2 activation in idiopathic Parkinson's disease. Sci. Transl. Med. 10, 1–12.

di Rita, A., D'Acunzo, P., Strappazzon, F., Cecconi, F., 2018. AMBRA1-mediated mitophagy counteracts oxidative stress and apoptosis induced by neurotoxicity in human neuroblastoma SH-SY5Y cells. Front. Cell. Neurosci. 12, 1–11. 92.

Dodiya, H.B., Forsyth, C.B., Voigt, R.M., Engen, P.A., Patel, J., Shaikh, M., GREEN, S.J., Naqib, A., Roy, A., Kordower, J.H., Pahan, K., Shannon, K.M., Keshavarzian, A., 2018. Chronic stress-induced gut dysfunction exacerbates Parkinson's disease phenotype and pathology in a rotenone-induced mouse model of Parkinson's disease. Neurobiol. Dis. 104352.

dos Santos, A.C., Castro, M.A., Jose, E.A., Delattre, A.M., Dombrowski, P.A., da Cunha, C., Ferraz, A.C., Lima, M.M., 2013. REM sleep deprivation generates cognitive and neurochemical disruptions in the intranigral rotenone model of Parkinson's disease. J. Neurosci. Res. 91, 1508–1516.

Drolet, R.E., Cannon, J.R., Montero, L., Greenamyre, J.T., 2009. Chronic rotenone exposure reproduces Parkinson's disease gastrointestinal neuropathology. Neurobiol. Dis. 36, 96–102.

Eleuteri, S., Monti, B., Brignani, S., Contestabile, A., 2009. Chronic dietary administration of valproic acid protects neurons of the rat nucleus basalis magnocellularis from ibotenic acid neurotoxicity. Neurotox. Res. 15, 127–132.

Elgayar, S.A.M., Abdel-HAFEZ, A.A.M., Gomaa, A.M.S., Elsherif, R., 2018. Vulnerability of glia and vessels of rat substantia nigra in rotenone Parkinson model. Ultrastruct. Pathol. 42, 181–192.

Emmrich, J.V., Hornik, T.C., Neher, J.J., Brown, G.C., 2013. Rotenone induces neuronal death by microglial phagocytosis of neurons. FEBS J. 280, 5030–5038.

Essa, M., Subash, S., Braidy, N., Al-Adawi, S., Al-Asmi, A., Al-Senawi, H., Vaishnav, R., Guillemin, G.J., 2013. Neuroprotective effects of pomegranate juice extracts on quinolinic acid; induced excitotoxicity in human neurons. Alzheimers Dement. 9, P800–P801.

Esteve-RUDD, J., Fernandez-Sanchez, L., Lax, P., de Juan, E., Martin-Nieto, J., Cuenca, N., 2011. Rotenone induces degeneration of photoreceptors and impairs the dopaminergic system in the rat retina. Neurobiol. Dis. 44, 102–115.

Ethell, D.W., Fei, Q.Y., 2009. Parkinson-linked genes and toxins that affect neuronal cell death through the Bcl-2 family. Antioxid. Redox Signal. 11, 529–540.

Fathalla, A.M., Soliman, A.M., Ali, M.H., Moustafa, A.A., 2016. Adenosine A2A receptor blockade prevents rotenone-induced motor impairment in a rat model of parkinsonism. Front. Behav. Neurosci. 10, 35.

Feng, J., 2006. Microtubule: a common target for parkin and Parkinson's disease toxins. Neuroscientist 12, 469–476.

Ferrante, R.J., Schulz, J.B., Kowall, N.W., Beal, M.F., 1997. Systemic administration of rotenone produces selective damage in the striatum and globus pallidus, but not in the substantia nigra. Brain Res. 753, 157–162.

Filomeni, G., Graziani, I., de Zio, D., Dini, L., Centonze, D., Rotilio, G., Ciriolo, M.R., 2012. Neuroprotection of kaempferol by autophagy in models of rotenone-mediated acute toxicity: possible implications for Parkinson's disease. Neurobiol. Aging 33, 767–785.

Finlayson, J., Siepmann, S., Trumbo, J., 2001. Chemical residues in surface and ground waters following rotenone application to California lakes and streams. In: Cailteux, R.L., DeMong, L., Finlayson, B.J., Horton, W., McClay, W., Schnick, R.A., Thompson, C. (Eds.), Rotenone in Fisheries: Are the Rewards Worth the Risks? American Fisheries Society, pp. 37–53.

Fleming, S.M., 2017. Mechanisms of gene-environment interactions in Parkinson's disease. Curr. Environ. Health Rep. 4, 192–199.

Fleming, S.M., Zhu, C.N., Fernagut, P.O., Mehta, A., Dicarlo, C.D., Seaman, R.L., Chesselet, M.F., 2004. Behavioral and immunohistochemical effects of chronic intravenous and subcutaneous infusions of varying doses of rotenone. Exp. Neurol. 187, 418–429.

Fullard, M.E., Tran, B., Xie, S.X., Toledo, J.B., Scordia, C., Linder, C., Purri, R., weintraub, D., Duda, J.E., Chahine, L.M., Morley, J.F., 2016. Olfactory impairment predicts cognitive decline in early Parkinson's disease. Parkinsonism Relat. Disord. 25, 45–51.

Gagne, J.J., Power, M.C., 2010. Anti-inflammatory drugs and risk of Parkinson disease: a meta-analysis. Neurology 74, 995–1002.

Gao, H.M., Hong, J.S., Zhang, W., Liu, B., 2002. Distinct role for microglia in rotenone-induced degeneration of dopaminergic neurons. J. Neurosci. 22, 782–790.

Gao, H.M., Liu, B., Hong, J.S., 2003. Critical role for microglial NADPH oxidase in rotenone-induced degeneration of dopaminergic neurons. J. Neurosci. 23, 6181–6187.

Gao, H., Yang, W., Qi, Z., Lu, L., Duan, C., Zhao, C., Yang, H., 2012. DJ-1 protects dopaminergic neurons against rotenone-induced apoptosis by enhancing ERK-dependent mitophagy. J. Mol. Biol. 423, 232–248.

Garcia-Dominguez, I., Vesela, K., Garcia-Revilla, J., Carrillo-Jimenez, A., Roca-Ceballos, M.A., Santiago, M., de Pablos, R.M., Venero, J.L., 2018. Peripheral inflammation enhances microglia response and Nigral dopaminergic cell death in an in vivo MPTP model of Parkinson's disease. Front. Cell. Neurosci. 12, 1–16. 398.

Garcia-Garcia, F., Ponce, S., Brown, R., Cussen, V., Krueger, J.M., 2005. Sleep disturbances in the rotenone animal model of Parkinson disease. Brain Res. 1042, 160–168.

George, S., Mok, S.S., Nurjono, M., Ayton, S., Finkelstein, D.I., Masters, C.L., Li, Q.X., Culvenor, J.G., 2010. Alpha-synuclein transgenic mice reveal compensatory increases in Parkinson's disease-associated proteins DJ-1 and parkin and have enhanced alpha-synuclein and PINK1 levels after rotenone treatment. J. Mol. Neurosci. 42, 243–254.

Goldman, S.M., Tanner, C.M., 2019. Rotenone and Parkinson's disease: reduced sensitivity in females. Toxicol. Sci. 170, 562–563.

Goswami, P., Gupta, S., Biswas, J., Joshi, N., Swarnkar, S., Nath, C., Singh, S., 2016a. Endoplasmic reticulum stress plays a key role in rotenone-induced apoptotic death of neurons. Mol. Neurobiol. 53, 285–298.

Goswami, P., Gupta, S., Biswas, J., Sharma, S., Singh, S., 2016b. Endoplasmic reticulum stress instigates the rotenone induced oxidative apoptotic neuronal death: a study in rat brain. Mol. Neurobiol. 53, 5384–5400.

Gotz, M.E., Freyberger, A., Riederer, P., 1990. Oxidative stress: a role in the pathogenesis of Parkinson's disease. J. Neural Transm. Suppl. 29, 241–249.

Gotz, M.E., Double, K., Gerlach, M., Youdim, M.B., Riederer, P., 2004. The relevance of iron in the pathogenesis of Parkinson's disease. Ann. N. Y. Acad. Sci. 1012, 193–208.

Greenamyre, J.T., Mackenzie, G., Peng, T.I., Stephans, S.E., 1999. Mitochondrial dysfunction in Parkinson's disease. Biochem. Soc. Symp. 66, 85–97. https://doi.org/10.1042/bss0660085.

Greenamyre, J.T., Cannon, J.R., Drolet, R., Mastroberardino, P.G., 2010. Lessons from the rotenone model of Parkinson's disease. Trends Pharmacol. Sci. 31, 141–142.

Greene, J.G., Noorian, A.R., Srinivasan, S., 2009. Delayed gastric emptying and enteric nervous system dysfunction in the rotenone model of Parkinson's disease. Exp. Neurol. 218, 154–161.

Guerreiro, R.J., Bras, J.M., Santana, I., Januario, C., Santiago, B., Morgadinho, A.S., Ribeiro, M.H., Hardy, J., Singleton, A., Oliveira, C., 2006. Association of HFE common mutations with Parkinson's disease, Alzheimer's disease and mild cognitive impairment in a Portuguese cohort. BMC Neurol. 6, 24.

Hajieva, P., Mocko, J.B., Moosmann, B., Behl, C., 2009. Novel imine antioxidants at low nanomolar concentrations protect dopaminergic cells from oxidative neurotoxicity. J. Neurochem. 110, 118–132.
Hansch, C., 1995. Comparative QSAR: understanding hydrophobic interactions. In: Hansch, C., Fujita, T. (Eds.), Classical and Three-Dimensional Qsar in Agrochemistry. In: 606, American Chemical Society, pp. 254–262.
Harischandra, D.S., Rokad, D., Ghaisas, S., Verma, S., Robertson, A., Jin, H., Anantharam, V., Kanthasamy, A., Kanthasamy, A.G., 2019. Enhanced differentiation of human dopaminergic neuronal cell model for preclinical translational research in Parkinson's disease. Biochim. Biophys. Acta, Mol. Basis Dis. 165533.
Hartman, R.E., Shah, A., Fagan, A.M., Schwetye, K.E., Parsadanian, M., Schulman, R.N., Finn, M.B., Holtzman, D.M., 2006. Pomegranate juice decreases amyloid load and improves behavior in a mouse model of Alzheimer's disease. Neurobiol. Dis. 24, 506–515.
He, Y., Imam, S.Z., Dong, Z.J., Jankovic, J., Ali, S.F., Appel, S.H., Le, W.D., 2003. Role of nitric oxide in rotenone-induced nigro-striatal injury. J. Neurochem. 86, 1338–1345.
Hedya, S.A., Safar, M.M., Bahgat, A.K., 2018. Cilostazol mediated Nurr1 and autophagy enhancement: neuroprotective activity in rat rotenone PD model. Mol. Neurobiol. 55, 7579–7587.
Heikkila, R.E., Nicklas, W.J., Vyas, I., Duvoisin, R.C., 1985. Dopaminergic toxicity of rotenone and the 1-methyl-4-phenylpyridinium ion after their stereotaxic administration to rats: implication for the mechanism of 1-methyl-4-phenyl-1,2,3,6-tetrahydropyridine toxicity. Neurosci. Lett. 62, 389–394.
Hoglinger, G.U., Feger, J., Prigent, A., Michel, P.P., Parain, K., Champy, P., Ruberg, M., Oertel, W.H., Hirsch, E.C., 2003. Chronic systemic complex I inhibition induces a hypokinetic multisystem degeneration in rats. J. Neurochem. 84, 491–502.
Hoglinger, G.U., Lannuzel, A., Khondiker, M.E., Michel, P.P., Duyckaerts, C., Feger, J., Champy, P., Prigent, A., Medja, F., Lombes, A., Oertel, W.H., Ruberg, M., Hirsch, E.C., 2005. The mitochondrial complex I inhibitor rotenone triggers a cerebral tauopathy. J. Neurochem. 95, 930–939.
Hosamani, R., Muralidhara, 2009. Neuroprotective efficacy of *Bacopa monnieri* against rotenone induced oxidative stress and neurotoxicity in *Drosophila melanogaster*. Neurotoxicology 30, 977–985.
Hsuan, S.L., Klintworth, H.M., Xia, Z.G., 2006. Basic fibroblast growth factor protects against rotenone-induced dopaminergic cell death through activation of extracellular signal-regulated kinases 1/2 and phosphatidylinositol-3 kinase pathways. J. Neurosci. 26, 4481–4491.
Hu, L.F., Lu, M., Tiong, C.X., Dawe, G.S., Hu, G., Bian, J.S., 2010. Neuroprotective effects of hydrogen sulfide on Parkinson's disease rat models. Aging Cell 9, 135–146.
Hwang, O., 2013. Role of oxidative stress in Parkinson's disease. Exp Neurobiol. 22, 11–17.
Imbeault, E., Mahvelati, T.M., Braun, R., Gris, P., Gris, D., 2014. Nlrx1 regulates neuronal cell death. Mol. Brain 7, 90.
Inden, M., Kitamura, Y., Takeuchi, H., Yanagida, T., Takata, K., Kobayashi, Y., Taniguchi, T., Yoshimoto, K., Kaneko, M., Okuma, Y., Taira, T., Ariga, H., Shimohama, S., 2007. Neurodegeneration of mouse nigrostriatal dopaminergic system induced by repeated oral administration of rotenone is prevented by 4-phenylbutyrate, a chemical chaperone. J. Neurochem. 101, 1491–1504.
Inden, M., Kitamura, Y., Tamaki, A., Yanagida, T., Shibaike, T., Yamamoto, A., Takata, K., Yasui, H., Taira, T., Ariga, H., Taniguchi, T., 2009. Neuroprotective effect of the antiparkinsonian drug pramipexole against nigrostriatal dopaminergic degeneration in rotenone-treated mice. Neurochem. Int. 55, 760–767.

Inden, M., Kitamura, Y., Abe, M., Tamaki, A., Takata, K., Taniguchi, T., 2011. Parkinsonian rotenone mouse model: reevaluation of long-term administration of rotenone in C57BL/6 mice. Biol. Pharm. Bull. 34, 92–96.

Iversen, S.D., Iversen, L.L., 2007. Dopamine: 50 years in perspective. Trends Neurosci. 30, 188–193.

Jang, S., Yu, L.R., Abdelmegeed, M.A., Gao, Y., Banerjee, A., Song, B.J., 2015. Critical role of c-Jun N-terminal protein kinase in promoting mitochondrial dysfunction and acute liver injury. Redox Biol. 6, 552–564.

Jia, F., Song, N., Zhao, C., Xie, J., Jiang, H., 2014. Unexpected improvements of spatial learning and memory abilities in chronic rotenone intoxicated mice. PLoS One 9, e91641.

Jiang, Q., Yan, Z., Feng, J., 2006. Neurotrophic factors stabilize microtubules and protect against rotenone toxicity on dopaminergic neurons. J. Biol. Chem. 281, 29391–29400.

Johnson, M.E., Lim, Y., Senthilkumaran, M., Zhou, X.F., Bobrovskaya, L., 2015. Investigation of tyrosine hydroxylase and BDNF in a low-dose rotenone model of Parkinson's disease. J. Chem. Neuroanat. 70, 33–41.

Johnson, M.E., Stringer, A., Bobrovskaya, L., 2018. Rotenone induces gastrointestinal pathology and microbiota alterations in a rat model of Parkinson's disease. Neurotoxicology 65, 174–185.

Kamel, F., Tanner, C.M., Umbach, D.M., Hoppin, J.A., Alavanja, M.C.R., Blair, A., Comyns, K., Goldman, S.M., Korell, M., Langston, J.W., Ross, G.W., Sandler, D.P., 2007. Pesticide exposure and self-reported Parkinson's disease in the agricultural health study. Am. J. Epidemiol. 165, 364–374.

Karuppagounder, S.S., Madathil, S.K., Pandey, M., Haobam, R., Rajamma, U., Mohanakumar, K.P., 2013. Quercetin up-regulates mitochondrial complex-I activity to protect against programmed cell death in rotenone model of Parkinson's disease in rats. Neuroscience 236, 136–148.

Kaur, D., Yantiri, F., Rajagopalan, S., Kumar, J., Mo, J.Q., Boonplueang, R., Viswanath, V., Jacobs, R., Yang, L., Beal, M.F., Dimonte, D., Volitaskis, I., Ellerby, L., Cherny, R.A., Bush, A.I., Andersen, J.K., 2003. Genetic or pharmacological iron chelation prevents MPTP-induced neurotoxicity in vivo: a novel therapy for Parkinson's disease. Neuron 37, 899–909.

Kaur, H., Chauhan, S., Sandhir, R., 2011. Protective effect of lycopene on oxidative stress and cognitive decline in rotenone induced model of Parkinson's disease. Neurochem. Res. 36, 1435–1443.

Kawasaki, A., Hayashi, T., Nakachi, K., Trosko, J.E., Sugihara, K., Kotake, Y., Ohta, S., 2009. Modulation of connexin 43 in rotenone-induced model of Parkinson's disease. Neuroscience 160, 61–68.

Khadrawy, Y.A., Salem, A.M., El-Shamy, K.A., Ahmed, E.K., Fadl, N.N., Hosny, E.N., 2017. Neuroprotective and therapeutic effect of caffeine on the rat model of Parkinson's disease induced by rotenone. J Diet Suppl. 14, 553–572.

Kilpatrick, K., Novoa, J.A., Hancock, T., Guerriero, C.J., Wipf, P., Brodsky, J.L., Segatori, L., 2013. Chemical induction of Hsp70 reduces alpha-synuclein aggregation in Neuroglioma cells. ACS Chem. Biol. 8, 1460–1468.

Klintworth, H., Newhouse, K., Li, T.T., Choi, W.S., Faigle, R., Xia, Z.G., 2007. Activation of c-Jun N-terminal protein kinase is a common mechanism underlying paraquat- and rotenone-induced dopaminergic cell apoptosis. Toxicol. Sci. 97, 149–162.

Koch, A., Lehmann-HORN, K., Daschsel, J.C., Gasser, T., Kahle, P.J., Lucking, C.B., 2009. Proteasomal inhibition reduces parkin mRNA in PC12 and SH-SY5Y cells. Parkinsonism Relat. Disord. 15, 220–225.

Kontopoulos, E., Parvin, J.D., Feany, M.B., 2006. Alpha-synuclein acts in the nucleus to inhibit histone acetylation and promote neurotoxicity. Hum. Mol. Genet. 15, 3012–3023.

Laforge, F.B., Haller, H.L., Smith, L.E., 1933. The determination of the structure of rotenone. Chem. Rev. 12, 181–213.

Lapointe, N., St-Hilaire, M., Martinoli, M.G., Blanchet, J., Gould, P., Rouillard, C., Cicchetti, F., 2004. Rotenone induces non-specific central nervous system and systemic toxicity. FASEB J. 18, 717–719.

Lawana, V., Singh, N., Sarkar, S., Charli, A., Jin, H., Anantharam, V., Kanthasamy, A.G., Kanthasamy, A., 2017. Involvement of c-Abl kinase in microglial activation of NLRP3 Inflammasome and impairment in autolysosomal system. J. Neuroimmune Pharmacol. 12, 624–660.

Leng, Y., Chuang, D.M., 2006. Endogenous alpha-synuclein is induced by valproic acid through histone deacetylase inhibition and participates in neuroprotection against glutamate-induced excitotoxicity. J. Neurosci. 26, 7502–7512.

Leng, A., Feldon, J., Ferger, B., 2003. Rotenone increases glutamate-induced dopamine release but does not affect hydroxyl-free radical formation in rat striatum. Synapse 50, 240–250.

Li, X., Liu, Z., Tamashiro, K., Shi, B., Rudnicki, D.D., Ross, C.A., Moran, T.H., Smith, W.W., 2010. Synphilin-1 exhibits trophic and protective effects against rotenone toxicity. Neuroscience 165, 455–462.

Li, C., Chen, X., Zhang, N., Song, Y., Mu, Y., 2012. Gastrodin inhibits neuroinflammation in rotenone-induced Parkinson's disease model rats. Neural Regen. Res. 7, 325–331.

Li, H., Jiang, H.B., Zhang, B.Y., Feng, J., 2018. Modeling Parkinson's disease using patient-specific induced pluripotent stem cells. J. Parkinsons Dis. 8, 479–493.

Liang, Y., Jing, X., Zeng, Z., Bi, W., Chen, Y., Wu, X., Yang, L., Liu, J., Xiao, S., Liu, S., Lin, D., Tao, E., 2015. Rifampicin attenuates rotenone-induced inflammation via suppressing NLRP3 inflammasome activation in microglia. Brain Res. 1622, 43–50.

Lima, M.M.S., 2013. Sleep disturbances in Parkinson's disease: the contribution of dopamine in REM sleep regulation. Sleep Med. Rev. 17, 367–375.

Lin, C.H., Huang, J.Y., Ching, C.H., Chuang, J.I., 2008. Melatonin reduces the neuronal loss, downregulation of dopamine transporter, and upregulation of D2 receptor in rotenone-induced parkinsonian rats. J. Pineal Res. 44, 205–213.

Lionnet, A., Leclair-Visonneau, L., Neunlist, M., Murayama, S., Takao, M., Adler, C.H., Derkinderen, P., Beach, T.G., 2018. Does Parkinson's disease start in the gut? Acta Neuropathol. 135, 1–12.

Liu, Y., Sun, J.D., Song, L.K., Li, J., Chu, S.F., Yuan, Y.H., Chen, N.H., 2015. Environment-contact administration of rotenone: a new rodent model of Parkinson's disease. Behav. Brain Res. 294, 149–161.

Lotharius, J., Brundin, P., 2002. Pathogenesis of Parkinson's disease: dopamine, vesicles and alpha-synuclein. Nat. Rev. Neurosci. 3, 932–942.

Lu, M., Hu, L.F., Hu, G., Bian, J.S., 2008. Hydrogen sulfide protects astrocytes against H(2)O(2)-induced neural injury via enhancing glutamate uptake. Free Radic. Biol. Med. 45, 1705–1713.

Luo, C.L., Huang, Y.C., Huang, B., Lu, T.S., Fu, Y.S., 2017. Caffeine may affect apoptosis and autophagy activity to promote dopaminergic cells survival under the rotenone induced Parkinson disease model. FASEB J. 31 (1).

Ma, C., Liu, Y., Neumann, S., Gao, X., 2017. Nicotine from cigarette smoking and diet and Parkinson disease: a review. Transl Neurodegener. 6, 18.

Madathil, S.K., Karuppagounder, S.S., Mohanakumar, K.P., 2013. Sodium salicylate protects against rotenone-induced Parkinsonism in rats. Synapse 67, 502–514.

Mansour, R.M., Ahmed, M.A.E., El-Sahar, A.E., El Sayed, N.S., 2018. Montelukast attenuates rotenone-induced microglial activation/p38 MAPK expression in rats: possible role of its antioxidant, anti-inflammatory and antiapoptotic effects. Toxicol. Appl. Pharmacol. 358, 76–85.
Maragos, W.F., Zhu, J., Chesnut, M.D., Dwoskin, L.P., 2002. Mitochondrial toxin inhibition of [H-3]dopamine uptake into rat striatal synaptosomes. Biochem. Pharmacol. 63, 1499–1505.
Marella, M., Seo, B.B., Nakamaru-Ogiso, E., Greenamyre, J.T., Matsuno-Yagi, A., Yagi, T., 2008. Protection by the NDI1 gene against neurodegeneration in a rotenone rat model of Parkinson's disease. PLoS One 3, e1433.
Martinez, T.N., Greenamyre, J.T., 2012. Toxin models of mitochondrial dysfunction in Parkinson's disease. Antioxid. Redox Signal. 16, 920–934.
Mastroberardino, P.G., Hoffman, E.K., Horowitz, M.P., betarbet, R., Taylor, G., Cheng, D.M., Na, H.M., Gutekunst, C.A., Gearing, M., Trojanowski, J.Q., Anderson, M., Chu, C.T., Peng, J.M., Greenamyre, J.T., 2009. A novel transferrin/TfR2-mediated mitochondrial iron transport system is disrupted in Parkinson's disease. Neurobiol. Dis. 34, 417–431.
Maulika, M., Mitra, S., Basmayor, A.M., Lu, B., Taylor, B.E., Bult-Ito, A., 2019. Genetic silencing of fatty acid desaturases modulates alpha-Synuclein toxicity and neuronal loss in Parkinson-like models of *C. elegans*. Front. Aging Neurosci. 11.
Metcalf, R.L., 1948. The Mode of Action of Organic Insecticides. Chemical-Biological Coordination Center- National Research Council, Washington D.C.
Meulener, M., Whitworth, A.J., Armstrong-Gold, C.E., Rizzu, P., Heutink, P., Wes, P.D., Pallanck, L.J., Bonini, N.M., 2005. Drosophila DJ-1 mutants are selectively sensitive to environmental toxins associated with Parkinson's disease. Curr. Biol. 15, 1572–1577.
Mili, A., Ben Charfeddine, I., Mamai, O., Abdelhak, S., Adala, L., Amara, A., Pagliarani, S., Lucchiarri, S., Ayadi, A., Tebib, N., Harbi, A., Bouguila, J., H'Mida, D., Saad, A., Limem, K., Comi, G.P., Gribaa, M., 2012. Molecular and biochemical characterization of Tunisian patients with glycogen storage disease type III. J. Hum. Genet. 57, 170–175.
Mocko, J.B., Kern, A., Moosmann, B., Behl, C., Hajieva, P., 2010. Phenothiazines interfere with dopaminergic neurodegeneration in Caenorhabditis elegans models of Parkinson's disease. Neurobiol. Dis. 40, 120–129.
Monti, B., Gatta, V., Piretti, F., Raffaelli, S.S., Virgili, M., Contestabile, A., 2010. Valproic acid is neuroprotective in the rotenone rat model of Parkinson's disease: involvement of alpha-Synuclein. Neurotox. Res. 17, 130–141.
Moon, Y., Lee, K.H., Park, J.H., Geum, D., Kim, K., 2005. Mitochondrial membrane depolarization and the selective death of dopaminergic neurons by rotenone: protective effect of coenzyme Q(10). J. Neurochem. 93, 1199–1208.
Morland, C., Boldingh, K.A., Iversen, E.G., Hassel, B., 2004. Valproate is neuroprotective against malonate toxicity in rat striatum: an association with augmentation of high-affinity glutamate uptake. J. Cereb. Blood Flow Metab. 24, 1226–1234.
Mulcahy, P., O'Doherty, A., Paucard, A., O'Brien, T., Kirik, D., Dowd, E., 2012. Development and characterisation of a novel rat model of Parkinson's disease induced by sequential Intranigral Administration of Aav-Alpha-Synuclein and the pesticide, rotenone. Neuroscience 203, 170–179.
Mulcahy, P., O'Doherty, A., Paucard, A., O'Brien, T., Kirik, D., Dowd, E., 2013. The behavioural and neuropathological impact of intranigral AAV-alpha-synuclein is exacerbated by systemic infusion of the Parkinson's disease-associated pesticide, rotenone, in rats. Behav. Brain Res. 243, 6–15.
Mullett, S.J., Hinkle, D.A., 2009. DJ-1 knock-down in astrocytes impairs astrocyte-mediated neuroprotection against rotenone. Neurobiol. Dis. 33, 28–36.

Natale, G., Pasquali, L., Ruggieri, S., Paparelli, A., Fornai, F., 2008. Parkinson's disease and the gut: a well known clinical association in need of an effective cure and explanation. Neurogastroenterol. Motil. 20, 741–749.

Neely, M.D., Davison, C.A., Aschner, M., Bowman, A.B., 2017. From the cover: manganese and rotenone-induced oxidative stress signatures differ in iPSC-derived human dopamine neurons. Toxicol. Sci. 159, 366–379.

Nehru, B., Verma, R., Khanna, P., Sharma, S.K., 2008. Behavioral alterations in rotenone model of Parkinson's disease: attenuation by co-treatment of centrophenoxine. Brain Res. 1201, 122–127.

Newhouse, K., Hsuan, S.L., Chang, S.H., Cai, B.B., Wang, Y.P., Xia, Z.G., 2004. Rotenone-induced apoptosis is mediated by p38 and JNK MAP kinases in human dopaminergic SH-SY5Y cells. Toxicol. Sci. 79, 137–146.

Ng, C.H., Mok, S.Z., Koh, C., Ouyang, X., Fivaz, M.L., Tan, E.K., Dawson, V.L., Dawson, T.M., Yu, F., Lim, K.L., 2009. Parkin protects against LRRK2 G2019S mutant-induced dopaminergic neurodegeneration in Drosophila. J. Neurosci. 29, 11257–11262.

Nistico, R., Mehdawy, B., Piccirilli, S., Mercuri, N., 2011. Paraquat- and rotenone-induced models of Parkinson's disease. Int. J. Immunopathol. Pharmacol. 24, 313–322.

Norazit, A., Meedeniya, A.C., Nguyen, M.N., Mackay-Sim, A., 2010. Progressive loss of dopaminergic neurons induced by unilateral rotenone infusion into the medial forebrain bundle. Brain Res. 1360, 119–129.

Norris, K.L., Hao, R., Chen, L.F., Lai, C.H., Kapur, M., Shaughnessy, P.J., Chou, D., Yan, J., Taylor, J.P., Engelender, S., West, A.E., Lim, K.L., Yao, T.P., 2015. Convergence of Parkin, PINK1, and alpha-Synuclein on stress-induced mitochondrial morphological remodeling. J. Biol. Chem. 290, 13862–13874.

Ohnishi, S.T., Ohnishi, T., Muranaka, S., Fujita, H., Kimura, H., Uemura, K., Yoshida, K., Utsumi, K., 2005. A possible site of superoxide generation in the complex I segment of rat heart mitochondria. J. Bioenerg. Biomembr. 37, 1–15.

Pahl, H.L., Baeuerle, P.A., 1996. Activation of NF-kappa B by ER stress requires both Ca2+ and reactive oxygen intermediates as messengers. FEBS Lett. 392, 129–136.

Pan, T.H., Li, X.Q., Xie, W.J., Jankovic, J., Le, W., 2005. Valproic acid-mediated Hsp70 induction and anti-apoptotic neuroprotection in SH-SY5Y cells. FEBS Lett. 579, 6716–6720.

Pan, T., Rawal, P., Wu, Y., Xie, W., Jankovic, J., Le, W., 2009a. Rapamycin protects against rotenone-induced apoptosis through autophagy induction. Neuroscience 164, 541–551.

Pan, T.H., Xie, W.J., Rawal, P., Jankovic, J., Le, W.D., 2009b. Rapamycin protects against rotenone-induced apoptosis through autophagy induction in SH-SY5Y cells. Neurology 72, A438–A439.

Pandareesh, M.D., Shrivash, M.K., Kumar, H.N.N., Misra, K., Bharath, M.M.S., 2016. Curcumin Monoglucoside shows improved bioavailability and mitigates rotenone induced neurotoxicity in cell and Drosophila models of Parkinson's disease. Neurochem. Res. 41, 3113–3128.

Pan-Montojo, F., Anichtchik, O., Dening, Y., Knels, L., Pursche, S., Jung, R., Jackson, S., Gille, G., Spillantini, M.G., Reichmann, H., Funk, R.H.W., 2010. Progression of Parkinson's Disease Pathology Is Reproduced by Intragastric Administration of Rotenone in Mice. PLoS One 5, e8762.

Parker Jr., W.D., Boyson, S.J., Parks, J.K., 1989. Abnormalities of the electron transport chain in idiopathic Parkinson's disease. Ann. Neurol. 26, 719–723.

Peng, K., Tao, Y., Zhang, J., Wang, J., Ye, F., Dan, G., Zhao, Y., Cai, Y., Zhao, J., Wu, Q., Zou, Z., Cao, J., Sai, Y., 2016. Resveratrol regulates mitochondrial biogenesis and fission/fusion to attenuate rotenone-induced neurotoxicity. Oxid. Med. Cell. Longev. 2016, 6705621.

Peng, K.G., Yang, L.K., Wang, J., Ye, F., Dan, G.R., Zhao, Y.P., Cai, Y., Cui, Z.H., Ao, L., Liu, J.Y., Zou, Z.M., Sai, Y., Cao, J., 2017. The interaction of mitochondrial biogenesis and fission/fusion mediated by PGC-1 alpha regulates rotenone-induced dopaminergic neurotoxicity. Mol. Neurobiol. 54, 3783–3797.
Peng, K.G., Xiao, J.S., Yang, L.K., Ye, F., Cao, J., Sai, Y., 2019. Mutual antagonism of PINK1/Parkin and PGC-1 contributes to maintenance of mitochondrial homeostasis in rotenone-induced neurotoxicity. Neurotox. Res. 35, 331–343.
Pfeiffer, R.F., 2016. Non-motor symptoms in Parkinson's disease. Parkinsonism Relat. Disord. 22, S119–S122.
Pfeiffer, R.F., 2018. Gastrointestinal dysfunction in Parkinson's disease. Curr. Treat. Options. Neurol. 20, 1–12. 54.
Pistollato, F., Canovas-Jorda, D., Zagoura, D., Bal-Price, A., 2017. Nrf2 pathway activation upon rotenone treatment in human iPSC-derived neural stem cells undergoing differentiation towards neurons and astrocytes. Neurochem. Int. 108, 457–471.
Radad, K., Moldzio, R., Taha, M., Rausch, W.D., 2009. Thymoquinone protects dopaminergic neurons against MPP+ and rotenone. Phytother. Res. 23, 696–700.
Rahimmi, A., Khosrobakhsh, F., Izadpanah, E., Moloudi, M.R., Hassanzadeh, K., 2015. N-acetylcysteine prevents rotenone-induced Parkinson's disease in rat: an investigation into the interaction of parkin and Drp1 proteins. Brain Res. Bull. 113, 34–40.
Ravenstijn, P.G., Merlini, M., Hameetman, M., Murray, T.K., Ward, M.A., Lewis, H., Ball, G., Mottart, C., de ville de Goyet, C., Lemarchand, T., van Belle, K., O'Neill, M.J., Danhof, M., de Lange, E.C., 2008. The exploration of rotenone as a toxin for inducing Parkinson's disease in rats, for application in BBB transport and PK-PD experiments. J. Pharmacol. Toxicol. Methods 57, 114–130.
Rcom-H'Cheo-Gauthier, A.N., Davis, A., Meedeniya, A.C.B., Pountney, D.L., 2016. Alpha-synuclein aggregates are excluded from calbindin-D28k-positive neurons in dementia with Lewy bodies and a unilateral rotenone mouse model. Mol. Cell. Neurosci. 77, 65–75.
Ren, Y., Liu, W., Jiang, H., Jiang, Q., Feng, J., 2005. Selective vulnerability of dopaminergic neurons to microtubule depolymerization. J. Biol. Chem. 280, 34105–34112.
Richter, F., Hamann, M., Richter, A., 2007. Chronic rotenone treatment induces behavioral effects but no pathological signs of parkinsonism in mice. J. Neurosci. Res. 85, 681–691.
Robertson, D.R., Smith-Vaniz, W.F., 2008. Rotenone: an essential but demonized tool for assessing marine fish diversity. Bioscience 58, 165–170.
Rodrigues, L.S., Targa, A.D.S., Noseda, A.C.D., Aurich, M.F., da Cunha, C., Lima, M.M.S., 2014. Olfactory impairment in the rotenone model of Parkinson's disease is associated with bulbar dopaminergic D2 activity after REM sleep deprivation. Front Cell Neurosci. 8, 1–11. 383.
Rojo, A.I., Cavada, C., de Sagarra, M.R., Cuadrado, A., 2007. Chronic inhalation of rotenone or paraquat does not induce Parkinson's disease symptoms in mice or rats. Exp. Neurol. 208, 120–126.
Ruan, Q.M., Harrington, A.J., Caldwell, K.A., Caldwell, G.A., Standaert, D.G., 2010. VPS41, a protein involved in lysosomal trafficking, is protective in Caenorhabditis elegans and mammalian cellular models of Parkinson's disease. Neurobiol. Dis. 37, 330–338.
Saha, S., Guillily, M.D., Ferree, A., Lanceta, J., Chan, D., Ghosh, J., Hsu, C.H., Segal, L., Raghavan, K., Matsumoto, K., Hisamoto, N., Kuwahara, T., Iwatsubo, T., Moore, L., Goldstein, L., Cookson, M., Wolozin, B., 2009. LRRK2 modulates vulnerability to mitochondrial dysfunction in Caenorhabditis elegans. J. Neurosci. 29, 9210–9218.
Sai, Y., Wu, Q., Le, W.D., Ye, F., Li, Y.P., Dong, Z.J., 2008. Rotenone-induced PC12 cell toxicity is caused by oxidative stress resulting from altered dopamine metabolism. Toxicol. In Vitro 22, 1461–1468.

Samantaray, S., Knaryan, V.H., Guyton, M.K., Matzelle, D.D., Ray, S.K., Banik, N.L., 2007. The parkinsonian neurotoxin rotenone activates calpain and caspase-3 leading to motoneuron degeneration in spinal cord of Lewis rats. Neuroscience 146, 741–755.
Sammi, S.R., Agim, Z.S., Cannon, J.R., 2018. Harmane-induced selective dopaminergic neurotoxicity in Caenorhabditis elegans. Toxicol. Sci. 161, 335–348.
Sanders, L.H., Greenamyre, J.T., 2013. Oxidative damage to macromolecules in human Parkinson disease and the rotenone model. Free Radic. Biol. Med. 62, 111–120.
Santiago, R.M., Barbieiro, J., Lima, M.M., Dombrowski, P.A., Andreatini, R., Vital, M.A., 2010. Depressive-like behaviors alterations induced by intranigral MPTP, 6-OHDA, LPS and rotenone models of Parkinson's disease are predominantly associated with serotonin and dopamine. Prog. Neuropsychopharmacol. Biol. Psychiatry 34, 1104–1114.
Saravanan, K.S., Sindhu, K.M., Senthilkumar, K.S., Mohanakumar, K.P., 2006. L-deprenyl protects against rotenone-induced, oxidative stress-mediated dopaminergic neurodegeneration in rats. Neurochem. Int. 49, 28–40.
Saravanan, K.S., Sindhu, K.M., Mohanakumar, K.P., 2007. Melatonin protects against rotenone-induced oxidative stress in a hemiparkinsonian rat model. J. Pineal Res. 42, 247–253.
Sarkar, S., Malovic, E., Harishchandra, D.S., Ghaisas, S., Panicker, N., Charli, A., Palanisamy, B.N., Rokad, D., Jin, H.J., Anantharam, V., Kanthasamy, A., Kanthasamy, A.G., 2017. Mitochondrial impairment in microglia amplifies NLRP3 inflammasome proinflammatory signaling in cell culture and animal models of Parkinson's disease. NPJ Parkinsons Dis. 3, 1–15. 30.
Saybasili, H., Yuksel, M., Haklar, G., Yalcin, A.S., 2001. Effect of mitochondrial electron transport chain inhibitors on superoxide radical generation in rat hippocampal and striatal slices. Antioxid. Redox Signal. 3, 1099–1104.
Schapira, A.H., Cooper, J.M., Dexter, D., Jenner, P., Clark, J.B., Marsden, C.D., 1989. Mitochondrial complex I deficiency in Parkinson's disease. Lancet 1, 1269.
Services, S., 1933. Rotenone as an insecticide. Nature 132, 167.
Sherer, T.B., Betarbet, R., Kim, J.H., Greenamyre, J.T., 2003a. Selective microglial activation in the rat rotenone model of Parkinson's disease. Neurosci. Lett. 341, 87–90.
Sherer, T.B., Betarbet, R., Testa, C.M., Seo, B.B., Richardson, J.R., Kim, J.H., Miller, G.W., Yagi, T., Matsuno-Yagi, A., Greenamyre, J.T., 2003b. Mechanism of toxicity in rotenone models of Parkinson's disease. J. Neurosci. 23, 10756–10764.
Sherer, T.B., Kim, J.H., Betarbet, R., Greenamyre, J.T., 2003c. Subcutaneous rotenone exposure causes highly selective dopaminergic degeneration and alpha-synuclein aggregation. Exp. Neurol. 179, 9–16.
Shinomol, G.K., Mythri, R.B., Srinivas Bharath, M.M., Muralidhara, 2012. *Bacopa monnieri* extract offsets rotenone-induced cytotoxicity in dopaminergic cells and oxidative impairments in mice brain. Cell. Mol. Neurobiol. 32, 455–465.
Smirnova, L., Harris, G., Delp, J., Valadares, M., Pamies, D., Hogberg, H.T., Waldmann, T., Leist, M., Hartung, T., 2016. A LUHMES 3D dopaminergic neuronal model for neurotoxicity testing allowing long-term exposure and cellular resilience analysis. Arch. Toxicol. 90, 2725–2743.
Sonia Angeline, M., Chaterjee, P., Anand, K., Ambasta, R.K., Kumar, P., 2012. Rotenone-induced parkinsonism elicits behavioral impairments and differential expression of parkin, heat shock proteins and caspases in the rat. Neuroscience 220, 291–301.
Sonia Angeline, M., Sarkar, A., Anand, K., Ambasta, R.K., Kumar, P., 2013. Sesamol and naringenin reverse the effect of rotenone-induced PD rat model. Neuroscience 254, 379–394.
Spillantini, M.G., Schmidt, M.L., Lee, V.M.Y., Trojanowski, J.Q., Jakes, R., Goedert, M., 1997. Alpha-synuclein in Lewy bodies. Nature 388, 839–840.

St Laurent, R., O'Brien, L.M., Ahmad, S.T., 2013. Sodium butyrate improves locomotor impairment and early mortality in a rotenone-induced Drosophila model of Parkinson's disease. Neuroscience 246, 382–390.

Stefanis, L., 2012. Alpha-synuclein in Parkinson's disease. Cold Spring Harb. Perspect. Med. 2, a009399.

Stephano, F., Nolte, S., Hoffmann, J., El-Kholy, S., Von Frieling, J., Bruchhaus, I., Fink, C., Roeder, T., 2018. Impaired Wnt signaling in dopamine containing neurons is associated with pathogenesis in a rotenone triggered Drosophila Parkinson's disease model. Sci. Rep. 8, 1–11. 2372.

Takeuchi, H., Yanagida, T., Inden, M., Takata, K., Kitamura, Y., Yamakawa, K., Sawada, H., Izumi, Y., Yamamoto, N., Kihara, T., Uemura, K., Inoue, H., Taniguchi, T., Akaike, A., Takahashi, R., Shimohama, S., 2009. Nicotinic receptor stimulation protects nigral dopaminergic neurons in rotenone-induced Parkinson's disease models. J. Neurosci. Res. 87, 576–585.

Tanner, C.M., Kamel, F., Ross, G.W., Hoppin, J.A., Goldman, S.M., Korell, M., Marras, C., Bhudhikanok, G.S., Kasten, M., Chade, A.R., Comyns, K., Richards, M.B., Meng, C., Priestley, B., Fernandez, H.H., Cambi, F., Umbach, D.M., Blair, A., Sandler, D.P., Langston, J.W., 2011. Rotenone, paraquat, and Parkinson's disease. Environ. Health Perspect. 119, 866–872.

Tansey, M.G., Goldberg, M.S., 2010. Neuroinflammation in Parkinson's disease: its role in neuronal death and implications for therapeutic intervention. Neurobiol. Dis. 37, 510–518.

Tapias, V., Cannon, J.R., Greenamyre, J.T., 2010. Melatonin treatment potentiates neurodegeneration in a rat rotenone Parkinson's disease model. J. Neurosci. Res. 88, 420–427.

Tapias, V., Cannon, J.R., Greenamyre, J.T., 2014. Pomegranate juice exacerbates oxidative stress and nigrostriatal degeneration in Parkinson's disease. Neurobiol. Aging 35, 1162–1176.

Tasselli, M., Chaumette, T., Paillusson, S., Monnet, Y., Lafoux, A., Huchet-Cadiou, C., Aubert, P., Hunot, S., Derkinderen, P., Neunlist, M., 2013. Effects of oral administration of rotenone on gastrointestinal functions in mice. Neurogastroenterol. Motil. 25, e183–e193.

Thakur, P., Nehru, B., 2014. Long-term heat shock proteins (HSPs) induction by carbenoxolone improves hallmark features of Parkinson's disease in a rotenone-based model. Neuropharmacology 79, 190–200.

Thakur, P., Nehru, B., 2015. Inhibition of neuroinflammation and mitochondrial dysfunctions by carbenoxolone in the rotenone model of Parkinson's disease. Mol. Neurobiol. 51, 209–219.

Thiffault, C., Langston, J.W., di Monte, D.A., 2000. Increased striatal dopamine turnover following acute administration of rotenone to mice. Brain Res. 885, 283–288.

Tong, Q., Wu, L., Gao, Q., Ou, Z., Zhu, D.Y., Zhang, Y.D., 2016. PPAR beta/delta agonist provides neuroprotection by suppression of IRE1 alpha-caspase-12-mediated endoplasmic reticulum stress pathway in the rotenone rat model of Parkinson's disease. Mol. Neurobiol. 53, 3822–3831.

Turner, N., 1932. Notes on rotenone as an insecticide. J. Econ. Entomol. 25, 1228–1237.

USEPA, 2007. In: Edwards, D. (Ed.), Reregistration Eligibility Decision for Rotenone. United States Environmental Protection Agency.

Uversky, V.N., 2004. Neurotoxicant-induced animal models of Parkinson's disease: understanding the role of rotenone, maneb and paraquat in neurodegeneration. Cell Tissue Res. 318, 225–241.

Ved, R., Saha, S., Westlund, B., Perier, C., Burnam, L., Sluder, A., Hoener, M., Rodrigues, C.M., Alfonso, A., Steer, C., Liu, L., Przedborski, S., Wolozin, B., 2005. Similar patterns of mitochondrial vulnerability and rescue induced by genetic modification of alpha-synuclein, parkin, and DJ-1 in Caenorhabditis elegans. J. Biol. Chem. 280, 42655–42668.

Venderova, K., Kabbach, G., Abdel-Messih, E., Zhang, Y., Parks, R.J., Imai, Y., Gehrke, S., Ngsee, J., Lavoie, M.J., Slack, R.S., Rao, Y., Zhang, Z., Lu, B., Haque, M.E., Park, D.S., 2009. Leucine-rich repeat kinase 2 interacts with Parkin, DJ-1 and PINK-1 in a *Drosophila melanogaster* model of Parkinson's disease. Hum. Mol. Genet. 18, 4390–4404.

Virga, D.M., Capps, J., Vohra, B.P.S., 2018. Enteric neurodegeneration is mediated through independent neuritic and somal mechanisms in rotenone and MPP plus toxicity. Neurochem. Res. 43, 2288–2303.

von Wrangel, C., Schwabe, K., John, N., Krauss, J.K., Alam, M., 2015. The rotenone-induced rat model of Parkinson's disease: behavioral and electrophysiological findings. Behav. Brain Res. 279, 52–61.

Wang, D., Tang, B., Zhao, G., Pan, Q., Xia, K., Bodmer, R., Zhang, Z., 2008. Dispensable role of Drosophila ortholog of LRRK2 kinase activity in survival of dopaminergic neurons. Mol. Neurodegener. 3, 3.

Wang, Y.L., Liu, W.W., Yang, J., Wang, F., Sima, Y.Z., Zhong, Z.M., Wang, H., Hu, L.F., Liu, C.F., 2017. Parkinson's disease-like motor and non-motor symptoms in rotenone-treated zebrafish. Neurotoxicology 58, 103–109.

Watabe, M., Nakaki, T., 2004. Rotenone induces apoptosis via activation of bad in human dopaminergic SH-SY5Y cells. J. Pharmacol. Exp. Ther. 311, 948–953.

Westerink, R.H.S., Ewing, A.G., 2008. The PC12 cell as model for neurosecretion. Acta Physiol. 192, 273–285.

Wise Jr., J.P., Cannon, J., 2016. From the cover: alterations in optineurin expression and localization in pre-clinical Parkinson's disease models. Toxicol. Sci. 153, 372–381.

Wise Jr., J.P., Price, C.G., Amaro, J.A., Cannon, J.R., 2018. Autophagy disruptions associated with altered optineurin expression in extranigral regions in a rotenone model of Parkinson's disease. Front. Neurosci. 12, 289.

Wolozin, B., Saha, S., Guillily, M., Ferree, A., Riley, M., 2008. Investigating convergent actions of genes linked to familial Parkinson's disease. Neurodegener Dis 5, 182–185.

WHO-IPCS, 1992. Rotenone: Health and Safety Guide [Online]. World Health Organization. Available: https://apps.who.int/iris/handle/10665/39581. [73].

Wu, F., Wang, Z., Gu, J.H., Ge, J.B., Liang, Z.Q., Qin, Z.H., 2013a. p38(MAPK)/p53-Mediated Bax induction contributes to neurons degeneration in rotenone-induced cellular and rat models of Parkinson's disease. Neurochem. Int. 63, 133–140.

Wu, L., Tian, Y.Y., Shi, J.P., Xie, W., Shi, J.Q., Lu, J., Zhang, Y.D., 2013b. Inhibition of endoplasmic reticulum stress is involved in the neuroprotective effects of candesartan cilexitil in the rotenone rat model of Parkinson's disease. Neurosci. Lett. 548, 50–55.

Wu, L., Luo, N., Zhao, H.R., Gao, Q., Lu, J., Pan, Y., Shi, J.P., Tian, Y.Y., Zhang, Y.D., 2014. Salubrinal protects against rotenone-induced SH-SY5Y cell death via ATF4-parkin pathway. Brain Res. 1549, 52–62.

Wu, F., Xu, H.D., Guan, J.J., Hou, Y.S., Gu, J.H., Zhen, X.C., Qin, Z.H., 2015. Rotenone impairs autophagic flux and lysosomal functions in Parkinson's disease. Neuroscience 284, 900–911.

Wu, X., Liang, Y., Jing, X., Lin, D., Chen, Y., Zhou, T., Peng, S., Zheng, D., Zeng, Z., Lei, M., Huang, K., Tao, E., 2018. Rifampicin prevents SH-SY5Y cells from rotenone-induced apoptosis via the PI3K/Akt/GSK-3beta/CREB signaling pathway. Neurochem. Res. 43, 886–893.

Xicoy, H., Wieringa, B., Martens, G.J.M., 2017. The SH-SY5Y cell line in Parkinson's disease research: a systematic review. Mol. Neurodegener. 12, 1–11. 10.

Xiong, N., Xiong, J., Khare, G., Chen, C., Huang, J., Zhao, Y., Zhang, Z., Qiao, X., Feng, Y., Reesaul, H., Zhang, Y., Sun, S., lin, Z., Wang, T., 2011. Edaravone guards dopamine neurons in a rotenone model for Parkinson's disease. PLoS One 6, e20677.

Xiong, N., Huang, J., Chen, C., Zhao, Y., Zhang, Z., Jia, M., Zhang, Z., Hou, L., Yang, H., Cao, X., Liang, Z., Zhang, Y., Sun, S., Lin, Z., Wang, T., 2012a. Dl-3-n-butylphthalide, a natural antioxidant, protects dopamine neurons in rotenone models for Parkinson's disease. Neurobiol. Aging 33, 1777–1791.

Xiong, N., Long, X., Xiong, J., Jia, M., Chen, C., Huang, J., Ghoorah, D., Kong, X., Lin, Z., Wang, T., 2012b. Mitochondrial complex I inhibitor rotenone-induced toxicity and its potential mechanisms in Parkinson's disease models. Crit. Rev. Toxicol. 42, 613–632.

Xiong, N., Xiong, J., Jia, M., Liu, L., Zhang, X.W., Chen, Z.Z., Huang, J.S., Zhang, Z.T., Hou, L.L., Luo, Z.J., Ghoorah, D., Lin, Z.C., Wang, T., 2013. The role of autophagy in Parkinson's disease: rotenone-based modeling. Behav. Brain Funct. 9, 1–12. 13.

Xiong, Z.K., Lang, J., Xu, G., Li, H.Y., Zhang, Y., Wang, L., Su, Y., Sun, A.J., 2015. Excessive levels of nitric oxide in rat model of Parkinson's disease induced by rotenone. Exp. Ther. Med. 9, 553–558.

Yang, F., Wolk, A., Hakansson, N., Pedersen, N.L., Wirdefeldt, K., 2017. Dietary antioxidants and risk of Parkinson's disease in two population-based cohorts. Mov. Disord. 32, 1631–1636.

Yang, B., Bai, Y., Yin, C., Qian, H., Xing, G., Wang, S., Li, F., Bian, J., Aschner, M., Lu, R., 2018. Activation of autophagic flux and the Nrf2/ARE signaling pathway by hydrogen sulfide protects against acrylonitrile-induced neurotoxicity in primary rat astrocytes. Arch. Toxicol. 92, 2093–2108.

Yi, P.L., Tsai, C.H., Lu, M.K., Liu, H.J., Chen, Y.C., Chang, F.C., 2007. Interleukin-1 beta mediates sleep alteration in rats with rotenone-induced parkinsonism. Sleep 30, 413–425.

Yy, S., Jiang, M., Xia, Y., Chen, Q.-Y., Wen, T.-Q., 2011. Rotenone induces more serious learning and memory impairment than α-synuclein A30P does in Drosophila. Journal of Shanghai University 15, 229–234.

Zaitone, S.A., Abo-Elmatty, D.M., Elshazly, S.M., 2012a. Piracetam and vinpocetine ameliorate rotenone-induced parkinsonism in rats. Indian J Pharmacol 44, 774–779.

Zaitone, S.A., Abo-Elmatty, D.M., Shaalan, A.A., 2012b. Acetyl-L-carnitine and alpha-lipoic acid affect rotenone-induced damage in nigral dopaminergic neurons of rat brain, implication for Parkinson's disease therapy. Pharmacol. Biochem. Behav. 100, 347–360.

Zaminelli, T., Gradowski, R.W., Bassani, T.B., Barbiero, J.K., Santiago, R.M., Maria-Ferreira, D., Baggio, C.H., Vital, M.A., 2014. Antidepressant and antioxidative effect of ibuprofen in the rotenone model of Parkinson's disease. Neurotox. Res. 26, 351–362.

Zhang, X.Y., Chen, L., Yang, Y., Xu, D.M., Zhang, S.R., Li, C.T., Zheng, W., Yu, S.Y., Wei, E.Q., Zhang, L.H., 2014. Regulation of rotenone-induced microglial activation by 5-lipoxygenase and cysteinyl leukotriene receptor 1. Brain Res. 1572, 59–71.

Zhang, Q., Chen, S., Yu, S., Qin, J., Zhang, J., Cheng, Q., Ke, K., Ding, F., 2016. Neuroprotective effects of pyrroloquinoline quinone against rotenone injury in primary cultured midbrain neurons and in a rat model of Parkinson's disease. Neuropharmacology 108, 238–251.

Zhang, Z.N., Zhang, J.S., Xiang, J., Yu, Z.H., Zhang, W., Cai, M., Li, X.T., Wu, T., Li, W.W., Cai, D.F., 2017. Subcutaneous rotenone rat model of Parkinson's disease: dose exploration study. Brain Res. 1655, 104–113.

Zharikov, A.D., Cannon, J.R., Tapias, V., Bai, Q., Horowitz, M.P., Shah, V., El Ayadi, A., Hastings, T.G., Greenamyre, J.T., Burton, E.A., 2015. shRNA targeting alpha-synuclein prevents neurodegeneration in a Parkinson's disease model. J. Clin. Investig. 125, 2721–2735.

Zhou, F., Wu, J.Y., Sun, X.L., Yao, H.H., Ding, J.H., Hu, G., 2007. Iptakalim alleviates rotenone-induced degeneration of dopaminergic neurons through inhibiting microglia-mediated neuroinflammation. Neuropsychopharmacology 32, 2570–2580.
Zhou, H., Zhang, F., Chen, S.H., Zhang, D., Wilson, B., Hong, J.S., Gao, H.M., 2012. Rotenone activates phagocyte NADPH oxidase by binding to its membrane subunit gp91(phox). Free Radic. Biol. Med. 52, 303–313.
Zhou, P., Hornberg, J.R., Fang, Q.Y., Wang, J.Q., Li, W.Z., Meng, X.Z., Shen, J.Q., Luan, Y., Liao, P., Swaab, D.F., Shan, L., Liu, C.Q., 2019. Histamine-4 receptor antagonist JNJ7777120 inhibits pro-inflammatory microglia and prevents the progression of Parkinson-like pathology and behaviour in a rat model. Brain Behav. Immun. 76, 61–73.

CHAPTER SIX

Neurotoxicity of amitraz, a formamidine pesticide

Lucio G. Costa*
Department of Environmental and Occupational Health Sciences, University of Washington, Seattle, WA, United States
Department of Medicine and Surgery, University of Parma, Parma, Italy
*Corresponding author: e-mail address: lgcosta@uw.edu

Contents

1. Introduction

Formamidines, such as chlordimeform [*N*′-(4-chloro-*o*-tolyl)-*N*,*N*-dimethylformamidine] or amitraz [*N*′-2,4-(dimethylphenyl)-*N*,*N*-((2,4-dimethylphenyl)imino)methyl-*N*-methanimidamide] were developed in the early 1960s for use in agriculture and in veterinary medicine as insecticides/acaricides (Hollingworth, 1976). Chlordimeform was first commercialized in 1966 (ICPS, 1998), while amitraz was registered in 1975. The structures of chlordimeform and amitraz are shown in Fig. 1, together with the closely related structure of the mammalian neurotransmitter norepinephrine and the insect neurotransmitter octopamine. The latter two are the endogenous neurotransmitters for the targets of formamidines in mammals and insects, respectively. All insecticides target the nervous system of insects (Costa, 2018); as most structural components of the nervous system are similar or identical in insects and mammals, insecticides also affect the nervous system of mammals including humans. Important classes

Advances in Neurotoxicology, Volume 4
ISSN 2468-7480
https://doi.org/10.1016/bs.ant.2019.11.003

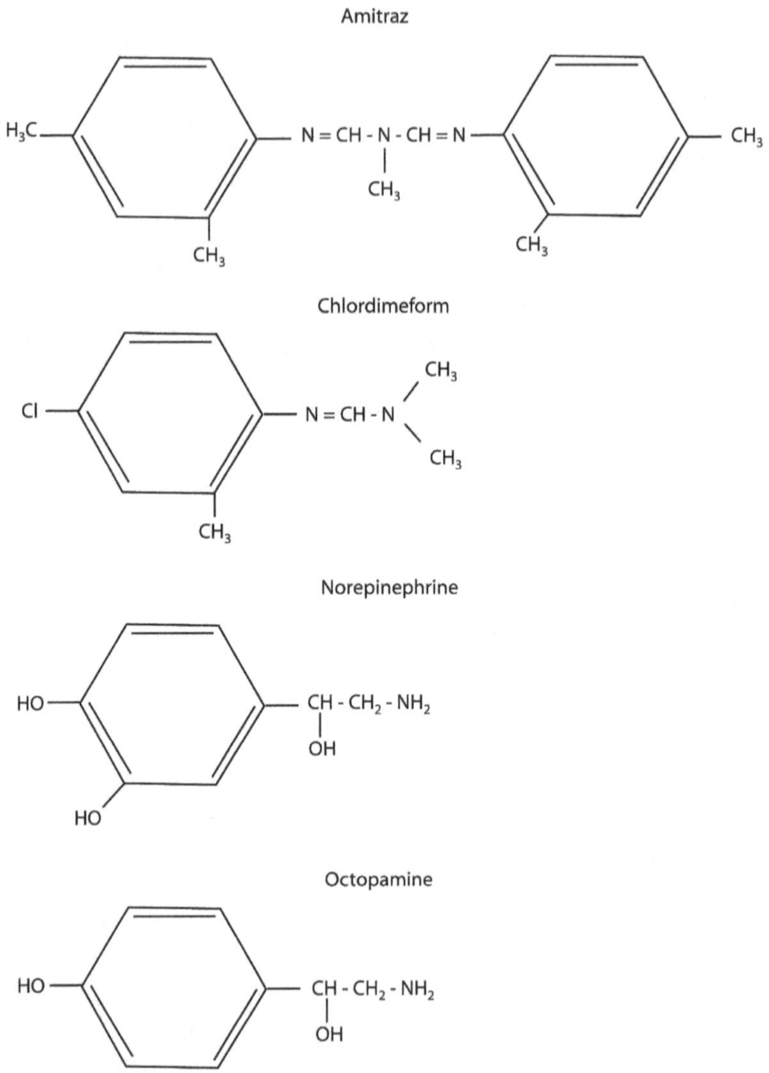

Fig. 1 Structures of amitraz, chlordimeform, norepinephrine and octopamine. *From Costa, L.G., 2018. Toxic effects of pesticides. In: Klaassen, C.D. (Ed.), Casarett and Doull's Toxicology: The Basic Science of Poisons. McGraw-Hill, New York, pp. 1055–1106, with permission.*

of insecticides target the cholinergic system (e.g., organophosphates and carbamates inhibit acetylcholinesterase, neonicotinoids are agonists at nicotinic receptors), the sodium channels (e.g., DDT, pyrethroids prolong the opening of the sodium channel) or GABA-linked chloride channels (e.g., cyclodienes block the chloride channel) (Costa, 2018). Formamidines are unique among insecticides as they exert their toxicity in invertebrates by activating an octopamine-dependent adenylate cyclase (Nathanson, 1985). In mammals, symptoms of formamidine poisoning are sympathomimetic in nature, and include hypotension, bradycardia, and sedation (Beeman and Matsumura, 1973; Del Pino et al., 2015b). The similarity between insect octopamine receptors and mammalian alpha$_2$-adrenergic receptors (alpha$_2$-adrenoceptors) suggested the latter as a possible target for formamidines. *In vivo* and *in vitro* studies have indeed shown that formamidines act as rather selective agonists for alpha$_2$–adrenoceptors (Altobelli et al., 2001; Costa et al., 1988; Hsu and Kakuk, 1984; Hsu and Lu, 1984).

For several years chlordimeform has been the most important compound of this class of insecticides. The N-demethylated metabolite of chlordimeform (desmethylchlordimeform) is more acutely toxic than chlordimeform and displays a >400-fold higher potency toward alpha$_2$-adrenoceptors (Costa and Murphy, 1987; Ghali and Hollingworth, 1985; Table 1). In contrast, two other metabolites of chlordimeform, 4-chloro-toluidine and *N*-formyl-4-chloro-*o*-toluidine, are believed to be responsible for the observed hemangioendotheliomas in mice observed in carcinogenicity studies (IPCS, 1998), which has led to the classification of chlordimeform as a probable human carcinogen (Group 2A) by IARC in 1990. Given the increasing evidence of an association between exposure to chlordimeform and 4-chloro-*o*-toluidine and bladder cancer (Popp et al., 1992), chlordimeform was withdrawn from the market in 1992.

In contrast, amitraz is still widely used for the control of ectoparasites in farm animals and crops. Though its acute toxicity is moderate

Table 1 *In vitro* interactions of amitraz, its metabolite DPMF, and chlordimeform's metabolite DCDM with octopamine alpha- and beta-adrenergic-like receptors.

	Octopamine alpha (1) receptors		**Octopamine beta (2) receptors**
Compound	**Ca^{2+}**	**cAMP**	**cAMP**
Octopamine	23.4	3030	5.2
Amitraz	9.6	708	2.0
DPMF	1.2	181	0.08
DCDM	5.3	ND	ND

Results show EC_{50} values (nM) for increases in Ca^{2+} and of cAMP levels in alpha (1)- and beta (2)-like adrenergic octopamine receptors cloned from silkworms and stably expressed in HEK-293 cells.
Adapted from Kita, T., Hayashi, T., Ohtani, T., Takao, H., Hiroshi, T., Liu, G., Ohta, H., Ozoe, F., Ozoe, Y., 2016. Amitraz and its metabolite differentially activate α- and β-adrenergic like octopamine receptors. Pest Manag. Sci. 73, 984–990.

(oral LD_{50} = 100 to >1000 mg/kg, depending on the animal species), several cases of acute amitraz poisoning in humans have been reported (see a following section). Signs and symptoms of poisoning with amitraz are in general adrenergic in nature, and mimic those of alpha$_2$-adrenoceptor agonists such as clonidine. Amitraz is not genotoxic but was classified as a possible human carcinogen by the EPA due to some evidence of carcinogenicity (lung, liver) in female mice at high doses (Del Pino et al., 2015b). However, because of its lack on carcinogenicity in rats of both sexes and in male mice, it was later indicated as unclassifiable (or nonquantifiable) as to human carcinogenicity (Del Pino et al., 2015b). This chapter will focus primarily on amitraz, and information on chlordimeform are provided solely for comparison purposes. Focus of the following sections are the interactions of amitraz with alpha$_2$-adrenoceptors in *in vitro* systems, and its effects *in vivo* in animals and in humans, which, as said, are adrenergic in nature, and rather unique among all insecticides.

2. Interaction with adrenoceptors: *In vitro* studies

The first reports of insecticidal/acaricidal effects of formamidines date back the early-mid 1960s (Hollingworth and Lund, 1982). The major difference noted between these compounds and more conventional insecticides (e.g., the organophosphates) was the relative importance of sublethal and behavioral effects in insects, as compared to direct lethality. These effects include increased excitability, increased motor activity and decreased feeding behavior (Beeman and Matsumura, 1978; Hollingworth and Lund, 1982; Lund et al., 1979). Chlordimeform and amitraz were the first and most relevant compounds of this new class. Both compounds are extensively metabolized *in vivo*, with *N*-monomethylformamidine analogs being the two main metabolites. These are desmethylchlordimeform (DCDM) for chlordimeform, and N^2-(2,4-dimethylphenyl)-N^1-methylformamidine (DPMF; also known as BTS-27271 or U-40481) for amitraz.

The mechanism(s) of insecticidal action of formamidines eluded researchers for some time. Inhibition of monoamine oxidase activity, uncoupling of oxidative phosphorylation, and blockade of neuromuscular transmission had been suggested and observed, but could not fully explain the action of these compounds (Hollingworth and Lund, 1982; Lund et al., 1979). The specific target of formamidines in insects was identified by Evans and Gee (1980) as the octopamine receptors. Octopamine, a biogenic phenolamine, is an important neurotransmitter and hormone in invertebrates (Axelrod and Saavedra, 1977), and although present in mammals, it is not playing an important role in the latter (David and Coulon, 1985). In invertebrates, octopamine affects a wide variety of physiological functions, including laying eggs, learning and memory, locomotion, flight, aggression, courtship, sleep, arousal, and feeding behavior (Ohta and Ozoe, 2014). Initial studies showed that both chlordimeform and its metabolite DCDM were potent activators of the octopamine receptors in the locust neuromuscular junction, with DCDM in the same order of potency as octopamine itself (Evans and Gee, 1980). A sub-class of octopamine receptors, the octopamine-2 receptors coupled to adenylate cyclase whose activation increases intracellular cAMP, was initially characterized by Nathanson (1985), who found that formamidines such as chlordimeform and DCDM act as partial agonists at these receptors (Nathanson and Hunnicut, 1981). Octopamine receptors were further characterized in subsequent years and are now subdivided into alpha adrenergic-like octopamine receptors (whose activation causes mobilization of calcium and secondarily the production of cAMP), and beta-adrenergic-like octopamine receptors, whose activation selectively increases cAMP levels (Evans and Maqueira, 2005; Ohta and Ozoe, 2014). As shown in Table 1, amitraz and its metabolite DPMF, as well as DCDM, were potent activators of these receptors, even more potent than octopamine itself (Kita et al., 2016).

Based on their pharmacological profile, octopamine receptors are similar, in mammals, to alpha$_2$-adrenoceptors, which are metabotropic cell membrane receptors coupled to GTP-binding proteins (G proteins). The G protein coupled to alpha$_2$-adrenoceptors is of the inhibitory type (Gi); thus, activation of alpha$_2$-adrenoceptors by agonists causes inhibition of the enzyme adenylate cyclase (which converts ATP to cAMP), and a decrease in the production of cAMP (Giovannitti et al., 2015; Maze and Tranquilli, 1991). Alpha$_2$-adrenoceptors can be further subdivided into the alpha$_{2A}$, alpha$_{2B}$, alpha$_{2C}$ and alpha$_{2D}$ subtypes. While the endogenous neurotransmitter agonists norepinephrine and epinephrine have similar affinities for all three subtypes, several pharmacological agonists and antagonists display selectivity for subtypes of alpha$_2$-adrenoceptors (Maze and Tranquilli, 1991). The alpha$_{2A}$ and alpha$_{2C}$ subtypes are found mainly, but not exclusively in the CNS, while alpha$_{2B}$-adrenoceptors are present in vascular tissue (Giovannitti et al., 2015). At the synaptic level, alpha$_2$-adrenoceptors can be located either pre-synaptically or post-synaptically, and activation of pre-synaptic receptors causes a decrease of norepinephrine release with a general decrease of sympathetic (noradrenergic) tone. Alpha$_2$-adrenoceptors agonists such as clonidine, are used primarily as anti-hypertensive drugs, particularly in anesthesia; additional effects of alpha$_2$-adrenoceptors activation include sedation, mydriasis, bradycardia, decreased insulin release, and decreased intestinal motility (Giovannitti et al., 2015; Maze and Tranquilli, 1991).

A series of *in vitro* studies by Costa et al. (Costa and Murphy, 1987; Costa et al., 1988) established the ability of formamidines to interact with alpha$_2$-adrenoceptors (Table 2). Chlordimeform, its active metabolite DCDM and amitraz were examined in these studies, while the potent amitraz metabolite DPMF was unfortunately unavailable. The findings clearly show that amitraz and DCDM interact with relatively high potency (IC$_{50}$s in the nanomolar range) with alpha$_2$-adrenoceptors labeled with either an agonist (clonidine) or an antagonist (yohimbine). Formamidines showed a high degree of specificity, as they did not interact with several other neurotransmitter receptors, except for the alpha$_1$-adrenoceptors toward which DCDM displayed an IC$_{50}$ in the low micromolar range (Table 2).

Table 2 *In vitro* interactions of amitraz, chlordimeform (CDM) and desmethylchlordimeform (DCDM) with neurotransmitter receptors.

Receptor type	Ligand	Amitraz	CDM (IC$_{50}$, μM)	DCDM
Alpha-2	Clonidine	0.13, 0.05	13.1, 18.2, 33.9	0.03, 0.04
Alpha-2	Yohimbine	0.22, 0.06	7.4, 13.1, 37.0	0.16
Alpha-1	Prazosin	>10	418, 87	2.33, 1.03
Beta	DHA	>10	909, 867	202, 134
GABA-A	Muscimol	>10	>1000, >1000	>1000, >1000
Benzodiazepine	Diazepam	>10	>1000, >1000	>1000, >1000
Muscarinic	QNB	>10	310, 413	161, 324
Histamine-1	Mepyramine	>10	ND	ND
Dopamine-2	Spiperone	9.4	10.2	34.1
Opiate mu	DHM	>10	434	164

Results show IC$_{50}$ values (μM) obtained in receptor binding assays in mouse or rat brain. Abbreviations: DHA, dihydroalprenolol; QNB, quinuclidinyl benzilate; DHM, dihydromorphine.

Adapted from: Costa, L.G., Murphy, S.D., 1987. Interaction of the pesticide chlordimeform with adrenergic receptors in mouse brain: an in vitro study. Arch. Toxicol. 59, 323–327; Costa, L.G., Olibet, G., Murphy, S.D., 1988. Alpha2-adrenoceptors as a target for formamidine pesticides: in vitro and in vivo studies in mice. Toxicol. Appl. Pharmacol. 93, 319–328; Costa, L.G., Olibet, G., Wu, D., Murphy, S.D., 1989. Acute and chronic effects of the pesticide amitraz on α2-adrenoceptors in the mouse brain. Toxicol. Lett. 47, 135–143; Costa, L.G., Wu, D., Olibet, G., Murphy, S.D., 1989. Formamidine pesticides and alpha2-adrenoceptors: studies with amitraz and chlordimeform in rats and development of a radioreceptor binding assay. Neurotoxicol. Teratol. 11, 405–411.

Additional evidence obtained in *in vitro* systems supports a relevant role of alpha$_2$-adrenoceptors activation in the effects of formamidines. In the isolated guinea pig ileum, amitraz caused a dose-dependent inhibition of transmurally-stimulated contractions and this effect was antagonized by an alpha$_2$-antagonist (idazoxan) but not by an alpha$_1$-antagonist (prazosin) (Hsu et al., 1987). Amitraz and its metabolite BTS-27271 (DPMF) also caused a dose-dependent increase in porcine myometrium contractility; this effect was antagonized by the alpha$_2$-antagonist yohimbine but not by prazosin and was due to an increase in extracellular calcium entry through voltage-dependent calcium channels (Shin and Hsu, 1994). In a rat beta cell line amitraz and its metabolite DPMF inhibited insulin release, as did medetomidine, a potent alpha$_2$-adrenoceptor agonist, and their effect was blocked by yohimbine but not by alpha$_1$-antagonists (Chen and Hsu, 1994). DPMF was slightly more potent than amitraz in this regard (Chen and Hsu, 1994). Both amitraz and chlordimeform were reported to inhibit the binding of [^{3}H]-yohimbine to canine platelets; the two formamidine insecticides, as well as clonidine, inhibited platelet aggregation induced by epinephrine in the presence of ADP (Koeiman and Hsu, 1991). In isolated rat synaptosomes, amitraz induced a reduction of the release of noradrenaline which mimicked that caused by clonidine (Altobelli et al., 2001). In human luteinized granulosa cells, amitraz was found to reduce the production of progesterone, and effect that was prevented by yohimbine (Young et al., 2005). Using rat primary hippocampal neurons as an *in vitro* model system, Del Pino et al. (2015a) found that amitraz disrupted GABAergic and glutamatergic transmission, and these effects were partially antagonized by idazoxan, and alpha$_2$-adrenoceptor blocker.

3. Role of alpha-2 adrenoceptors in the *in vivo* toxicity of formamidines

The direct interaction of formamidines with alpha$_2$-adrenoceptors *in vitro*, and the results of several studies utilizing *in vitro* systems evidencing an alpha$_2$-adrenoceptor-mediated mode of action for these compounds, suggested that these receptors may also be responsible for the effects observed upon their *in vivo* administration. Strong indications in this regard derive from the fact that alpha$_2$-adrenergic agonists elicit similar effects as amitraz or chlordimeform, and that alpha$_2$-adrenoceptor antagonists block the effects of formamidines. Effects of formamidines like those elicited by the partial alpha$_2$-adrenoceptor agonist clonidine include, for example in the case of chlordimeform, a reduction of motor activity (Landauer et al., 1984), analgesia (Costa and Murphy, 1987; Pfister and Yim, 1977), hypotension (Lund et al., 1979), and alterations in visual functions (Boyes and Moser, 1988). The latter two effects are also antagonized by the alpha$_2$-adrenoceptor blocker yohimbine (Boyes and Moser, 1988; Costa and Murphy, 1987).

Several *in vivo* studies with chlordimeform and amitraz have confirmed that alpha$_2$-adrenoceptors are targeted by these two compounds (and by their active metabolites DCDM and DPMF). Occupation of alpha$_2$-adrenoceptors in mouse brain by formamidines significantly decreased [^{3}H]-clonidine binding, particularly in cerebral cortex, hippocampus, and midbrain areas (including thalamus and hypothalamus) (Costa et al., 1988). This agrees with alterations in hippocampal evoked potentials and visual cortical evoked potential (both antagonized by yohimbine) reported following treatment with formamidines (Boyes and Moser, 1987; Gilbert and Dyer, 1988). Time-course studies indicated that recovery of [^{3}H]-clonidine binding was much faster in case of chlordimeform (full recovery at 5 h), and long-lasting following administration of amitraz (almost no recovery by 24 h and full recover by 72–96 h; Costa et al., 1988, 1989a,b). This finding supports the results of behavioral and of pharmacokinetic studies. For example, in case of amitraz, the effects on motor activity and on hippocampal evoked potentials lasted for 72–96 h (Crofton et al., 1987; Gilbert and Dyer, 1988). In addition, while an earlier study had indicated that amitraz and DPMF are rapidly eliminated (Knowles and Benezet, 1981), a more recent study indicates that DPMF, which binds very little to plasma proteins, can persist in brain tissue for long periods and may be the chemical entity responsible (more than amitraz itself) for the alpha$_2$-adrenergic neurotoxicity (Hu et al., 2019).

Table 3 summarizes the effects observed upon administration of amitraz or chlordimeform to various animal species, which are believed to be mediated for the most part by activation of alpha$_2$-adrenoceptors. While many effects of formamidines are believed to involve activation of CNS alpha$_2$-adrenoceptors, the distribution of alpha$_2$-adrenoceptors in the peripheral nervous system also allows for multiple peripheral actions of formamidines. For example, both chlordimeform and amitraz caused a dose-dependent decrease in hepatic glutathione (GSH) in mice, which was of longer duration with amitraz, and was antagonized by yohimbine but not by prazosin or propranolol, antagonists of alpha$_1$- and beta-adrenoceptors, respectively (Costa et al., 1991). The effects mimicked those seen with clonidine

Table 3 *In vivo* effects of formamidines in animals.

Animal species	Effect	Alpha-2 antagonism	Reference
Amitraz			
Mouse	Delayed gastrointestinal transit	YOH	Hsu and Lu (1984)
Rat	Mydriasis, bradycardia	YOH, Phentolamine	Hsu and Kakuk (1984)
Dog	Delayed gastrointestinal transit	YOH	Hsu and McNeel (1985)
Dog	Bradycardia	YOH	Hsu and McNeel (1985)
Dog	CNS depression, bradycardia	YOH	Hsu and Hopper (1986)
Dog	Bradycardia, hypertension	YOH	Hsu et al. (1986)
Rat	Decreased operant performance	ND	Moser et al. (1987)
Rat	Altered visual evoked potential	ND	Boyes and Moser (1987)
Dog	Bradycardia, depressed respiration	ND	Cullen and Reynoldson (1987)
Rat	Decreased motor activity, altered acoustic startle response, hypothermia	ND	Crofton et al. (1987)
Dog	Hypertension, bradycardia	Phentolamine	Cullen and Reynoldson (1988)
Rat	Increased susceptibility to kindled seizures	ND	Gilbert (1988)
Rat	Increased hippocampal excitability	ND[a]	Gilbert and Dyer (1988)
Rat	Decreased motor activity	ND	Moser and MacPhail (1989)
Rat	Glucose-induced hyperglycemia	YOH	Smith et al. (1990)
Dog	Hypotension, bradycardia	YOH	Cullen and Reynoldson (1990)

Continued

Table 3 *In vivo* effects of formamidines in animals.—Cont'd

Animal species	Effect	Alpha-2 antagonism	Reference
Dog	Depression of shuttle avoidance	YOH	Schaffer et al. (1990)
Mouse	Decreased hepatic glutathione	YOH	Costa et al. (1991)
Rat	Alterations in F.O.B.[b]	ND	Moser (1991)
Sheep	Decreased intestinal motility	YOH	Pass and Mogg (1991)
Dog	Sedation, bradycardia, hypothermia, hyperglycemia	Atipamezole	Hugnet et al. (1996)
Horse	Analgesia, sedation	YOH	Queiroz-Neto et al. (1998)
Rat[c]	Inhibition of insulin secretion, decreased glucagon secretion	Idoxazan	Abu-Basha et al. (1999)
Dog	Sedation, hypothermia, bradycardia, mydriasis, hypotension	Atipamezole	Andrade and Sakate (2003)
Chlordimeform			
Rat	Decreased operant performance	ND	Moser et al. (1987)
Rat	Altered visual evoked potentials	ND	Boyes and Moser (1987)
Rat	Altered visual evoke potentials	YOH	Boyes and Moser (1988)
Rat	Increased susceptibility to kindled seizures	ND	Gilbert (1988)
Mouse	Decreased hepatic glutathione	YOH	Costa et al. (1991)

[a]Antagonism not determined, but the same effects were seen with clonidine.
[b]Depressed arousal and rearing, hypothermia, body weight loss, decreased defecation, altered gait, loss of pupil reflex, handling hypersensitivity.
[c]Perfused rat pancreas.
Abbreviations: YOH, yohimbine; ND, not determined; F.O.B., Functional Observational Battery.

(e.g., only about 40% of the hepatic pool of GSH is under alpha$_2$-adrenergic control), and the effects of clonidine and formamidines were not additive (Costa et al., 1991). Similar results with clonidine on hepatic GSH levels have been obtained in another study (Harbison et al., 1991). The results with amitraz were later confirmed in rats, where amitraz decreased hepatic GSH levels, particularly when co-administered with the pyrethroid insecticide cypermethrin (Kanbur et al., 2016).

Table 3 provides a list of effects seen in animals upon administration of amitraz at various dose levels, but below the lethal ones; it also indicates whether an alpha$_2$-antagonist was used to antagonize the observed effect. Several effects are recurrent in multiple studies in different animal species. In addition to the effect on hepatic GSH levels discussed in the preceding paragraph, another peripheral effect is on the gastrointestinal system, where amitraz causes a delay in gastrointestinal transit (Hsu and Lu, 1984; Hsu and McNeel, 1985; Hsu et al., 1987; Moser, 1991; Pass and Mogg, 1991). This effect is antagonized by an alpha$_2$-antagonist such as yohimbine, and is also seen upon *in vivo* administration of clonidine (Maze and Tranquilli, 1991). Furthermore, as indicated earlier, it is also seen *in vitro* in the isolated guinea pig ileum where it is antagonized by another alpha$_2$-antagonist, idazoxan (Hsu et al., 1987). It is suggested that amitraz decreases intestinal contractions by activating alpha$_2$-adrenoceptors in the myenteric (Auerbach's) plexus, thus inhibiting the parasympathetic tone (Hsu et al., 1987). In the perfused pancreas of anesthetized rats, Abu-Basha et al. (1999) measured the ability of amitraz to modulate the secretion of insulin and glucagon. By activating alpha$_{2D}$-adrenoceptors amitraz inhibited the secretion of insulin, thereby confirming the action of amitraz in a rat beta cell line (Chen and Hsu, 1994). Amitraz also stimulated the secretion of glucagon, and both effects were antagonized by the alpha$_2$-adrenoceptor antagonist idazoxan (Abu-Basha et al., 1999).

Amitraz also causes an alpha$_2$-adrenoceptor-mediated bradycardia in several animal species (Andrade and Sakate, 2003; Cullen and Reynoldson, 1987, 1988; Hsu and Hopper, 1986; Hsu and Kakuk, 1984; Hsu and McNeel, 1985; Hsu et al., 1986; Hugnet et al., 1996), which may result from a diminished sympathetic tone (CNS alpha-2 effect) and an increased vascular resistance (peripheral alpha-2 effect). A prolonged bradycardia is one of the main effects of clonidine overdose in humans (Isbister et al., 2017), as well as one of the main effects of amitraz intoxication in humans (see following section). Mydriasis (dilated pupils) is another typical alpha$_2$-adrenoceptor-mediated effects of amitraz in animals (Andrade and Sakate, 2003; Hsu and Kakuk, 1984), and is also a main effect of clonidine (Heal et al., 1989; Koss, 1986). However, findings in human also indicate possible miosis following amitraz intoxication (see following section). Clonidine has been used for over 40 years to treat hypertension

(Pettinger, 1975), and some studies in animals have reported hypotension following administration of amitraz (Andrade and Sakate, 2003); however, hypertension has also been found (Cullen and Reynoldson, 1988; Hsu et al., 1986), and both opposite effects on blood pressure appear to be mediated by alpha$_2$-adrenoceptors (Table 3). CNS effects are among the primary effects observed after administration of amitraz (Table 3), with the most relevant one being sedation, which also represents a main effect of clonidine and other alpha-2 adrenoceptor agonists (Hayes et al., 1986; Sinclair, 2003).

The effects of amitraz and chlordimeform shown in Table 3 are seen at dose levels below the lethal ones. An issue that needs to be addressed is whether the acute lethal toxicity of formamidines is due to activation of alpha$_2$-adrenoceptors. Formamidines act as partial agonists at octopamine and alpha$_2$-adrenergic receptors *in vitro* (Nathanson and Hunnicut, 1981; Robinson and Bittle, 1979). If the same would occur *in vivo*, one would expect formamidines to act as agonists at low doses and as antagonists at higher doses. This hypothesis is partially supported by the findings that animals pre-treated with phentolamine (a mixed alpha$_1$/alpha$_2$ antagonist) died more quickly than controls and showed a slightly lower LD_{50} value for chlordimeform (Robinson et al., 1975), as would be expected if two antagonists (i.e., phentolamine and high dose chlordimeform and DCDM) had an additive effect. Furthermore, the alpha$_2$-antagonist yohimbine was found to potentiate the toxicity of an LD_{50} dose of chlordimeform in mice (Costa and Murphy, 1987). In contrast to these findings with chlordimeform, yohimbine was found to attenuate the delayed lethality induced by amitraz in mice (Moser and MacPhail, 1985). Another way to look at this issue is to compare the potency of the parent compounds and their metabolites at alpha$_2$-adrenoceptors with their acute toxicity. For example, DCDM is more potent than chlordimeform toward alpha$_2$-adrenoceptors (Costa and Murphy, 1987; Costa et al., 1988) and has a higher acute toxicity *in vivo* (Ghali and Hollingworth, 1985); however, the latter has been attributed to pharmacokinetic factors, i.e., higher levels of DCDM in brain tissue (Ghali and Hollingworth, 1985) rather than higher potency toward alpha$_2$-adrenoceptors. In case of amitraz, its metabolite DPMF is slightly more potent than the parent compound toward alpha$_2$-adrenoceptors (Abu-Basha et al., 1999; Chen and Hsu, 1994), but has a much higher acute lethality (lower LD_{50}) (Del Pino et al., 2015b). Pharmacokinetic issues may explain these findings as well, since DPMF binds minimally to plasma proteins and gain easy access to the brain, in contrast to amitraz (Hu et al., 2019). Indeed, after the administration of amitraz, most effects may be due to the active metabolite rather than amitraz itself (Hu et al., 2019).

4. Human poisoning

There are several reports of human poisoning with amitraz, primarily in Turkey, but also in South Africa, India or Sri Lanka, with many cases involving children (Proudfoot, 2003). Most cases involved oral exposures, in which amitraz was accidentally ingested or was self-administered for suicidal purposes. One case involved the intravenous administration of amitraz by a 22-year-old female in Turkey (Gursoy et al., 2005). Poisonings in adults have been published as single case reports (Basetty et al., 2017; Dhooria et al., 2015; Elinav et al., 2005; Gursoy et al., 2005; Herath et al., 2017; Jorens et al., 1997; Saseedharan et al., 2018; Wilson and Murty, 2018), or summaries of multiple poisonings (Avsarogullari et al., 2006; Demirel et al., 2006; Doganay et al., 2002; Kennel et al., 1996; Ulukaya et al., 2001; Veale et al., 2011). Most reports of children poisoning originate from Turkey and involved 8 cases (age 2–4) (Aydin et al., 1997), 11 cases (age 2.5–6) (Yaramis et al., 2000), 21 cases (age 0.5–9) (Ertekin et al., 2002), 43 cases (age 0.8–13) (Kalyoncu et al., 2002), 14 cases (age 2–5) (Atabek et al., 2002), 9 cases (age 0.8–5) (Yilmaz and Yildizdas, 2003), and 8 cases (age 1–4) (Caksen et al., 2003). Additional cases were reported in Sri Lanka (1 case, age 13; Rajesh et al., 2014), and in India (2 cases, age 2 and 14; Shitole et al., 2010; 1 case, age 15; Sweta and Agarwal, 2013; 1 case, age 3; Shashidhara et al., 2017), and a few others have been reported by Dhooria and Agarwal (2016). The acute toxicity of amitraz is in the moderate range, with LD_{50} values upon oral administration of 600 mg/kg (rat), 400–600 mg/kg (guinea pig), ~1000 mg/kg (mouse), 100 mg/kg (dog and rabbit), and 100–250 mg/kg (baboon) (Del Pino et al., 2015b). A few grams of amitraz should thus be lethal in humans. The amount of amitraz consumed was reported only in about 30% of all cases and ranged between 63 mg and 100 g (Dhooria and Agarwal, 2016). However, only one fatality due to amitraz poisoning has been described, a 53-old female from Taiwan who ingested ~100 g of amitraz for suicidal purposes (Hu et al., 2010). She was found unconscious 1 h after ingesting amitraz and died within 24–36 h; the cause of death was identified as a refractory torsade de pointes (Hu et al., 2010).

Table 4 Main signs and symptoms of amitraz poisoning in humans.

CNS effects: lethargy, sedation, unconsciousness
Bradycardia
Mydriasis
Miosis
Respiratory depression
Hypotension
Hypothermia
Hyperglycemia

Most signs and symptoms seen in humans upon amitraz poisoning reflect those seen in animals (Tables 3 and 4). The most common ones are sedation or unconsciousness, bradycardia, hypotension, hyperglycemia and respiratory depression (Dhooria and Agarwal, 2016). Activation of CNS alpha$_2$-adrenoceptors by amitraz and/or its metabolite DPMF is responsible for the altered mental status (sedation and unconsciousness) and for the lowered blood pressure and bradycardia, because of a diminished peripheral sympathetic tone. Hyperglycemia is due to an action of amitraz on alpha$_2$-adrenoceptors in the pancreas, whose activation inhibits the release of insulin. Respiratory depression, in the form of bradypnea, can be also ascribed to activation of alpha$_2$-adrenoceptors, as it is seen also with clonidine (Maze and Tranquilli, 1991), though the exact mechanism is uncertain. Animal studies have indicated the ability of amitraz to induce mydriasis (Andrade and Sakate, 2003; Hsu and Kakuk, 1984), an effect that has also been seen with clonidine, and that has been ascribed to activation of CNS alpha$_2$-adrenoceptors (Heal et al., 1989; Koss, 1986). In contrast, reports of human amitraz poisoning indicate that in several cases miosis rather than mydriasis is present (e.g., Aydin et al., 1997; Shitole et al., 2010; Ulukaya et al., 2001; Veale et al., 2011). This is being explained by the fact that low doses of amitraz may preferentially activate pre-synaptic alpha$_2$-adrenoceptors, thus reducing the sympathetic tone and causing miosis; higher doses of amitraz would instead activate post-synaptic alpha$_2$-adrenoceptors and cause mydriasis.

An important issue to consider is that in human poisoning, exposure occurs not to the relatively pure active ingredient (e.g., amitraz), but to a pesticide formulation containing other chemical compounds. These so called "inert ingredients" often have, however, biological activities that can either aggravate some of the effects of amitraz or cause additional effects not seen in animal studies (Jones, 1990). For example, several amitraz formulations contain xylene, a solvent displaying a variety of adverse health effects (Niaz et al., 2015), some of which have been seen in human amitraz poisoning. Vomiting has been observed in more than 30% of individuals poisoned with pesticides containing amitraz (Dhooria and Agarwal, 2016); in none of the animal species (those that can vomit) has such effect of amitraz been reported, but vomiting has been associated with exposure to xylene (Niaz et al., 2015). Xylene has an inhibitory effect on the CNS, which can be additive to that of amitraz, as sedation to unconsciousness are among the most reported effects in human amitraz poisoning (Dhooria and Agarwal, 2016). Some unusual cardiovascular effects, for example ventricular arrhythmias, can also be ascribed to xylene (Jones, 1990).

There is no specific antidote for amitraz poisoning, as alpha$_2$-adrenoceptor blockers such as yohimbine which have been shown to antagonize clinical signs of poisoning in animals (see Table 3), have not been used in human poisoning. Management of amitraz poisoning is thus supportive and symptomatic. It is important to monitor respiratory, cardiovascular, and CNS functions. Use of gastric lavage or activated charcoal has not proven to be effective (Dhooria and Agarwal, 2016). At times, amitraz poisoning is confused with organophosphate poisoning, and atropine and/or oximes are administered; they are not effective, except perhaps atropine in cases with symptomatic bradycardia and miosis (Yilmaz and Yildizdas, 2003). Fortunately, amitraz poisoning carries a good prognosis with an extremely low mortality rate and no reported long-term effect.

5. Other mechanisms of formamidine neurotoxicity

While most effects of formamidine in the CNS and the periphery can be attributed to activation of pre- and/or post-synaptic alpha$_2$-adrenoceptors, there are other targets of formamidines that have been reported, and in some cases such additional mechanisms may be involved in effects seen in animals and/or humans as a result of exposure to amitraz or chlordimeform. Yim et al. (1978) found that both amitraz (IC_{50} = 880 μM) and chlordimeform (IC_{50} = 34 μM) could inhibit the synthesis of prostaglandin E2 in bovine seminal microsomes. This effect is believed to be responsible for the observed hypothermic and anti-inflammatory effect of formamidines, observed by several investigators (Andrade and Sakate, 2003; Crofton et al., 1987; Hugnet et al., 1996; Moser, 1991; Yim et al., 1978). Hypothermia is also an effect seen in about 25% of humans poisoned with amitraz (Dhooria and Agarwal, 2016).

Early on, it was evident that symptoms of formamidine poisoning in mammals were sympathomimetic in nature (Beeman and Matsumura, 1973), and inhibition of monoamine oxidase (MAO) both *in vitro* (amitraz $IC_{50} = 50\ \mu M$) and *in vivo* had been an earlier reported action of these compounds (Benezet et al., 1978; Moser and MacPhail, 1989). Inhibition of MAO does not appear to be involved in the acute effects of formamidines (e.g., lethality, Robinson and Smith, 1977; changes in visual evoked potentials, Boyes et al., 1985), but given the long duration of inhibition (Bailey et al., 1982) may be involved in some of their delayed effects, such as diminished pain sensitivity (Costa et al., 1988). Inhibition of MAO (both MAO-A and MAO-B) has been suggested as a mechanism underlying the changes in the levels of monoamine (norepinephrine, dopamine, 5-hydroxytryptamine) seen after administration of amitraz in adult or developing rats (Del Pino et al., 2011, 2013, 2017).

Amitraz was found to antagonize (in part) the contractions of the guinea pig ileum induced by histamine and other H1-receptor agonists, suggesting that it may have H1 antihistaminic activity (Pass and Seawright, 1982). However, formamidines had no effect on [^{3}H]-mepyramine binding to H1-histamine receptors *in vitro* (Table 2). Other reported effects of formamidines (mainly chlordimeform) include inhibition of oxidative phosphorylation (Abo-Khatwa and Hollingworth, 1974), local anesthetic like action (Pfister et al., 1978), block of calcium channels (Candura et al., 1992), and neuromuscular blockage (Wang et al., 1975).

6. Summary and conclusion

Amitraz, and other previously used formamidines such as chlordimeform, are unique among insecticides/acaricides as their target of toxicity is the monoaminergic system. In the target species, activation of octopamine receptors leads to functional and behavioral changes rather than strict lethality, but formamidines are nevertheless very effective pesticides. In mammals, the main targets for their toxicity are the alpha$_2$-adrenoceptors which are located pre- and post-synaptically in the CNS and the peripheral nervous system. Though other possible targets have been suggested, most adverse effects of amitraz can be ascribed to alpha$_2$-adrenoceptors activation. Effects seen *in vivo* in various animal species and in humans are similar, and include primarily CNS sedation, hypotension, and bradycardia. These are antagonized by alpha$_2$-blockers in animals, but the latter are not indicated as antidote in humans, and treatment is always symptomatic. Cases of human poisoning continue to occur, particularly in children (with symptoms similar as in adults), particularly in some countries.

Acknowledgments

Research by the author is supported by the grants from NIEHS (R01ES022949, R01ES028273, P30ES07033, P42ES04696).

References

Abo-Khatwa, N., Hollingworth, R.M., 1974. Pesticidal chemicals affecting some energy-linked functions in rat liver mitochondria in vivo. Bull. Environ. Contam. Toxicol. 12, 446–454.

Abu-Basha, E.A., Yibchok-Anun, S., Hopper, D.L., Hsu, W.H., 1999. Effects of the pesticide amitraz and its metabolite BTS 27271 on insulin and glucagon secretion from the perfused rat pancreas: involvement of alpha$_{2D}$-adrenergic receptors. Metabolism 48, 1461–1469.

Altobelli, D., Martire, M., Maurizi, S., Preziosi, P., 2001. Interaction of formamidine pesticides with the presynaptic α_2-adrenoceptor regulating [^{3}H]-noradrenaline release from rat hypothalamic synaptosomes. Toxicol. Appl. Pharmacol. 172, 179–185.

Andrade, S.F., Sakate, M., 2003. The comparative efficacy of yohimbine and atipamezole to treat amitraz intoxication in dogs. Vet. Hum. Toxicol. 45, 124–127.

Atabek, M.E., Aydin, K., Erkul, I., 2002. Different clinical features of amitraz poisoning in children. Hum. Exp. Toxicol. 21, 13–16.

Avsarogullari, L., Ikizceli, I., Sangur, M., Sozuer, E., Akdur, O., Yucei, M., 2006. Acute amitraz poisoning in adults: clinical features, laboratory findings, and management. Clin. Toxicol. 44, 19–23.

Axelrod, J., Saavedra, J.M., 1977. Octopamine. Nature 265, 501–504.

Aydin, K., Kurtoglu, S., Poyrazoglu, M.H., Uzum, K., Ustunbas, H.B., Hallac, I.K., 1997. Amitraz poisoning in children: clinical and laboratory findings of eight cases. Hum. Exp. Toxicol. 16, 680–682.

Bailey, B.A., Martin, R.J., Downer, R.G.H., 1982. Monoamine oxidase inhibition and brain catecholamine levels in the rat following treatment with chlordimeform. Pestic. Biochem. Physiol. 17, 293–300.

Basetty, S., Mishra, A.K., Sathyendra, S., 2017. Neurological effects of an unusual insecticide poison: amitraz. J. Family Med. Prim. Care 6, 686–687.

Beeman, R.W., Matsumura, G., 1973. Chlordimeform; a pesticide acting upon amine regulatory mechanisms. Nature 242, 273–274.

Beeman, R.W., Matsumura, G., 1978. Formamidine pesticides—actions in insects and acarines. In: Shankland, D.L., Hollingworth, R.M., Smith, T. (Eds.), Pesticide and Venom Neurotoxicity. Plenum Press, New York, pp. 179–188.

Benezet, J.J., Chang, K.M., Knowles, C.O., 1978. Formamidine pesticides: metabolic aspects of neurotoxicity. In: Shankland, D.L., Hollingworth, R.M., Smith, T. (Eds.), Pesticide and Venom Neurotoxicity. Plenum Press, New York, pp. 189–206.

Boyes, W.K., Moser, V.C., 1987. Investigations of amitraz neurotoxicity in rats. II. Effects on visual evoked potentials. Fundam. Appl. Toxicol. 9, 140–153.
Boyes, W.K., Moser, V.C., 1988. An α_2-adrenergic mode of action of chlordimeform on rat visual function. Toxicol. Appl. Pharmacol. 92, 402–418.
Boyes, W.K., Moser, V.C., MacPhail, R.C., Dyer, R.S., 1985. Monoamine oxidase inhibition cannot account for changes in visual evoked potentials produced by chlordimeform. Neuropharmacology 24, 853–860.
Caksen, H., Obadas, D., Arslan, S., Akgun, C., Atas, B., Akbayram, S., Tuncer, O., 2003. Report of eight children with amitraz intoxication. Hum. Exp. Toxicol. 22, 95–97.
Candura, S.M., Marraccini, P., Costa, L.G., Manzo, L., Rossi, A., Coccini, T., Tonini, M., 1992. Calcium entry blockade as a mechanism for chlordimeform-induced inhibition of motor activity in the isolated guinea-pig ileum. Pharmacol. Toxicol. 71, 426–433.
Chen, T.S., Hsu, W.H., 1994. Inhibition of insulin release by a formamidine pesticide amitraz and its metabolites in a rat beta-cell line: an action mediated by alpha-2 adrenoceptors, a GTP-binding protein and a decrease in cyclic AMP. J. Pharmacol. Exp. Ther. 271, 1240–1245.
Costa, L.G., 2018. Toxic effects of pesticides. In: Klaassen, C.D. (Ed.), Casarett and Doull's Toxicology: The Basic Science of Poisons. McGraw-Hill, New York, pp. 1055–1106.
Costa, L.G., Murphy, S.D., 1987. Interaction of the pesticide chlordimeform with adrenergic receptors in mouse brain: an in vitro study. Arch. Toxicol. 59, 323–327.
Costa, L.G., Olibet, G., Murphy, S.D., 1988. Alpha$_2$-adrenoceptors as a target for formamidine pesticides: in vitro and in vivo studies in mice. Toxicol. Appl. Pharmacol. 93, 319–328.
Costa, L.G., Olibet, G., Wu, D., Murphy, S.D., 1989a. Acute and chronic effects of the pesticide amitraz on α_2-adrenoceptors in the mouse brain. Toxicol. Lett. 47, 135–143.
Costa, L.G., Wu, D., Olibet, G., Murphy, S.D., 1989b. Formamidine pesticides and alpha2-adrenoceptors: studies with amitraz and chlordimeform in rats and development of a radioreceptor binding assay. Neurotoxicol. Teratol. 11, 405–411.
Costa, L.G., Gastel, J., Murphy, S.D., 1991. The formamidine pesticides chlordimeform and amitraz decrease hepatic glutathione in mice through an interaction with alpha$_2$-adrenoceptors. J. Toxicol. Environ. Health 33, 349–358.
Crofton, K.M., Boncek, V.M., Reiter, L.W., 1987. The effects of amitraz (AMZ) administration on locomotor activity and the acoustic startle response: dosage and time dependent effects in the rat. Toxicologist 7, 254.
Cullen, L.K., Reynoldson, J.A., 1987. Cardiovascular and respiratory effects of the acaricide amitraz. J. Vet. Pharmacol. Ther. 10, 134–143.
Cullen, L.K., Reynoldson, J.A., 1988. Cardiovascular responses to amitraz in the presence of autonomic antagonists and agonists. Arch. Int. Pharmacodyn. Ther. 296, 45–56.
Cullen, L.K., Reynoldson, J.A., 1990. Central and peripheral alpha-adrenoceptor actions of amitraz in the dog. J. Vet. Pharmacol. Ther. 13, 86–92.
David, J.C., Coulon, J.F., 1985. Octopamine in invertebrates and vertebrates. A review. Prog. Neurobiol. 24, 141–185.
Del Pino, J., Martinez, M.A., Castellano, V.J., Ramos, E., Martinez-Larranaga, M.R., Anadon, A., 2011. Effects of prenatal and postnatal exposure to amitraz on norepinephrine, serotonin and dopamine levels in brain regions of male and female rats. Toxicology 287, 145–152.
Del Pino, J., Martinez, M.A., Castellano, V., Ramos, E., Martinez-Larranaga, M.R., Anadon, A., 2013. Effects of exposure to amitraz on noradrenaline, serotonin, and dopamine levels in brain regions of 30 and 60 days old male rats. Toxicology 308, 88–95.
Del Pino, J., Frejo, M.T., Anadon Baselga, M.J., Moyano, P., Diaz, M.J., 2015a. Impaired glutamatergic and GABAergic transmission by amitraz in primary hippocampal cells. Neurotoxicol. Teratol. 50, 82–87.

Del Pino, J., Moyano-Cires, P.V., Anadon, M.J., Diaz, M.J., Lobo, M., Capo, M.A., Frejo, M.T., 2015b. Molecular mechanisms of amitraz mammalian toxicity: a comprehensive review of existing data. Chem. Res. Toxicol. 28, 1073–1094.

Del Pino, J., Moyano, P., Ruiz, M., Anadon, M.J., Diaz, M.J., Garcia, J.M., Labajo-Gonzalez, E., Frejo, M.T., 2017. Amitraz changes NE, DA and 5-HT biosynthesis and metabolism mediated by alterations in estradiol content in CNS of male rats. Chemosphere 181, 518–529.

Demirel, Y., Yilmaz, A., Gursoy, S., Kaygusuz, K., Mimaroglu, C., 2006. Acute amitraz intoxication: retrospective analysis of 45 cases. Hum. Exp. Toxicol. 25, 613–617.

Dhooria, S., Agarwal, R., 2016. Amitraz, an unrecognized poison: a systematic review. Indian J. Med. Res. 144, 348–358.

Dhooria, S., Behera, D., Agarwal, R., 2015. Amitraz: a mimicker of organophosphate poisoning. BMJ Case Rep. 2015. https://doi.org/10.1136/bcr-2015-210296.

Doganay, Z., Aygun, D., Altintop, L., Guven, H., Bildik, F., 2002. Basic toxicological approach has been effective in two poisoned patients with amitraz ingestion: case reports. Hum. Exp. Toxicol. 21, 55–57.

Elinav, E., Shapira, Y., Ofran, Y., Hassin, T., Ben-Dov, I.Z., 2005. Near-fatal amitraz intoxication: the overlooked pesticide. Basic Clin. Pharmacol. Toxicol. 97, 185–187.

Ertekin, V., Alp, H., Selimoglu, M.A., Karacan, M., 2002. Amitraz poisoning in children: retrospective analysis of 21 cases. J. Int. Med. Res. 30, 203–205.

Evans, P.D., Gee, J.D., 1980. Action of formamidine pesticides on octopamine receptors. Nature 287, 60–62.

Evans, P.D., Maqueira, B., 2005. Insect octopamine receptors: a new classification scheme based on studies of cloned *Drosophila* G-protein coupled receptors. Invert. Neurosci. 5, 111–118.

Ghali, T.G., Hollingworth, R.M., 1985. Influence of mixed function oxygenase metabolism on the acute neurotoxicity of the pesticide chlordimeform in mice. Neurotoxicology 6, 215–238.

Gilbert, M.E., 1988. Formamidine pesticides enhance susceptibility to kindled seizures in amygdala and hippocampus of the rat. Neurotoxicol. Teratol. 10, 221–227.

Gilbert, M.E., Dyer, R.S., 1988. Increased hippocampal excitability produced by amitraz. Neurotoxicol. Teratol. 10, 229–235.

Giovannitti, J.A., Thoms, S.M., Crawford, J.J., 2015. Alpha-2 adrenergic receptor agonists: a review of current clinical applications. Anesth. Prog. 62, 31–38.

Gursoy, S., Kunt, N., Kaygusuz, K., Kafali, H., 2005. Intravenous amitraz poisoning. Clin. Toxicol. 43, 113–116.

Harbison, R.D., James, R.C., Roberts, S.M., 1991. Hepatic glutathione suppression by the α-adrenoceptor stimulating agent phenylephrine and clonidine. Toxicology 69, 279–290.

Hayes, A.G., Skingle, M., Tyers, M.B., 1986. Alpha-adrenoceptor-mediated antinociception and sedation in the rat and dog. Neuropharmacology 25, 391–396.

Heal, D.J., Prow, M.R., Buckett, W.R., 1989. Clonidine produces mydriasis in conscious mice by activating central α_2-adrenoceptors. Eur. J. Pharmacol. 170, 11–18.

Herath, H.M.M.T.B., Pahalagamage, S.P., Yogendranathan, N., Wijayabandara, M.D.M.S., 2017. Amitraz poisoning: a case report of an unusual pesticide poisoning in Sri Lanka and literature review. BMC Pharmacol. Toxicol. 18, 6.

Hollingworth, R.M., 1976. Chemistry, biological activity and uses of formamidine pesticides. Environ. Health Perspect. 14, 57–69.

Hollingworth, R.M., Lund, A.E., 1982. Biological and neurotoxic effects of amidine pesticides. In: Coats, J.R. (Ed.), Insecticide Mode of Action. Academic Press, San Diego, pp. 189–227.

Hsu, W.H., Hopper, D.L., 1986. Effect of yohimbine on amitraz-induced CNS depression and bradycardia in dogs. J. Toxicol. Environ. Health 18, 423–429.

Hsu, W.H., Kakuk, T.J., 1984. Effect of amitraz and chlordimeform on heart rate and pupil diameter in rats: mediated by α_2-adrenoceptors. Toxicol. Appl. Pharmacol. 73, 411–415.

Hsu, W.H., Lu, Z.X., 1984. Amitraz-induced delay of gastrointestinal transit in mice: mediated by α_2-adrenergic receptors. Drug Dev. Res. 4, 655–660.

Hsu, W.H., McNeel, S.V., 1985. Amitraz-induced prolongation of gastrointestinal transit and bradycardia in dogs and their antagonism by yohimbine: preliminary study. Drug Chem. Toxicol. 8, 239–253.

Hsu, W.H., Lu, Z.X., Hembrough, F.B., 1986. Effect of amitraz on heart rate and aortic blood pressure in conscious dogs: influence of atropine, prazosin, tolazoline, and yohimbine. Toxicol. Appl. Pharmacol. 84, 418–422.

Hsu, W.H., Shaw, R.N., Schaffer, D.D., Crump, M.H., Greer, M.H., 1987. Further evidence to support the α_2-adrenergic nature of amitraz-induced decrease in intestinal motility. Arch. Int. Pharmacodyn. Ther. 286, 145–151.

Hu, S.Y., Hsu, C.L., Tsan, Y.T., Hung, D.Z., Hu, W.H., Li, H.P., 2010. Torsades de pointes in amitraz poisoning. Resuscitation 81, 366–367.

Hu, S.X., Benner, C.P., White, J.A., Martin, R.A., Feenstra, K.L., 2019. Pharmacokinetics and brain distribution of amitraz and its metabolites in rats. Environ. Toxicol. Pharmacol. 65, 40–45.

Hugnet, C., Buronrosse, F., Pineau, X., Cadore, J.L., Lorgue, G., Berny, P.J., 1996. Toxicity and kinetics of amitraz in dogs. Am. J. Vet. Res. 57, 1506–1510.

IPCS (International Programme on Chemical Safety), 1998. Chlordimeform. Environmental Health Criteria 199. World Health Organization, Geneva, p. 159.

Isbister, G.K., Heppell, S.P., Page, C.P., Ryan, N.M., 2017. Adult clonidine overdose: prolonged bradycardia and central nervous system depression, but not severe toxicity. Clin. Toxicol. 55, 187–192.

Jones, R.D., 1990. Xylene/amitraz: a pharmacologic review and profile. Hum. Exp. Toxicol. 32, 446–448.

Jorens, P.G., Zandijk, E., Belmans, L., Schepens, P.J.C., Bossaert, L.L., 1997. An unusual poisoning the unusual pesticide amitraz. Hum. Exp. Toxicol. 16, 600–601.

Kalyoncu, M., Dilber, E., Okten, A., 2002. Amitraz intoxication in children in the rural Black Sea region: analysis of forty-three patients. Hum. Exp. Toxicol. 21, 269–272.

Kanbur, M., Silig, Y., Eraslan, G., Karabacak, M., Sarica, Z.S., Sahin, S., 2016. The toxic effect of cypermethrin, amitraz and combinations of cypermethrin-amitraz in rats. Environ. Sci. Pollut. Res. 23, 5232–5242.

Kennel, O., Prince, C., Garnier, R., 1996. Four cases of amitraz poisoning in humans. Vet. Hum. Toxicol. 38, 28–30.

Kita, T., Hayashi, T., Ohtani, T., Takao, H., Hiroshi, T., Liu, G., Ohta, H., Ozoe, F., Ozoe, Y., 2016. Amitraz and its metabolite differentially activate α- and β-adrenergic like octopamine receptors. Pest Manag. Sci. 73, 984–990.

Knowles, C.O., Benezet, H.J., 1981. Excretion balance, metabolic fate and tissue residues following treatment of rats with amitraz and N^1-(2,4-dimethyl-phenyl)-N-methylformamidine. J. Environ. Sci. Health B 16, 547–555.

Koeiman, N.R., Hsu, W.H., 1991. Interaction between amitraz and alpha 2-adrenoceptors inhibits epinephrine-induced canine platelet aggregation. Arch. Int. Pharmacodyn. Ther. 310, 56–65.

Koss, M.C., 1986. Pupillary dilation as an index of central nervous system α_2-adrenoceptor activation. J. Pharmacol. Methods 15, 1–19.

Landauer, M.R., Tomlinson, W.T., Balster, R.L., MacPhail, R.C., 1984. Some effects of the formamidine pesticide chlordimeform on the behavior of mice. Neurotoxicology 5, 91–100.

Lund, A.E., Hollingworth, R.M., Yim, G.K.W., 1979. The comparative neurotoxicity of formamidine pesticides. In: Narahashi, T. (Ed.), Neurotoxicology of Insecticides and Pheromones. Plenum Press, New York, pp. 119–137.

Maze, M., Tranquilli, W., 1991. Alpha-2 adrenoceptor agonists: defining the role in clinical anesthesia. Anesthesiology 74, 581–605.

Moser, V.C., 1991. Investigations of amitraz neurotoxicity in rats. IV. Assessment of toxicity syndrome using a functional observational battery. Fundam. Appl. Toxicol. 17, 7–16.

Moser, V.C., MacPhail, R.C., 1985. Yohimbine attenuates the delayed lethality induced in mice by amitraz, a formamidine pesticide. Toxicol. Lett. 28, 99–104.

Moser, V.C., MacPhail, R.C., 1989. Investigations of amitraz neurotoxicity in rats. III. Effects on motor activity and inhibition of monoamine oxidase. Fundam. Appl. Toxicol. 12, 12–22.

Moser, V.C., Boyes, W.K., MacPhail, R.C., 1987. Investigations of amitraz neurotoxicity in rats. I. Effects on operant performance. Fundam. Appl. Toxicol. 9, 131–139.

Nathanson, J.A., 1985. Characterization of octopamine-sensitive adenylate cyclase: elucidation of a class of potent and selective octopamine-2 receptor agonists with toxic effects in insects. Proc. Natl. Acad. Sci. U. S. A. 82, 599–603.

Nathanson, J.A., Hunnicut, E.J., 1981. N-demethylchlordimeform. A potent partial agonist of octopamine-sensitive adenylate cyclase. Mol. Pharmacol. 20, 68–75.

Niaz, K., Bahadar, H., Maqbool, F., Abdollahi, M., 2015. A review of environmental and occupational exposure to xylene and its health concerns. EXCLI J. 14, 1167–1186.

Ohta, H., Ozoe, Y., 2014. Molecular signaling, pharmacology, and physiology of octopamine and tyramine receptors as potential insect pest control targets. Adv. Insect Physiol. 46, 73–166.

Pass, M.A., Mogg, T.D., 1991. Effect of amitraz and its metabolites on intestinal motility. Comp. Biochem. Physiol. 99C, 169–172.

Pass, M.A., Seawright, A.A., 1982. Effect of amitraz on contractions of the guinea-pig ileum. Comp. Biochem. Physiol. 73C, 419–422.

Pettinger, W.A., 1975. Drug therapy: clonidine, a new antihypertensive drug. N. Engl. J. Med. 293, 1179–1180.

Pfister, W.R., Yim, G.K.W., 1977. Antipyretic-analgesic action of the formamidine pesticide chlordimeform. Pharmacologist 19, 216.

Pfister, W., Chin, C., Noland, V., Yim, G.K.W., 1978. Similar pharmacological actions of chlordimeform and local anesthetics. Pestic. Biochem. Physiol. 9, 148–156.

Popp, W., Schmieding, W., Speck, M., Vahrenholz, C., Norpoth, K., 1992. Incidence of bladder cancer in a cohort of workers exposed to 4-chloro-o-toluidine while synthesizing chlordimeform. Br. J. Ind. Med. 49, 529–531.

Proudfoot, A.T., 2003. Poisoning with amitraz. Toxicol. Rev. 22, 71–74.

Queiroz-Neto, A., Zamur, G., Goncalves, S.C., Carregaro, A.B., Mataqueiro, M.I., Harkins, J.D., Tobin, T., 1998. Characterization of the antinociceptive and sedative effect of amitraz in horses. J. Vet. Pharmacol. Ther. 21, 400–405.

Rajesh, N.T., Pandian, N.S., Mathai, J., 2014. Poisoning with amitraz, a veterinary drug, in an adolescent boy. Sri Lanka J. Child Health 43, 55–56.

Robinson, C.P., Bittle, I., 1979. Vascular effects of demethylchlordimeform, a metabolite of chlordimeform. Pestic. Biochem. Physiol. 11, 46–55.

Robinson, C.P., Smith, P.W., 1977. Lack of involvement of monoamine oxidase inhibition in the lethality of acute poisoning by chlordimeform. J. Toxicol. Environ. Health 3, 565–568.

Robinson, C.P., Smith, P.W., Zelenski, J.D., Endecott, B.R., 1975. Lack of an effect of interference with amine mechanisms on the lethality of chlordimeform in the rat. Toxicol. Appl. Pharmacol. 33, 380–383.

Saseedharan, S., Pathrose, E.J., Madhav, B.V., 2018. A clinical conundrum called amitraz poisoning—a case report. Indian J. Crit. Care Med. 22, 195–196.
Schaffer, D.D., Hsu, W.H., Hopper, D.L., 1990. The effects of yohimbine and four other antagonists on amitraz-induced depression of shuttle avoidance responses in dogs. Toxicol. Appl. Pharmacol. 104, 543–547.
Shashidhara, S., Mundkur, S.C., Hebbar, S.A., Kumar, A., Sambasivan, V., 2017. Amitraz. A perpetual silent menace among children in developing countries—a case report. Pediatr. Emerg. Care 35, e169–e171.
Shin, D.H., Hsu, W.H., 1994. Influence of the formamidine pesticide amitraz and its metabolites on porcine myometrial contractility: involvement of α_2-adrenoceptors and Ca^{2+} channels. Toxicol. Appl. Pharmacol. 128, 45–49.
Shitole, D.G., Kulkarni, R.S., Sathe, S.S., Rahate, P.R., 2010. Amitraz poisoning—an unusual pesticide poisoning. J. Assoc. Physicians India 58, 317–319.
Sinclair, M.D., 2003. A review of the physiological effects of α_2-agonists related to the clinical use of medetomidine in small animal practice. Can. Vet. J. 44, 885–897.
Smith, B.E., Hsu, W.H., Yang, P.C., 1990. Amitraz-induced glucose intolerance in rats: antagonism by yohimbine but not prazosin. Arch. Toxicol. 64, 680–683.
Sweta, U.S., Agarwal, A., 2013. Amitraz: an unfamiliar poisoning with familiar pesticide. J. Anaesthesiol. Clin. Pharmacol. 29, 420–421.
Ulukaya, S., Demirag, K., Moral, A.R., 2001. Acute amitraz intoxication in human. Intensive Care Med. 27, 930–933.
Veale, D.J., Wium, C.A., Muller, G.J., 2011. Amitraz poisoning in South Africa: a two-year survey (2008-2009). Clin. Toxicol. 49, 40–44.
Wang, C.M., Narahashi, T., Fukami, J.I., 1975. Mechanism of neuromuscular block by chlordimeform. Pestic. Biochem. Physiol. 5, 119–125.
Wilson, W., Murty, S., 2018. Amitraz poisoning: the (un)common poisoning. J. Emerg. Trauma Shock 11, 140–142.
Yaramis, A., Soker, M., Bilici, M., 2000. Amitraz poisoning in children. Hum. Exp. Toxicol. 19, 1–3.
Yilmaz, H.L., Yildizdas, D.R., 2003. Amitraz poisoning, an emerging problem: epidemiology, clinical features, management, and preventive strategies. Arch. Dis. Child. 88, 130–134.
Yim, G.K.W., Holsapple, M.P., Pfister, W.R., Hollingworth, R.M., 1978. Prostaglandin synthesis inhibited by formamidine pesticides. Life Sci. 23, 2509–2516.
Young, F.M., Menadue, M.F., Lavranos, T.C., 2005. Effects of the insecticide amitraz, an α_2-adrenergic receptor agonist, on human luteinized granulosa cells. Hum. Reprod. 20, 3018–3025.

FSC
www.fsc.org
MIX
Papier | Fördert gute Waldnutzung
FSC® C083411

Kolibri 360 GmbH
Ferdinand-Jühlke-Straße 7
99095 Erfurt, Deutschland
produktsicherheit@kolibri360.de